Multivariable and Optimal Systems

D. H. OWENS

Department of Control Engineering, University of Sheffield, Sheffield, U.K.

1981

ACADEMIC PRESS
London New York Toronto Sydney San Francisco
A Subsidiary of Harcourt Brace Jovanovich, Publishers

ACADEMIC PRESS INC. (LONDON) LTD.
24/28 Oval Road
London NW1

United States Edition published by
ACADEMIC PRESS INC.
111 Fifth Avenue
New York, New York 10003

British Library Cataloguing in Publication Data

Owens, D.H.
Multivariable and optimal systems.
1. Control theory
I. Title
629.8'312 QA402.3
ISBN Hardback: 0 12 531720 4
ISBN Paperback: 0 12 531722 0

LCCCN: 81–67886

Printed in Great Britain at the Alden Press
Oxford London and Northampton

ML: ADDIT

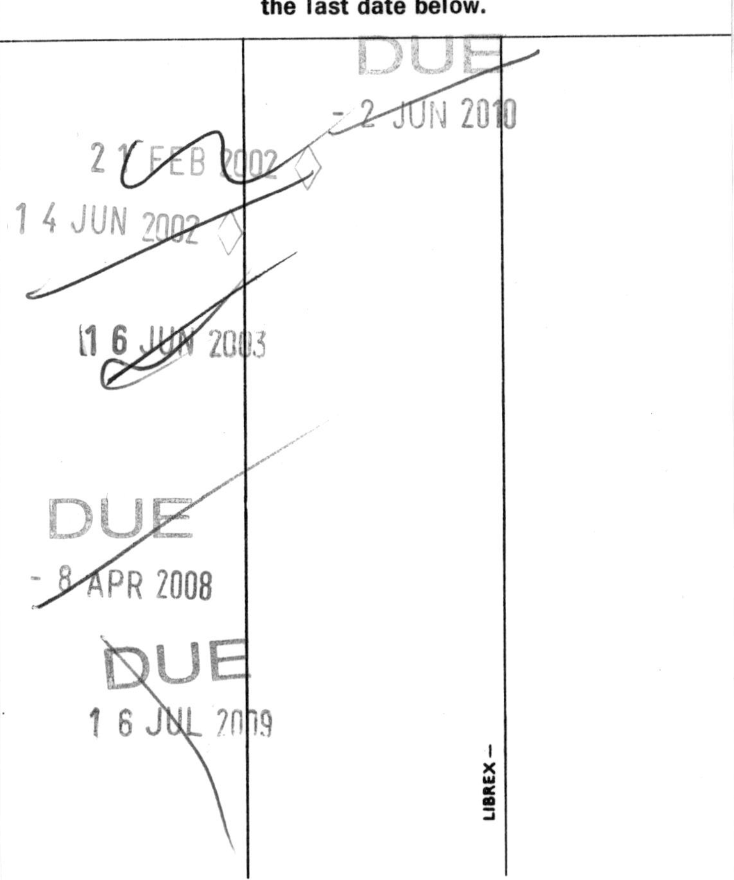

Multivariable and Optimal Systems

Preface

Multivariable and optimal systems are now an established part of systems science and control engineering degree courses. This text aims to provide a course and self study textbook for undergraduate and Master's degree control engineering students in universities and polytechnics covering the conceptual basis of multivariable systems control theory and optimal control and illustrating its application to simple multivariable process plant. It is motivated by the observation that available texts either lie at the research level demanding a degree of mathematical sophistication that few undergraduates can cope with, or are fairly general control texts containing some information on multivariable and optimal systems but, due to lack of space, tending to make the treatment rather superficial. The proposed text attempts to bring depth whilst using, in its simplest form, mathematics normally included in undergraduate engineering courses. In this way the reader can obtain a (relatively) painless and firm basis for future specialist studies.

In my experience, the major obstacles met by the typical student are the steps from classical control design methods to "thinking multivariable-style" and from there to expressing multivariable concepts in matrix form. The approach taken, therefore, is to introduce concepts in the context of dynamics studies of simple process plant to illustrate the natural source of the techniques. I have also found that engineering students, in general, respond best if the engineering design applications aspects of the material are emphasized rather than the general systems theoretical topics. Thus the text takes the view that design principles and design practice are the most important part of the armoury of the future control engineer in the sense that design decisions are made by human judgement based on experience of synthesis procedures seen to work in elementary cases.

An important comment is that, despite the elementary style of the text, the standard of mathematical rigour is, on the whole, high and based on the maxim, "He who does not know the limitations of the theory is lost" (anon.). In this sense the reader is presented with a rigorous development of most of the essential results forming the cornerstones of applications studies and techniques described in more advanced texts. The validity and potential of the concepts are also illustrated by simple but meaningful physical examples. The subtleties of more advanced computer-aided design and synthesis procedures are, however, left for further study in more advanced texts.

The text is divided naturally into three parts. Chapters 1 and 2 provide a sound introduction to the basic ideas of modelling systems behaviour by continuous or discrete state variable models and the effect of a general form of state feedback. Chapters 3 and 4 lay a firm foundation for control studies using Laplace transforms, poles, zeros and design criteria of systems with more than one output. A number of essentially multivariable design concepts are illustrated by a detailed consideration of multivariable generalizations of a first order lag. Finally, Chapters 5 to 7 return to the time-domain and formulate and solve a number of continuous and discrete optimal control problems. Emphasis is placed on the linear quadratic optimal control problem and minimum energy and minimum fuel problems to illustrate principles. A statement of the Euler-Lagrange equations and the Minimum Principle of Pontryagin are also included for completeness.

This text is the product of several courses given by the author at the University of Sheffield since 1973 and the comments and questions of the undergraduates and postgraduates who attended them. Many thanks go to Mrs P. Turner and Mrs J. Stubbs for typing the draft manuscript and to my wife and family for enduring its preparation.

D.H. Owens
August 1981

Contents

For BEN and PENNY

1. Systems and Dynamics

This chapter introduces the fundamental language of systems models, systems dynamics, systems structure and simulation used in the multivariable and optimal control studies described in the remaining chapters. The material is frequently initiated by illustrating how it naturally arises in practice and its distinctive differences from, and similarities to, classical ordinary differential equation methods are highlighted to emphasize fundamentally new concepts and to reassure the timid reader that he is not too far from familiar territory.

1.1 Introduction and Review of Basic Concepts

The schematic diagram shown in Fig. 1 is the familiar block diagram representation of a classical dynamic system with manipulable *input* u driving the system from given initial conditions to produce the consequent measured *output* y. Either or both could be continuously varying with time or be represented by discrete/sampled values at defined points of time. For simplicity at this stage we will assume that all signals are continuous. The only significant exclusions from this general picture are the possibilities of other measurable or stochastic (i.e. noise) inputs to the process from the environment.

The mathematical building blocks used to express, analyse and predict the relationships between input, output and initial conditions are surprisingly few in number. The most fundamental branch of knowledge is the theory of nth

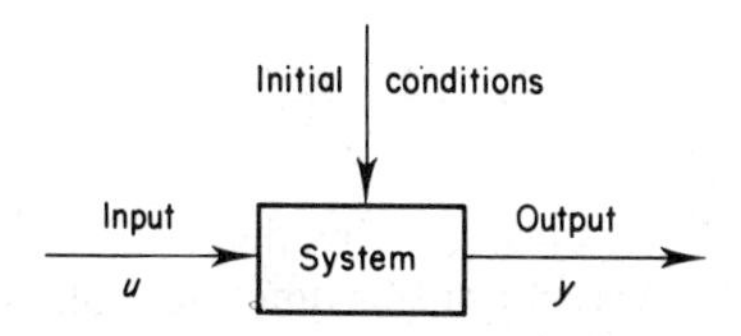

Fig. 1. Deterministic representation of system dynamics.

order ordinary differential equations. More precisely the most commonly used method of expressing the relationship between the input and output uses an nth order ordinary differential equation of the general (nonlinear) form

$$\frac{\mathrm{d}^n y(t)}{\mathrm{d}t^n} = \psi\left(y(t), \frac{\mathrm{d}y(t)}{\mathrm{d}t}, \ldots, \frac{\mathrm{d}^{n-1}y(t)}{\mathrm{d}t^{n-1}}, u(t), \ldots, \frac{\mathrm{d}^n u(t)}{\mathrm{d}t^n}, t\right) \tag{1.1}$$

with initial conditions of the form

$$y(0) = \mathrm{d}_0, \left.\frac{\mathrm{d}y}{\mathrm{d}t}\right|_{t=0} = \mathrm{d}_1, \ldots, \left.\frac{\mathrm{d}^{n-1}y(t)}{\mathrm{d}t^{n-1}}\right|_{t=0} = d_{n-1} \tag{1.2}$$

These models are normally obtained either by "analytical modelling" of the system dynamics directly from the fundamental and empirical laws governing the process or by "indirect" model-fitting techniques based on available plant transient or frequency response data.

In the case of equation (1.1) being linear of the form (after a little reorganization)

$$a_0 \frac{\mathrm{d}^n y}{\mathrm{d}t^n} + a_1 \frac{\mathrm{d}^{n-1} y}{\mathrm{d}t^{n-1}} + \ldots + a_{n-1} \frac{\mathrm{d}y}{\mathrm{d}t} + a_n y = b_0 \frac{\mathrm{d}^n u}{\mathrm{d}t^n}$$

$$+ b_1 \frac{\mathrm{d}^{n-1} u}{\mathrm{d}t^{n-1}} + \ldots + b_{n-1} \frac{\mathrm{d}u}{\mathrm{d}t} + b_n u, \qquad (a_0 \neq 0) \tag{1.3}$$

the mathematical machinery of analysis is extremely highly developed. In particular the use of Laplace transform techniques has long been established as a powerful tool. Perhaps the most significant idea is that of the system transfer function. Introducing the differential operator $\mathrm{D} = \mathrm{d}/\mathrm{d}t$ and the polynomials

$$P(\lambda) = a_0\lambda^n + a_1\lambda^{n-1} + \ldots + a_{n-1}\lambda + a_n \tag{1.4}$$

$$Q(\lambda) = b_0\lambda^n + b_1\lambda^{n-1} + \ldots + b_{n-1}\lambda + b_n \tag{1.5}$$

then the system of equation (1.3) has the compact form

$$P(\mathrm{D})y(t) = Q(\mathrm{D})u(t) \tag{1.6}$$

The system *transfer function* is the rational function of the complex variable s defined by

$$g(s) = \frac{Q(s)}{P(s)} \tag{1.7}$$

and the system is frequently represented by the block diagram of Fig. 2. A fundamental property of the system transfer function is obtained for the case of zero initial conditions, $d_{k-1} = 0$, $1 \leqslant k \leqslant n$, and the case of inputs possessing

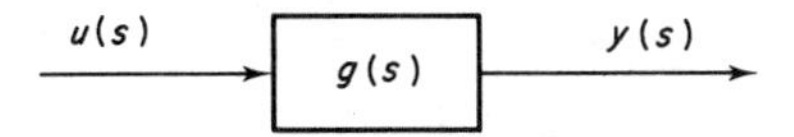

Fig. 2. Transfer function representation of system dynamics.

the property that $D^{k-1}u(t)|_{t=0} = 0, 1 \leqslant k \leqslant n$, namely

$$y(s) = g(s)u(s) \tag{1.8}$$

where, for notational simplicity, the Laplace transform of $u(t)$ and $y(t)$ are denoted $u(s)$ and $y(s)$ respectively. If $\mathscr{L}$ and $\mathscr{L}^{-1}$ denote the operations of taking the Laplace transform and taking inverse Laplace transforms respectively, then (1.8) takes the form

$$y(t) = \mathscr{L}^{-1}[g(s)u(s)] = \int_0^t h(t-t')u(t')\mathrm{d}t' \tag{1.9}$$

where $h(t) = \mathscr{L}^{-1}[g(s)]$ is the system *impulse response* and the final integral is termed the *convolution* of h and u.

The power of the transfer function representation of system dynamics originates in the simplicity of representation of the dynamics of composite systems. Consider the parallel, series and feedback configurations illustrated in Fig. 3., the reader will easily verify that the transfer functions in each case are

(a) $$g(s) = g_1(s) + g_2(s) + \ldots + g_m(s)$$

(b) $$g(s) = g_1(s)g_2(s)\ldots g_m(s)$$

(c) $$g(s) = \frac{g_1(s)}{1+g_1(s)g_2(s)} \tag{1.10}$$

The major impact of transfer function methods are felt, however, in the design of the forward path and minor loop elements in the general scalar *feedback system* illustrated in Fig. 4., by the use of the *frequency response* $g(j\omega)$ ($j^2 = -1$) and the *Nyquist stability criterion.* Alternatively, the factorization

$$g(s) = \frac{g_0(s-z_1)(s-z_2)\ldots(s-z_{n_z})}{(s-p_1)(s-p_2)\ldots(s-p_n)} = g_0 \prod_{l=1}^{n_z}(s-z_l) \Big/ \prod_{l=1}^{n}(s-p_l) \tag{1.11}$$

and consideration of the poles $\{p_1, p_2, \ldots, p_n\}$, zeros $\{z_1, z_2, \ldots, z_{n_z}\}$, order n, rank $n-n_z$ and gain g_0 of the transfer function (a) provide valuable information on open-loop stability and transient performance and (b) can be used as the basis for the choice of control elements using the well known *root-locus method.*

The following chapters can be regarded as a self-contained introduction to the

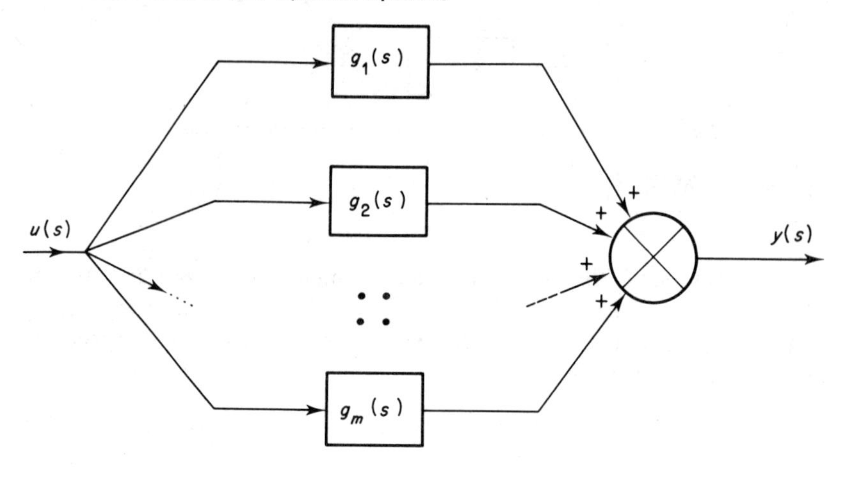

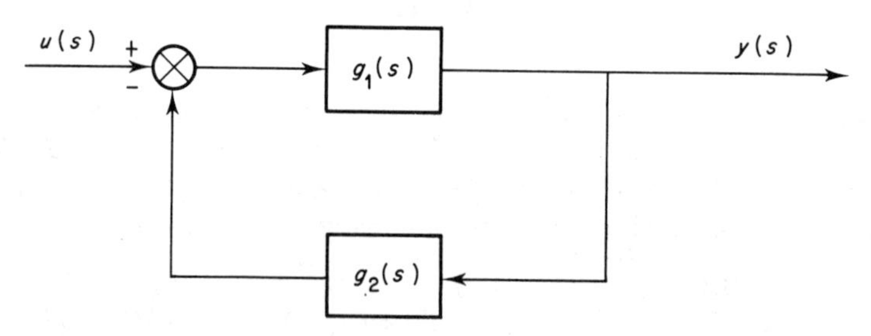

Fig. 3. Parallel, series and feedback configurations.

methodology required for the extension of the above ideas to cover the modelling, analysis and design of control systems for engineering systems with more than one input and more than one output. Such systems are commonly called multi-input/multi-output or MULTIVARIABLE systems and are represented schematically as shown in Fig. 5. They reduce to the single-input/single-output case wherever $m = l = 1$. An important general property of such systems (and a

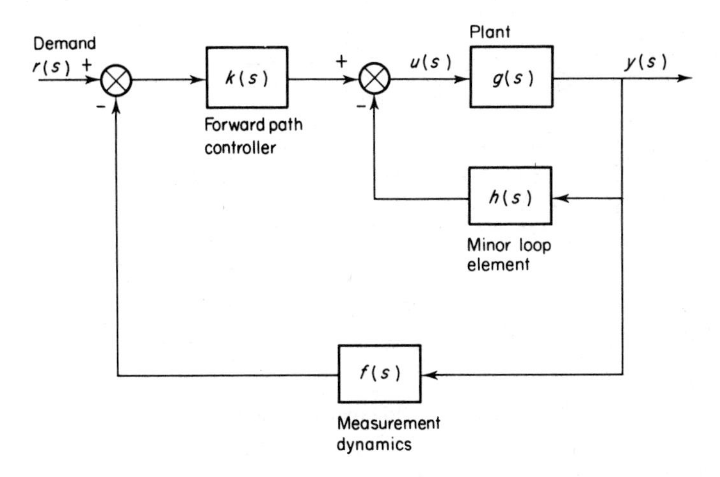

Fig. 4. Feedback system with minor loop and transducer dynamics.

major source of design difficulties) is INTERACTION between inputs and outputs in the sense that any input u_k (say) will have a dynamic effect on all outputs y_i, $1 \leqslant i \leqslant m$. In the case of $m = l > 1$ (i.e. we have equal numbers of inputs and outputs) and each input u_k only has a dynamic effect on y_k, $1 \leqslant k \leqslant m$, the system is said to be NON-INTERACTING. Examples of interacting and non-interacting two-input/two-output systems are illustrated in Figs 6 and 7 respectively. The source of the interaction is easily identified by noting that Fig. 6 reduces to Fig. 7 in the case of $\alpha = \beta = 0$. Note that non-interacting systems can be regarded as a collection of distinct single-input/single-output systems, each of which can be controlled independently of the others.

The success of the classical ideas of feedback and transfer functions in the analysis and design of single-input/single-output systems is a major motivation for the attempt to generalize them to cover the case of multi-input/multi-output

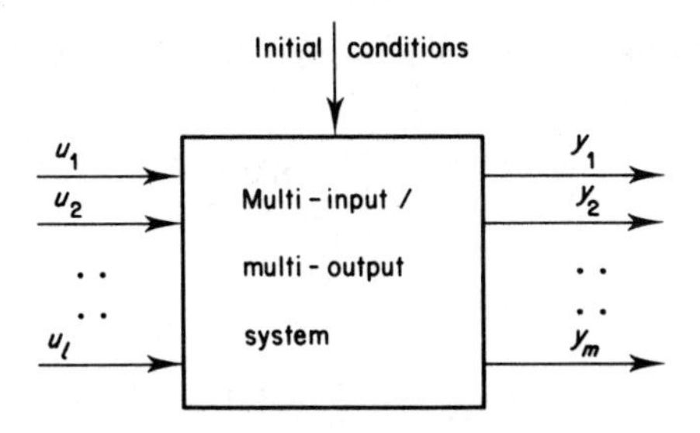

Fig. 5. Deterministic representation of a multi-input/multi-output system.

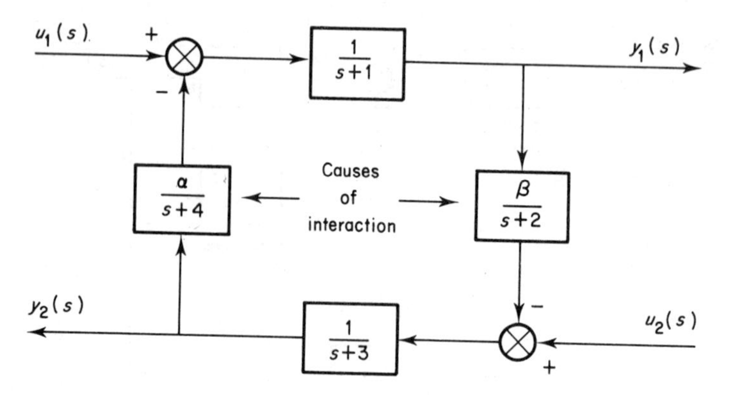

Fig. 6. An interacting multi-input/multi-output system.

systems. Such an attempt must formulate and provide precise answers to questions such as

(i) What form of mathematical model is most suitable for multi-input/multi-output studies?

(ii) What is the generalization of the notion of transfer function?

(iii) What do we mean by the ideas of poles, zeros and frequency response?

(iv) What is feedback in this general case?

(v) What is the generalization of the notion of stability?

(vi) Are there useful generalizations of the ideas of Nyquist diagrams and root-loci?

Of course the answers to these questions will apply to single-input/single-output systems also.

In many cases the required generalizations take their most natural form when expressed in the language of matrix theory. This formal simplicity is simultaneously of great value in removing mathematical clutter and revealing the basic structure of the problem and in the conversion of design relationships into a form suitable for evaluation by a digital computer. The reader should beware,

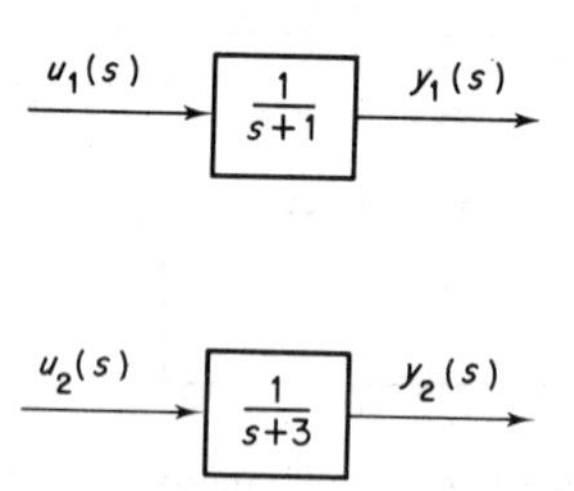

Fig. 7. A non-interacting multi-input/multi-output system.

however, of its tendency to mask the difficulties inherent in multi-input/multi-output control problems. In other cases, such as the development of optimal controllers and a consideration of the integrity of the design, the development has no counterpart in classical theory.

A final word on the use of matrix theory: there is no doubt that it is a powerful tool for analysis and a natural setting for converting design relations into a form suitable for digital computer calculations. It can even be agreed that matrix methods are a major contributor to the rapid rate of advance of control science over the last two decades. The methods do, however, take a little getting used to and, for this reason, the text restricts its attention to design methods that can be formulated using only simple matrix operations, such as addition, multiplication by scalars, inversion and calculation of eigenvalues and eigenvectors. Subroutines performing such operations are commonly available as library software on digital computers.

1.2 State Variable Models of Process Dynamics

Although the scalar nth order differential equations of (1.3) and (1.6) have been used frequently in classical design studies, it is true that modelling of process dynamics directly from known physical laws rarely produces such equations. In contrast, the system model frequently takes the form of a set of ordinary differential equations, each obtained from modelling individual system compo-

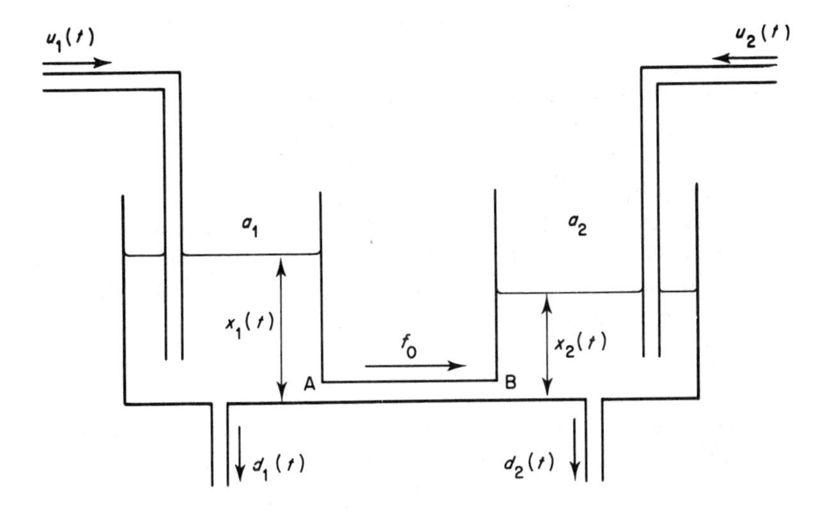

Fig. 8. Two-input liquid level system.

nents and their interconnections. Only in the case of single-input/single-output ($m = l = 1$) systems where all such equations are linear with constant coefficients can these equations be reduced to a single ordinary differential equation. In many cases, this can require considerable computational effort. For these, and other reasons, the mathematical models used in the study of the dynamics and control of multi-input/multi-output systems take a general form that bears little formal similarity to (1.3) and (1.6). The following examples will be used to motivate the general form to be used.

EXAMPLE 1.2.1 (nonlinear model of a two-vessel liquid level system). Consider the elementary dynamic system illustrated in Fig. 8 consisting of two interconnected vessels of uniform cross-section and areas a_1, a_2 respectively. Let $x_i(t)$, $i = 1,2$, denote the height of liquid in vessel i, $u_i(t)$, $i = 1,2$, denote the input flow rate into vessel i (m^3/s), and let $d_i(t)$, $i = 1,2$, denote known disturbance outlet flow rates (m^3/s) from the bottom of the vessels. The intervessel flow (m^3/s) is assumed to be a function only of the pressure drop between A and B and is hence taken as a function $f_0(x_1(t) - x_2(t))$ of the height difference $x_1(t) - x_2(t)$.

Elementary considerations of the mass balance in each vessel yields the following (nonlinear) differential equations describing the dynamics of the process:

$$a_1 \frac{dx_1(t)}{dt} = u_1(t) - f_0(x_1(t) - x_2(t)) - d_1(t)$$

$$a_2 \frac{dx_2(t)}{dt} = u_2(t) + f_0(x_1(t) - x_2(t)) - d_2(t) \tag{1.12}$$

The system initial conditions can be taken as the known values of $x_1(t)$ and $x_2(t)$ at a given time $t = t_0$. Note that each equation cannot be solved independently of the other, i.e. the vessels dynamics are coupled by the intervessel flow.

The model will be complete when the system ouputs are defined. There are a number of possibilities here. For example, if separate measurements of the levels $x_1(t)$ and $x_2(t)$ are available, the outputs could be defined as

$$y_1(t) = x_1(t), \qquad y_2(t) = x_2(t) \tag{1.13}$$

Alternatively, the outputs could be taken to be the total volume of liquids in both vessels and the difference in head,

$$y_1(t) = a_1x_1(t) + a_2x_2(t), \qquad y_2(t) = x_1(t) - x_2(t) \tag{1.14}$$

EXERCISE 1.2.1. In the case of $a_1 = a_2 = a$ with outputs defined by (1.14) show that the model (1.12) reduces to the equations

$$\frac{dy_1(t)}{dt} = (u_1(t) + u_2(t)) - (d_1(t) + d_2(t))$$

$$\frac{dy_2(t)}{dt} = a^{-1}(u_1(t) - u_2(t)) - 2a^{-1}f(y_2(t)) - a^{-1}(d_1(t) - d_2(t)) \tag{1.15}$$

Note that each equation can be solved independently of the other if the initial conditions and inputs are known. (This is not the case for the outputs defined by (1.13).) Note also that, if we used the input variables $\hat{u}_1 = u_1 + u_2$ and $\hat{u}_2 = u_1 - u_2$, equation (1.15) indicates that the resulting system is non-interacting.

EXAMPLE 1.2.2 (model of a double spring-mass system). Consider the mechanical system of Fig. 9 consisting of two masses m_1, m_2 connected by lossless springs with linear displacement/force characteristics. The system has one input $u_1(t)$ equal to the vertical displacement of the supporting platform from an equilibrium position. Applying the normal laws of motion, the following linear equations are obtained

$$m_1 \frac{d^2x_1(t)}{dt^2} = k_1(u_1(t) - x_1(t)) - k_2(x_1(t) - x_2(t))$$

$$m_2 \frac{d^2x_2(t)}{dt^2} = k_2(x_1(t) - x_2(t)) \tag{1.16}$$

i.e. two coupled second order ordinary differential equations. An equivalent set of four first order ordinary differential equations are obtained by defining auxiliary variables $x_3(t) = dx_1(t)/dt$ and $x_4(t) = dx_2(t)/dt$,

$$\frac{dx_1(t)}{dt} = x_3(t)$$

$$\frac{dx_2(t)}{dt} = x_4(t)$$

$$\frac{dx_3(t)}{dt} = \frac{k_1}{m_1}(u_1(t) - x_1(t)) - \frac{k_2}{m_1}(x_1(t) - x_2(t))$$

$$\frac{dx_4(t)}{dt} = \frac{k_2}{m_2}(x_1(t) - x_2(t)) \tag{1.17}$$

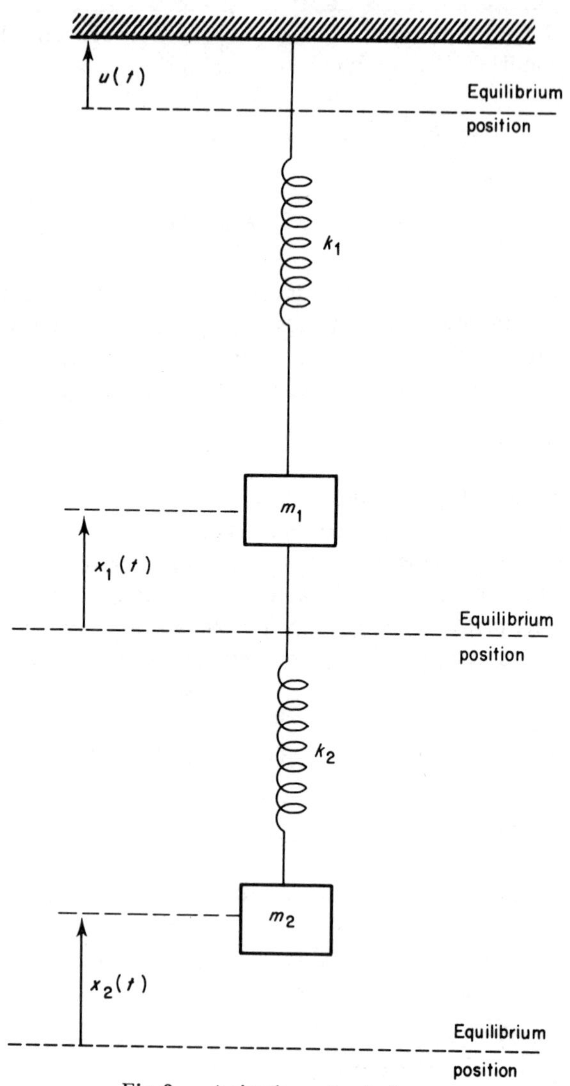

Fig. 9. A simple mechanical system.

The system initial conditions can be taken to be known values of $x_i(t)$, $1 \leqslant i \leqslant 4$ (i.e. known positions and velocities of the masses), at a given time $t = t_0$.

The model is completed when the system outputs are defined. For example, if separate measurements of the displacements of both masses are available, the output equations are

$$y_1(t) = x_1(t), \qquad y_2(t) = x_2(t) \tag{1.18}$$

The overall structure of the general process model is illustrated by the above examples, namely a set of n first order ordinary differential equations of the form

$$\frac{\mathrm{d}x_i(t)}{\mathrm{d}t} = f_i(x_1(t), \ldots, x_n(t), u_1(t), \ldots, u_l(t), t), \qquad 1 \leqslant i \leqslant n \tag{1.19}$$

(where the f_i are defined functions of their arguments) with the initial conditions

$$x_i(t_0) = x_{i0}, \qquad 1 \leqslant i \leqslant n \tag{1.20}$$

at a time $t = t_0$ and a *set of output equations*

$$y_i(t) = g_i(x_1(t), \ldots, x_n(t), u_1(t), \ldots, u_l(t), t), \qquad 1 \leqslant i \leqslant m \tag{1.21}$$

defining the measured *output variables* $y_i(t)$, $1 \leqslant i \leqslant m$, in terms of the n system *state variables* $x_i(t)$, $1 \leqslant i \leqslant n$, and the system *input variables* $u_i(t)$, $1 \leqslant i \leqslant l$. In most applications there are many more state variables than inputs and outputs in the sense that $n \gg \max(m, l)$. It is also true that, in most applications, the functions f_i, $1 \leqslant i \leqslant n$, and g_i, $1 \leqslant i \leqslant m$, do not depend explicitly on time t. In such cases the last argument in equations (1.19) and (1.21) is excluded. Matrix versions of (1.19), (1.20) and (1.21) are obtained by defining the column matrices (commonly called vectors)

$$x(t) = \begin{bmatrix} x_1(t) \\ x_2(t) \\ \vdots \\ x_n(t) \end{bmatrix}, \quad y(t) = \begin{bmatrix} y_1(t) \\ y_2(t) \\ \vdots \\ y_m(t) \end{bmatrix}, \quad u(t) = \begin{bmatrix} u_1(t) \\ u_2(t) \\ \vdots \\ u_l(t) \end{bmatrix}, \quad x_0 = \begin{bmatrix} x_{10} \\ x_{20} \\ \vdots \\ x_{n0} \end{bmatrix} \tag{1.22}$$

and the vector functions

$$f(x(t), u(t), t) = \begin{bmatrix} f_1(x_1(t), x_2(t), \ldots, x_n(t), u_1(t), \ldots, u_l(t), t) \\ f_2(x_1(t), \quad \ldots \ldots \ldots \quad , t) \\ \vdots \\ f_n(x_1(t), \quad \ldots \ldots \quad , x_n(t), u_1(t), \ldots, u_l(t), t) \end{bmatrix} \tag{1.23}$$

$$g(x(t), u(t), t) = \begin{bmatrix} g_1(x_1(t), \quad \ldots \quad , u_l(t), t) \\ \vdots \\ g_m(x_1(t), \quad \ldots \quad , u_l(t), t) \end{bmatrix} \tag{1.24}$$

Defining the derivative of a matrix with respect to time as the matrix of derivatives, then

$$\frac{dx(t)}{dt} = \begin{bmatrix} \frac{dx_1(t)}{dt} \\ \vdots \\ \frac{dx_n(t)}{dt} \end{bmatrix} \tag{1.25}$$

and the general process equations (1.19)–(1.21) take the matrix form

$$\frac{dx(t)}{dt} = f(x(t), u(t), t) \tag{1.26}$$

$$x(t_0) = x_0 \tag{1.27}$$

$$y(t) = g(x(t), u(t), t) \tag{1.28}$$

The time-dependent vectors $x(t)$, $u(t)$ and $y(t)$ are termed the $n \times 1$ *state vector*, the $l \times 1$ *input vector* and the $m \times 1$ *output vector* respectively. The integer numbers n, l, m are termed the *state, input* and *output dimensions* respectively. The three equations (1.26)–(1.28) are said to define a *state-variable model* of the system input-output relationships. Alternative descriptions are "state-vector model" or "state-space model".

The use of matrix notation in the formulation of the model most certainly provides a formal simplicity "on paper" that removes much of the confusion generated by the multiple arguments and indices required for (1.19)–(1.21). The system can, in fact, be regarded as possessing a single vector input producing a single vector output and represented by the *vector block diagram* of Fig. 10. Note the formal similarity to Fig. 1.

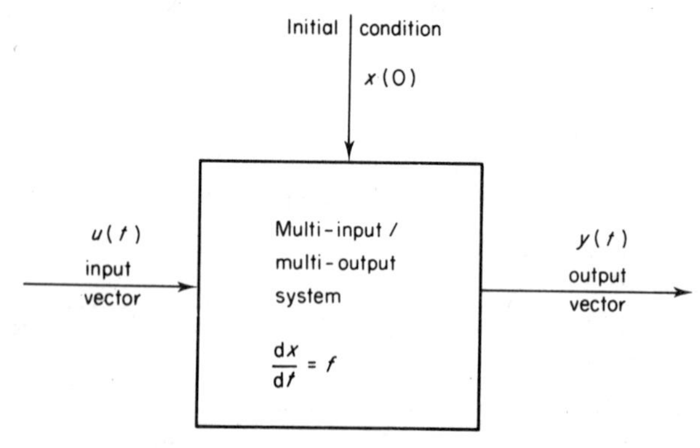

Fig. 10. State variable representation of a multi-input/multi-output system.

EXAMPLE 1.2.3. Consider the construction of a state-variable model of a single-input/single-output system described by the following special form of (1.1) and (1.2)

$$\frac{d^n y(t)}{dt^n} = Q\left(y(t), \frac{dy(t)}{dt}, \ldots, \frac{d^{n-1}y(t)}{dt^{n-1}}, u(t), t\right)$$

$$y(0) = d_0, \left.\frac{dy(t)}{dt}\right|_{t=0} = d_1, \ldots, \left.\frac{d^{n-1}y(t)}{dt^{n-1}},\right|_{t=0} = d_{n-1} \quad (1.29)$$

This can be achieved by setting $x_1(t) = y(t)$, $x_2(t) = dy(t)/dt, \ldots$, $x_n(t) = d^{n-1}y(t)/dt^{n-1}$ when the system model takes the form

$$\frac{dx_1(t)}{dt} = x_2(t), \frac{dx_2(t)}{dt} = x_3(t), \ldots, \frac{dx_{n-1}(t)}{dt} = x_n(t)$$

$$\frac{dx_n(t)}{dt} = Q(x_1(t), x_2(t), \ldots, x_n(t), u(t), t)$$

$$y(t) = x_1(t) \quad (1.30)$$

with initial conditions $x_i(0) = d_{i-1}$, $1 \leqslant i \leqslant n$. In vector form, this is simply

$$\frac{dx(t)}{dt} = \begin{bmatrix} x_2(t) \\ x_3(t) \\ \vdots \\ x_{n-1}(t) \\ Q(x_1(t), \ldots, x_n(t), u(t), t \end{bmatrix}, \quad x(0) = \begin{bmatrix} d_0 \\ d_1 \\ \vdots \\ d_{n-1} \end{bmatrix} \quad (1.31)$$

EXERCISE 1.2.2. Use the procedure of Example 1.2.3 to convert the single-input/single-output system $(D^2 + 2D + 1)y(t) = u(t)$, $y(0) = 1$, $Dy(0) = 0$, into the state-variable model

$$\frac{dx(t)}{dt} = \begin{bmatrix} x_2(t) \\ \\ -x_1(t) - 2x_2(t) + u(t) \end{bmatrix}, \quad x(0) = \begin{bmatrix} 1 \\ \\ 0 \end{bmatrix}$$

$$y(t) = x_1(t) \quad (1.32)$$

A case of particular importance in this book (and in practice) is the situation when the f_i, $1 \leqslant i \leqslant n$, and g_i, $1 \leqslant i \leqslant m$, have the following linear form

$$f_i(x_1, \ldots, x_n, u_1, \ldots, u_l, t) = A_{i1}(t)x_1 + A_{i2}(t)x_2 + \ldots + A_{in}(t)x_n + B_{i1}(t)u_1 + \ldots + B_{il}(t)u_l \qquad 1 \leqslant i \leqslant n \tag{1.33}$$

$$g_i(x_1, \ldots, u_l, t) = C_{i1}(t)x_1 + \ldots + C_{in}(t)x_n + D_{i1}(t)u_1 + \ldots + D_{il}(t)u_l, \qquad 1 \leqslant i \leqslant m \tag{1.34}$$

Defining the following *time-varying* matrices

$$A(t) = \begin{bmatrix} A_{11}(t) & \ldots & A_{1n}(t) \\ \vdots & & \vdots \\ A_{n1}(t) & \ldots & A_{nn}(t) \end{bmatrix}_{n \times n}, \qquad B(t) = \begin{bmatrix} B_{11}(t) & \ldots & B_{1l}(t) \\ \vdots & & \vdots \\ B_{n1}(t) & \ldots & B_{nl}(t) \end{bmatrix}_{n \times l}$$

$$C(t) = \begin{bmatrix} C_{11}(t) & \ldots & C_{1n}(t) \\ \vdots & & \vdots \\ C_{m1}(t) & \ldots & C_{mn}(t) \end{bmatrix}_{m \times n}, \qquad D(t) = \begin{bmatrix} D_{11}(t) & \ldots & D_{1l}(t) \\ \vdots & & \vdots \\ D_{m1}(t) & \ldots & D_{ml}(t) \end{bmatrix}_{m \times} \tag{1.35}$$

then a simple application of the rules of matrix multiplication indicates that

$$f(x, u, t) = A(t)x + B(t)u, \qquad g(x, u, t) = C(t)x + D(t)u \tag{1.36}$$

and hence that the system model has the form

$$\frac{dx(t)}{dt} = A(t)x(t) + B(t)u(t), \qquad x(t_0) = x_0$$

$$y(t) = C(t)x(t) + D(t)u(t) \tag{1.37}$$

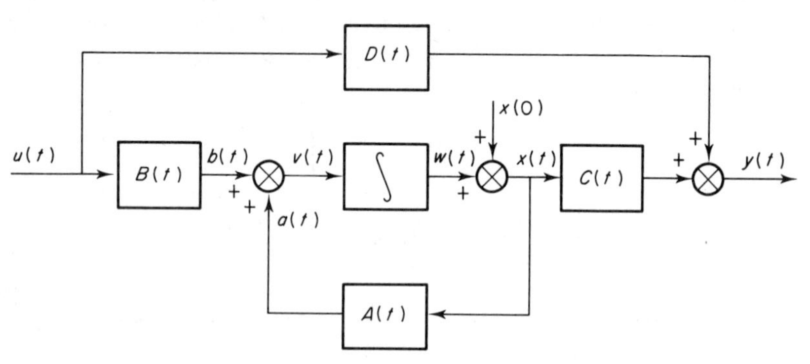

Fig. 11. Block representation of a linear, time-varying state-variable model.

These equations are said to define a linear state-variable (vector/space) model of system dynamics. If any of the matrices $A(t), B(t), C(t), D(t)$ have elements that vary with time, the system is said to be time varying. If all matrices A, B, C, D are constant, the system is said to be time-invariant. This second case is the one most frequently encountered in practice and the one of primary concern in this text.

EXERCISE 1.2.3. Show that the mechanical system of Example 1.2.2 has a linear time-invariant state variable model defined by the matrices

$$A = \begin{bmatrix} 0 & 0 & 1 & 0 \\ 0 & 0 & 0 & 1 \\ \dfrac{-(k_1+k_2)}{m_1} & \dfrac{k_2}{m_1} & 0 & 0 \\ \dfrac{k_2}{m_2} & \dfrac{-k_2}{m_2} & 0 & 0 \end{bmatrix}, \quad B = \begin{bmatrix} 0 \\ 0 \\ \dfrac{k_1}{m_1} \\ 0 \end{bmatrix}$$

$$C = \begin{bmatrix} 1 & 0 & 0 & 0 \\ 0 & 1 & 0 & 0 \end{bmatrix}, \quad D = \begin{bmatrix} 0 \\ 0 \end{bmatrix} \tag{1.38}$$

EXERCISE 1.2.4. Show that the system of exercise 1.2.2 has a linear time-invariant state vector model

$$\frac{dx(t)}{dt} = \begin{bmatrix} 0 & 1 \\ -1 & -2 \end{bmatrix} x(t) + \begin{bmatrix} 0 \\ 1 \end{bmatrix} u(t), \qquad x(0) = \begin{bmatrix} 1 \\ 0 \end{bmatrix}$$

$$y(t) = [1 \quad 0]\, x(t) \tag{1.39}$$

Hence identify A, B, C, D by comparison with (1.37).

A useful intuitive representation of the system structure in the linear case is obtained by integrating to yield

$$x(t) = x_0 + \int_0^t \{A(t')x(t') + B(t')u(t')\}\,dt' \tag{1.40}$$

(Note: the integral of a matrix function of time is simply the matrix of the integral of the elements). The system can hence be represented by the vector block diagram shown in Fig. 11 and suggests that

(a) B represents the way that the states are affected by the inputs.
(b) C represents the way that the states affect the outputs.
(c) A has a feedback position and hence (intuitively) will dominate the stability of the system.
(d) D represents a feedforward element in the system structure.

The representation can be misleading, however, unless the reader remembers that each summing junction is a vector summing junction representing the matrix addition operation $v(t) = b(t) + a(t)$ (for example). In element form this is equivalent to the n additions $v_i(t) = b_i(t) + a_i(t)$, $1 \leqslant i \leqslant n$. The vector summing junction hence requires n scalar summers for its realization. In a similar manner, the vector integrator represents the matrix integration $w(t) = \int_0^t v(t')\,dt'$ and is equivalent to the use of n scalar integrators $w_i(t) = \int_0^t v_i(t')\,dt'$, $1 \leqslant i \leqslant n$.

1.3 Equilibrium States and Linearization

In the case of a single-input/single-output system, the linear differential equation (1.3) is much easier to handle than the general nonlinear differential equation (1.1). A similar situation holds for multi-input/multi-output systems, namely that the linear state-variable model of equation (1.37) is much easier to handle than the general nonlinear model (1.26)–(1.28). It is natural therefore that most applicable control theory is based on linear models that approximate to a more accurate nonlinear model of system dynamics. The questions arise, of course, as to how one devises such models and in what situations do the two models have similar properties? Fortunately, these questions have fairly simple solutions in a case of great practical relevance.

1.3.1 System steady states

In many practical applications the ideal operating condition for the system is one where all system inputs, states and outputs are constant in time. Typical examples of such applications are base-load operation of power plant and temperature and humidity control in environmental systems. In such applications, the role of the control system is primarily that of a regulator to return the system variables to their ideal values following a disturbance. Suppose that the system is described by the nonlinear model

$$\frac{dx(t)}{dt} = f(x(t), u(t)), \qquad x(0) = x_0 \tag{1.41}$$

$$y(t) = g(x(t), u(t)) \tag{1.42}$$

where the vector functions f and g do not depend explicitly on time t. The mathematical representation of a constant ideal operating point is via the use of the notion of equilibrium points or steady-states.

Definition 1.3.1. *The* $n \times 1$ *vector* x_s *is an* equilibrium point, equilibrium state *or* steady state *corresponding to the* constant *input* $u(t) \equiv u_s$ *if, and only if,*

$$f(x_s, u_s) = 0 \tag{1.43}$$

The corresponding equilibrium (*or steady-state*) output *is the constant* $m \times 1$ *vector* y_s,

$$y_s = g(x_s, u_s) \tag{1.44}$$

(Note: equation (1.43) is equivalent to n algebraic equations obtained by setting each element of f equal to zero.)

A number of useful general properties of equilibrium states are illustrated in the following exercise and examples.

EXERCISE 1.3.1. Show that the solution of the state equations (1.41) with input $u(t) \equiv u_s$ and $x(0) = x_s$ is $x(t) \equiv x_s$. That is, x_s represents a constant solution of the state equations corresponding to u_s. In fact, u_s, x_s, y_s define a system condition where all system inputs, states and outputs are constant in time.

EXAMPLE 1.3.1. To compute the steady states of the single-input/single-output system

$$\frac{dx_1(t)}{dt} = x_1(t) + x_2(t) - 2u(t)$$

$$\frac{dx_2(t)}{dt} = x_1(t) - x_2(t) \tag{1.45}$$

$$y(t) = x_1(t) + \tfrac{1}{2}(x_2(t))^2 \tag{1.46}$$

corresponding to the constant input $u(t) \equiv u_s$, simply set the right-hand side of equation (1.45) to zero, i.e.

$$f_1 = x_1 + x_2 - 2u_s = 0, \qquad f_2 = x_1 - x_2 = 0 \tag{1.47}$$

The system hence has a unique equilibrium point $x_s = \begin{bmatrix} 1 \\ 1 \end{bmatrix} u_s$ corresponding to each u_s. The corresponding steady state output is $y_s = u_s(1 + \tfrac{1}{2}u_s)$.

EXERCISE 1.3.2. Show that the single-input/single-output system

$$\frac{dx_1(t)}{dt} = x_1(t) + x_2(t) - 2u(t)$$

$$\frac{dx_2(t)}{dt} = x_2(t)(x_1(t) - x_2(t)) \tag{1.48}$$

has two distinct equilibrium states corresponding to the constant input $u(t) \equiv u_s$, i.e.

$$x_s = \begin{bmatrix} 1 \\ 1 \end{bmatrix} u_s, \qquad x_s = \begin{bmatrix} 2 \\ 0 \end{bmatrix} u_s \tag{1.49}$$

(Note: Example 1.3.1 and Exercise 1.3.2 indicate that a dynamic system may have one or more equilibrium points. The following exercises provide illustrations of systems with an infinity of equilibrium points and systems with no equilibrium points at all.)

EXERCISE 1.3.3. Show that the nonlinear system

$$\frac{dx_1(t)}{dt} = x_1(t) + u(t)x_2(t), \qquad \frac{dx_2(t)}{dt} = x_1(t) + x_2(t) \tag{1.50}$$

has the unique equilibrium point $x_s = \begin{bmatrix} 0 \\ 0 \end{bmatrix}$ if $u_s \neq 1$. If, however, $u_s = 1$ the system has an infinite number of steady states $x_s = \lambda \begin{bmatrix} 1 \\ -1 \end{bmatrix}$ where λ is arbitrary.

EXAMPLE 1.3.2. Consider the nonlinear system

$$\frac{dx_1(t)}{dt} = x_1(t) - x_2(t),$$

$$\frac{dx_2(t)}{dt} = 1 + (x_1(t) - x_2(t))\,u(t) + (x_1(t))^2 \tag{1.51}$$

The system equilibrium points are obtained by solving the algebraic equations

$$x_1 - x_2 = 0, \qquad 1 + (x_1 - x_2)u_s + x_1^2 = 0 \tag{1.52}$$

The only solutions to these equations are $x_1 = x_2 = \pm j$ (j = "the square root of minus one"). In almost all physical applications the state-variables must be real. In this sense the above system has no equilibrium points at all.

EXAMPLE 1.3.3. Consider the linear, time-invariant equation

$$\frac{dx(t)}{dt} = Ax(t) + Bu(t) \qquad (= f(x(t), u(t))) \tag{1.53}$$

The evaluation of steady states corresponds to the solution of linear equations of the form

$$Ax_s = -Bu_s \tag{1.54}$$

There are two cases in particular,

(a) If A is non-singular then the unique solution of (1.54) is $x_s = -A^{-1}Bu_s$

(b) If A is singular, (1.54) either has no solution or an infinity of solutions. An example where no solution exists is

$$\begin{bmatrix} 1 & -1 \\ 1 & -1 \end{bmatrix} x_s = \begin{bmatrix} 1 & 0 \\ -1 & 1 \end{bmatrix} u_s, \qquad u_s = \begin{bmatrix} 1 \\ 0 \end{bmatrix} \tag{1.55}$$

If, however, a solution x_s exists, then the singularity of A indicates that it has a zero eigenvalue and hence a non-zero eigenvector w satisfying $Aw = 0$. It is left as an exercise for the reader to show that $x_s + \lambda w$ is then a solution of (1.54) independent of the value of λ and hence to infer that the system has an infinite number of equilibrium points.

1.3.2 *Linearization procedures*

The process of linearization of the general nonlinear model of equations (1.41)–(1.42) in the *vicinity* of a given steady state condition specified by the triple x_s, u_s, y_s consists of *approximation* of the vector functions f, g by linear functions

$$f(x, u) \simeq f(x_s, u_s) + A(x - x_s) + B(u - u_s) \tag{1.56}$$

$$g(x, u) \simeq g(x_s, u_s) + C(x - x_s) + D(u - u_s) \tag{1.57}$$

where A, B, C, D are suitable constant $n \times n$, $n \times l$, $m \times n$ and $m \times l$ matrices respectively. Defining the perturbation variables $\tilde{x}_i(t) = x_i(t) - (x_s)_i$, $1 \leqslant i \leqslant n$, $\tilde{y}_i(t) = y_i(t) - (y_s)_i$, $1 \leqslant i \leqslant m$, and $\tilde{u}_i(t) = u_i(t) - (u_s)_i$, $1 \leqslant i \leqslant l$, in the vector forms

$$\tilde{x}(t) = x(t) - x_s, \tilde{y}(t) = y(t) - y_s, \tilde{u}(t) = u(t) - u_s \tag{1.58}$$

then, using equations (1.43), (1.44), (1.56) and (1.57), we obtain

$$\frac{d\tilde{x}(t)}{dt} = \frac{d}{dt}(x(t)-x_s) = \frac{dx(t)}{dt} \tag{1.59}$$

$$y(t) \simeq y_s + C\tilde{x}(t) + D\tilde{u}(t) \tag{1.60}$$

$$f(x(t), u(t)) \simeq A\tilde{x}(t) + B\tilde{u}(t) \tag{1.61}$$

and hence the relations

$$\frac{d\tilde{x}(t)}{dt} \simeq A\tilde{x}(t) + B\tilde{u}(t)$$

$$\tilde{y}(t) \simeq C\tilde{x}(t) + D\tilde{u}(t) \tag{1.62}$$

These expressions suggest that the dynamic behaviour of the perturbations $\tilde{x}$ and $\tilde{y}$ in response to the perturbation input $\tilde{u}$ can be approximated by solutions of the linear time-invariant state-variable model obtained from (1.62) by replacing approximation signs by equalities. The question of the choice of A, B, C, D and the sense in which the approximations are valid are, of course, very important. Some indications of the form of the answers are given below.

(a) *Graphical Linearization*

Linearization can proceed by application of graphical or analytical methods. The graphical method can be illustrated by consideration of the liquid-level system described in Example 1.2.1., equations (1.12) and (1.15). Suppose that the disturbance flows $d_1(t) = d_1$ and $d_2(t) = d_2$ are constant and consider the steady state levels $x_1(t) \equiv (x_s)_1$ and $x_2(t) \equiv (x_s)_2$ corresponding to the constant inputs $u_1(t) \equiv (u_s)_1$ and $u_2(t) \equiv (u_s)_2$. These are defined by the relations

$$0 = (u_s)_1 - d_1 - f_0((x_s)_1 - (x_s)_2)$$

$$0 = (u_s)_2 - d_2 + f_0((x_s)_1 - (x_s)_2) \tag{1.63}$$

Adding these equations yields, after some manipulation,

$$(u_s)_1 + (u_s)_2 = d_1 + d_2 \tag{1.64}$$

which simply states that, in a steady state condition, the total volume of liquid in the two vessels is constant in time. It certainly provides no information on the steady state levels $(x_s)_i$, $i = 1, 2$. This is obtained by considering either of the equations in (1.63). In fact, taking, for example, the equation

$$f_0((x_s)_1 - (x_s)_2) = (u_s)_1 - d_1 \tag{1.65}$$

we see that it is only possible to solve for the *difference* in head $z_s = (x_s)_1 - (x_s)_2$ and hence that there are an infinity of possible steady states (corresponding to different total liquid volumes but constant difference in head z_s).

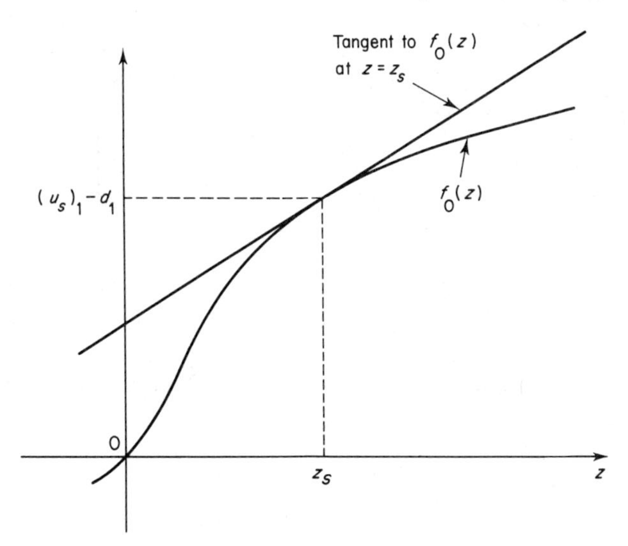

Fig. 12. Graphical linearization of the liquid level system.

Given a known steady state condition, consider the construction of the linear approximation

$$f_0(z) \simeq f_0(z_s) + \beta(z - z_s) \tag{1.66}$$

to $f_0(z)$ at $z = z_s$ by construction of the tangent to the flow characteristic as shown in Fig. 12. Such an approximation is quite obviously good provided $|z_s - z_s|$ is *small*, i.e. the difference in head does not differ too much from its equilibrium value. The approximation matrices A, B can be deduced by writing

$$f(x,u) = \begin{bmatrix} a_1^{-1}(u_1 - f_0(x_1 - x_2) - d_1) \\ a_2^{-1}(u_2 + f_0(x_1 - x_2) - d_2) \end{bmatrix}$$

$$\simeq \begin{bmatrix} a_1^{-1}(u_1 - f_0((x_s)_1 - (x_s)_2) - \beta((x_1 - (x_s)_1) - (x_2 - (x_s)_2)) - d_1) \\ a_2^{-1}(u_2 + f_0((x_s)_1 - (x_s)_2) + \beta((x_1 - (x_s)_1) - (x_2 - (x_s)_2)) - d_2) \end{bmatrix}$$

$$= \begin{bmatrix} a_1^{-1}((u_s)_1 - f_0((x_s)_1 - (x_s)_2) - d_1) \\ a_2^{-1}((u_s)_2 + f_0((x_s)_1 - (x_s)_2) - d_2) \end{bmatrix} +$$

$$+ \begin{bmatrix} -\dfrac{\beta}{a_1} & \dfrac{\beta}{a_1} \\ \dfrac{\beta}{a_2} & -\dfrac{\beta}{a_2} \end{bmatrix}(x - x_s) + \begin{bmatrix} \dfrac{1}{a_1} & 0 \\ 0 & \dfrac{1}{a_2} \end{bmatrix}(u - u_s) \tag{1.67}$$

indicating that

$$A = \begin{bmatrix} -\dfrac{\beta}{a_1} & \dfrac{\beta}{a_1} \\ \dfrac{\beta}{a_2} & -\dfrac{\beta}{a_2} \end{bmatrix}, \qquad B = \begin{bmatrix} \dfrac{1}{a_1} & 0 \\ 0 & \dfrac{1}{a_2} \end{bmatrix} \tag{1.68}$$

Also, by writing

$$g(x, u) = \begin{bmatrix} x_1 \\ x_2 \end{bmatrix} = x = x_s + (x - x_s) \tag{1.69}$$

it is seen that an exact linear "approximation" to the output equation is possible of the form of (1.57) with

$$C = \begin{bmatrix} 1 & 0 \\ 0 & 1 \end{bmatrix}, \qquad D = \begin{bmatrix} 0 & 0 \\ 0 & 0 \end{bmatrix} \tag{1.70}$$

The resulting linearized model describing the approximate dynamics of perturbation variables takes the form

$$\frac{\mathrm{d}\tilde{x}(t)}{\mathrm{d}t} = \begin{bmatrix} -\dfrac{\beta}{a_1} & \dfrac{\beta}{a_1} \\ \dfrac{\beta}{a_2} & -\dfrac{\beta}{a_2} \end{bmatrix} \tilde{x}(t) + \begin{bmatrix} \dfrac{1}{a_1} & 0 \\ 0 & \dfrac{1}{a_2} \end{bmatrix} \tilde{u}(t)$$

$$\tilde{y}(t) = \begin{bmatrix} 1 & 0 \\ 0 & 1 \end{bmatrix} \tilde{x}(t) + \begin{bmatrix} 0 & 0 \\ 0 & 0 \end{bmatrix} \tilde{u}(t) \tag{1.71}$$

and is valid provided the change in head $\tilde{x}_1(t) - \tilde{x}_2(t)$ is small.

(b) *Analytical Linearization*

Direct graphical linearization in the manner described above is not always possible. If, however, the analytical form of the nonlinearity is known, application of Taylor series methods yields a formally simple solution by neglecting second and higher order terms. More precisely, the Taylor series expansion of f_i at the equilibrium point

$$f_i((x_s)_1 + \tilde{x}_1, \ldots, (x_s)_n + \tilde{x}_n, (u_s)_1 + \tilde{u}_1, \ldots, (u_s)_l + \tilde{u}_l)$$

$$= f_i((x_s)_1, \ldots, (x_s)_n, \ldots, (u_s)_l) + \sum_{k=1}^{n} \left.\frac{\partial f_i}{\partial x_k}\right|_s \tilde{x}_k + \sum_{k=1}^{l} \left.\frac{\partial f_i}{\partial u_k}\right|_s \tilde{u}_k$$

$$+ \text{higher order terms}, \qquad 1 \leqslant i \leqslant n \tag{1.72}$$

(where the subscript s means that the derivatives are evaluated at the equilibrium points x_s, u_s) can be written in the form of (1.56) by neglecting the higher order terms. In fact, we obtain

$$A = \begin{bmatrix} \left.\frac{\partial f_1}{\partial x_1}\right|_s & \cdots & \left.\frac{\partial f_1}{\partial x_n}\right|_s \\ \vdots & & \vdots \\ \left.\frac{\partial f_n}{\partial x_1}\right|_s & \cdots & \left.\frac{\partial f_n}{\partial x_n}\right|_s \end{bmatrix}, \qquad B = \begin{bmatrix} \left.\frac{\partial f_1}{\partial u_1}\right|_s & \cdots & \left.\frac{\partial f_1}{\partial u_l}\right|_s \\ \vdots & & \vdots \\ \left.\frac{\partial f_n}{\partial u_1}\right|_s & \cdots & \left.\frac{\partial f_n}{\partial u_l}\right|_s \end{bmatrix} \tag{1.73}$$

as the (so-called) "Jacobian matrices". Similar considerations for the output equation yields (1.57) with

$$C = \begin{bmatrix} \left.\frac{\partial g_1}{\partial x_1}\right|_s & \cdots & \left.\frac{\partial g_1}{\partial x_n}\right|_s \\ \vdots & & \vdots \\ \left.\frac{\partial g_m}{\partial x_1}\right|_s & \cdots & \left.\frac{\partial g_m}{\partial x_n}\right|_s \end{bmatrix}, \qquad D = \begin{bmatrix} \left.\frac{\partial g_1}{\partial u_1}\right|_s & \cdots & \left.\frac{\partial g_1}{\partial u_l}\right|_s \\ \vdots & & \vdots \\ \left.\frac{\partial g_m}{\partial u_1}\right|_s & \cdots & \left.\frac{\partial g_m}{\partial u_l}\right|_s \end{bmatrix} \tag{1.74}$$

The resulting linearized model is valid whenever the neglected high order terms are "small enough", i.e. when the system state and input variables do not deviate too much from their equilibrium values.

Example 1.3.4. Consider the liquid level system of equations (1.12) and (1.13) with constant disturbance flows. We have

$$\begin{aligned} f_1(x_1, x_2, u_1, u_2) &= a_1^{-1}(u_1 - d_1 - f_0(x_1 - x_2)) \\ f_2(x_1, x_2, u_1, u_2) &= a_2^{-1}(u_2 - d_2 + f_0(x_1 - x_2)) \\ g_i(x_1, x_2, u_1, u_2) &= x_i, \qquad i = 1, 2 \end{aligned} \tag{1.75}$$

and hence

$$\frac{\partial f_1}{\partial x_1} = -\frac{\partial f_1}{\partial x_2} = -a_1^{-1} \left.\frac{\mathrm{d}f_0(z)}{\mathrm{d}z}\right|_{x_1 - x_2}$$

$$\frac{\partial f_2}{\partial x_1} = -\frac{\partial f_2}{\partial x_2} = a_2^{-1} \left.\frac{\mathrm{d}f_0(z)}{\mathrm{d}z}\right|_{x_1 - x_2}$$

$$\frac{\partial f_1}{\partial u_1} = a_1^{-1}, \qquad \frac{\partial f_1}{\partial u_2} = 0 = \frac{\partial f_2}{\partial u_1}, \qquad \frac{\partial f_2}{\partial u_2} = a_2^{-1}$$

$$\frac{\partial g_1}{\partial x_1} = \frac{\partial g_2}{\partial x_2} = 1, \qquad \frac{\partial g_1}{\partial x_2} = \frac{\partial g_2}{\partial x_1} = 0$$

$$\frac{\partial g_1}{\partial u_1} = \frac{\partial g_1}{\partial u_2} = \frac{\partial g_2}{\partial u_1} = \frac{\partial g_2}{\partial u_2} = 0 \tag{1.76}$$

Evaluating these at their steady state values and substituting into (1.73) and (1.74) yields the results (1.69) and (1.70) obtained by graphical linearization with

$$\beta = \left. \frac{\mathrm{d}f_0(z)}{\mathrm{d}z} \right|_{(x_s)_1 - (x_s)_2} \tag{1.77}$$

EXAMPLE 1.3.5. It is easily shown by substitution that the two-input/two-output system described by the nonlinear differential equations

$$\frac{\mathrm{d}x_1}{\mathrm{d}t} = -(x_1 - 1)^2 + x_2 + x_3 - 2u_1$$

$$\frac{\mathrm{d}x_2}{\mathrm{d}t} = x_1^2 - (x_2 - 1)^2 + x_1 x_2 - u_1^2 - u_2$$

$$\frac{\mathrm{d}x_3}{\mathrm{d}t} = x_1 + x_2 - x_3 - u_2$$

$$y_1 = x_1(1 + x_2) + u_1$$

$$y_2 = x_2 + x_3 - u_2 \tag{1.78}$$

has an equilibrium point at $x_1 = x_2 = x_3 = 1$ corresponding to the constant inputs $u_1 = u_2 = 1$. The corresponding A matrix of the linearization at this equilibrium point is obtained by evaluating the following derivatives at the equilibrium point

$$\frac{\partial f_1}{\partial x_1} = -2(x_1 - 1), \qquad \frac{\partial f_1}{\partial x_2} = \frac{\partial f_1}{\partial x_3} = 1$$

$$\frac{\partial f_2}{\partial x_1} = 2x_1 + x_2, \qquad \frac{\partial f_2}{\partial x_2} = -2(x_2 - 1) + x_1, \frac{\partial f_2}{\partial x_3} = 0$$

$$\frac{\partial f_3}{\partial x_1} = \frac{\partial f_3}{\partial x_2} = 1, \qquad \frac{\partial f_3}{\partial x_3} = -1 \tag{1.79}$$

i.e.

$$A = \begin{bmatrix} 0 & 1 & 1 \\ 3 & 1 & 0 \\ 1 & 1 & -1 \end{bmatrix} \tag{1.80}$$

The corresponding B, C and D matrices are obtained in a similar manner to be

$$B = \begin{bmatrix} -2 & 0 \\ -2 & -1 \\ 0 & -1 \end{bmatrix}, \qquad C = \begin{bmatrix} 2 & 1 & 0 \\ 0 & 1 & 1 \end{bmatrix}$$

$$D = \begin{bmatrix} 1 & 0 \\ 0 & -1 \end{bmatrix} \tag{1.81}$$

1.4 Numerical Solution of the State Equations

A primary use of the mathematical model of a dynamical system is in the prediction of system transient performance in situations of interest. For example, it may be used to investigate transient behaviour under fault conditions or to evaluate the nature of the system response to setpoint or input changes. In many cases, these problems can be replaced by the mathematical problem of finding a solution of the state-variable equations with a given input and a given initial state in the time interval $t_0 \leqslant t \leqslant t_f$. In general, analytical solution is impossible and the engineer must resort to the generation of approximate solutions by analog or digital computer. It is not the purpose of this text to dwell on the use of computers in this context or to go into the relative merits and mathematical intricacies of various numerical algorithms. It is, however, instructive to illustrate the simplifying effect of our matrix notation by description of a simple digital simulation method (the so-called Euler method).

Suppose that it is desirable to obtain the solution of the vector model $\mathrm{d}x(t)/\mathrm{d}t = f(x(t), u(t), t)$ in the time interval $t_0 \leqslant t \leqslant t_f$, given the input vector $u(t)$ and the initial condition $x(t_0) = x_0$. Divide the interval $t_0 \leqslant t \leqslant t_f$ into N equal intervals of width equal to the so-called *step length*

$$h = \frac{t_f - t_0}{N} \tag{1.82}$$

and set $t_k = kh + t_0$, $0 \leqslant k \leqslant N$, noting that $t_f = t_N$. The problem of solving the vector differential equations is now replaced by obtaining sufficiently accurate approximations to the state vector $x(t)$ at the times $t_1, t_2, t_3, \ldots, t_N$ as illustrated schematically in Fig. 13. The required accuracy of the approximation will vary from application to application.

We now approximate the derivatives $\mathrm{d}x_k/\mathrm{d}t$ at time t_i by the simple formula $(x_k(t_{i+1}) - x_k(t_i))/h$, $1 \leqslant k \leqslant n$, $0 \leqslant i \leqslant N-1$ and hence write

$$\frac{x(t_{i+1}) - x(t_i)}{h} \simeq \left.\frac{\mathrm{d}x(t)}{\mathrm{d}t}\right|_{t=t_i} = f(x(t_i), u(t_i), t_i) \tag{1.83}$$

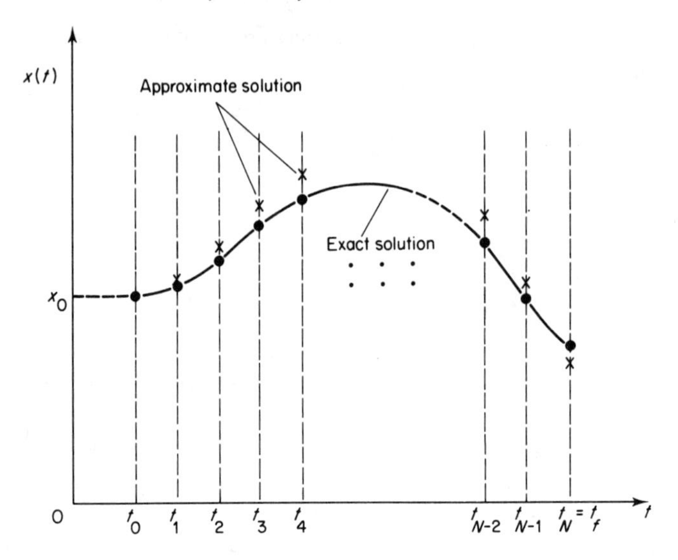

Fig. 13. Approximate solution of the state equations.

This approximation can be made arbitrarily good by making h small enough (or, equivalently, making N large enough). Rearranging (1.83) it is easily verified that approximations to $x(t_i)$, $i \geqslant 1$, can be obtained by solving the recursion relations

$$x(t_{i+1}) = x(t_i) + hf(x(t_i), u(t_i), t_i), \qquad 0 \leqslant i \leqslant N-1 \tag{1.84}$$

Given $x(t_0) = x_0$, these equations are easily solved sequentially

$$\begin{aligned} x(t_1) &= x(t_0) + hf(x(t_0), u(t_0), t_0) \\ x(t_2) &= x(t_1) + hf(x(t_1), u(t_1), t_0) \\ &\vdots \end{aligned} \tag{1.85}$$

using a digital computer.

The above relations have a particularly simple form in the case of a linear, time-invariant system where $f(x, u, t) = Ax + Bu$. In this case (1.84) takes the form

$$x(t_{i+1}) = (I + hA)x(t_i) + hBu(t_i), \qquad 0 \leqslant i \leqslant N-1 \tag{1.86}$$

Equivalently, writing

$$\Phi_h = I + hA, \qquad \Delta_h = hB, \tag{1.87}$$

we obtain

$$x(t_{i+1}) = \Phi_h x(t_i) + \Delta_h u(t_i), \qquad 0 \leqslant i \leqslant N-1 \tag{1.88}$$

which is an example of a discrete state-variable model (see Section 1.8). The realization of this algorithm in the form of a digital computer programme is a straightforward matter as it only requires the operations of matrix multiplication and addition commonly available as library subroutines. The general structure of the algorithm is as follows

Step 1: input $A, B, T = t_f - t_0, N, u_i = u(t_{i-1}), 1 \leqslant i \leqslant N$ and x_0
Step 2: compute $h = T/N$, Φ_h and Δ_h
Step 3: set $i = 0$ and $v_1 = x_0$
Step 4: set $i = i + 1$
Step 5: compute $v_2 = \Phi_h v_1 + \Delta_h u_i$
Step 6: set $x(t_i) = v_2$, $v_1 = v_2$
Step 7: if $i = N$, STOP! If $i < N$ go back to step 4.

Despite the usefulness of computers, a familiarity with analytic methods is a great help in providing checks for computational results and an invaluable aid to the development of insight into system behaviour. Some of the most important analytical methodology is introduced in the next section.

1.5 Exact Solution of the State Equations: Linear Time-invariant Systems

1.5.1 Change of variables

The major difficulty in analytic solution of the linear time-invariant equation

$$\frac{\mathrm{d}x(t)}{\mathrm{d}t} = Ax(t) + Bu(t)$$

$$x(0) = x_0 \tag{1.89}$$

is the general fact that the equation for $x_i(t)$

$$\frac{\mathrm{d}x_i}{\mathrm{d}t} = A_{i1}x_1 + A_{i2}x_2 + \ldots + A_{in}x_n$$

$$+ B_{i1}u_1 + \ldots + B_{il}u_l \tag{1.90}$$

contains contributions from other states x_k, $k \neq i$. This *cross-coupling* precludes the possibility of solving the equations independently! To illustrate this point consider the linearized liquid level model of (1.71) with $a_1 = a_2 = \beta = 1$,

$$\frac{\mathrm{d}x_1}{\mathrm{d}t} = -x_1 + x_2 + u_1 \tag{1.91}$$

$$\frac{\mathrm{d}x_2}{\mathrm{d}t} = x_1 - x_2 + u_2 \tag{1.92}$$

(where we have dropped the use of "tildas" for notational simplicity). We see that, in order to solve (1.91) for $x_1(t)$, we need to know $x_2(t)$. But, in order to solve (1.92) for $x_2(t)$, we need to know $x_1(t)$. An apparently impossible position!

A route out of this difficulty is obtained by adding (1.91) and (1.92) to obtain

$$\frac{d}{dt}(x_1+x_2) = u_1+u_2 \tag{1.93}$$

and then subtracting them to obtain

$$\frac{d}{dt}(x_1-x_2) = -2(x_1-x_2)+u_1-u_2 \tag{1.94}$$

Defining the new variables $z_1(t)=x_1(t)+x_2(t)$ and $z_2(t)=x_1(t)-x_2(t)$ it is easily verified that

$$\frac{dz_1(t)}{dt} = u_1(t)+u_2(t), \frac{dz_2(t)}{dt} = -2z_2(t)+u_1(t)-u_2(t) \tag{1.95}$$

and that there is no cross-coupling between these new equations. They can hence be solved independently.

A possible key to the analytical solution of (1.89) in the general case is hence the choice of new state variables defined as linear combinations of the old. The key to the general representation of this procedure is obtained by consideration of the liquid level example in matrix form, i.e.

$$\begin{bmatrix} z_1(t) \\ z_2(t) \end{bmatrix} = \begin{bmatrix} 1 & 1 \\ 1 & -1 \end{bmatrix} \begin{bmatrix} x_1(t) \\ x_2(t) \end{bmatrix} \tag{1.96}$$

This suggests that we write our state variable transformation in the form

$$z(t) = T^{-1}x(t) \tag{1.97}$$

where $z(t)$ is the new $n \times 1$ state vector and T is a *nonsingular,* constant $n \times n$ matrix. There are, of course, an infinite number of possible choices of T. Not all of them are useful, however.

As T is constant it is easily verified that $dz(t)/dt = T^{-1}dx(t)/dt = T^{-1}Ax(t)+T^{-1}Bu(t)$. That is, writing $x(t)=TT^{-1}x(t)=Tz(t)$, we see that the transformed state is the solution of the state equations

$$\frac{dz(t)}{dt} = A^*z(t)+B^*u(t)$$

$$z(0) = z_0 \tag{1.98}$$

where

$$A^* = T^{-1}AT, \qquad B^* = T^{-1}B, \qquad z_0 = T^{-1}x_0 \tag{1.99}$$

It is also easily verified that the output equation

$$y(t) = Cx(t) + Du(t) \tag{1.100}$$

is transformed to

$$y(t) = C^*z(t) + D^*u(t) \tag{1.101}$$

where

$$C^* = CT, \qquad D^* = D \tag{1.102}$$

The two models (1.89, (1.100) and (1.98), (1.101) do, of course, have identical input-output properties.

EXAMPLE 1.5.1. Consider the solution of the state equations

$$\frac{dx}{dt} = \begin{bmatrix} -\frac{1}{2} & \frac{1}{2} \\ -\frac{1}{2} & \frac{1}{2} \end{bmatrix} x + \begin{bmatrix} -1 \\ 1 \end{bmatrix} u, \qquad x(0) = \begin{bmatrix} -1 \\ 1 \end{bmatrix} \tag{1.103}$$

Choosing the transformation matrix $T = \begin{bmatrix} 1 & -1 \\ 1 & 1 \end{bmatrix}$, it is easily verified that

$$A^* = \begin{bmatrix} 0 & 1 \\ 0 & 0 \end{bmatrix}, \qquad B^* = \begin{bmatrix} 0 \\ 1 \end{bmatrix}, \qquad T^{-1}x(0) = \begin{bmatrix} 0 \\ 1 \end{bmatrix} \tag{1.104}$$

and hence that, with $z = T^{-1}x$,

$$\frac{dz}{dt} = \begin{bmatrix} 0 & 1 \\ 0 & 0 \end{bmatrix} z + \begin{bmatrix} 0 \\ 1 \end{bmatrix} u, \qquad z(0) = \begin{bmatrix} 0 \\ 1 \end{bmatrix} \tag{1.105}$$

or, equivalently, $dz_1/dt = z_2$, $z_1(0) = 0$, $dz_2/dt = u(t)$, $z_2(0) = 1$. Although these equations are coupled, they can be solved sequentially. For example, taking a step input $u(t) \equiv 1, t \geqslant 0$, it is easily verified that $z_2(t) = 1 + \int_0^t u(t')dt' = 1 + t$ and hence $z_1(t) = \int_0^t z_2(t')dt' = t + \frac{1}{2}t^2$. Substituting these solutions into the relations $x = Tz$ yields the required solutions

$$x_1(t) = \tfrac{1}{2}t^2 - 1, \qquad x_2(t) = 1 + 2t + \tfrac{1}{2}t^2 \tag{1.106}$$

which are easily checked by substituting back into (1.103).

A little reflection shows that this example was solved easily because the transformed matrix A^* was upper triangular and that, in general, the choice of T to make A^* triangular enables an analytic solution to be obtained. A special case of this methodology is described in the next section.

1.5.2 Eigenvector transformations and diagonal forms

The transformed state equations (1.98) are uncoupled if we can choose T so that A^* is diagonal, i.e.

$$A^* = T^{-1}AT = \begin{bmatrix} \lambda_1 & 0 & \cdots & 0 \\ 0 & \lambda_2 & & \vdots \\ \vdots & & \ddots & 0 \\ 0 & \cdots & 0 & \lambda_n \end{bmatrix} \tag{1.107}$$

when they take the form, $1 \leqslant i \leqslant n$,

$$\frac{dz_i(t)}{dt} = \lambda_i z_i(t) + B^*_{i1}u_1(t) + \ldots + B^*_{il}u_l(t), \quad z_i(0) = (z_0)_i \tag{1.108}$$

In fact it is easily verified that the solution has the explicit form

$$z_i(t) = e^{\lambda_i t}z_i(0) + \sum_{k=1}^{l} B^*_{ik} \int_0^t e^{\lambda_i(t-t')}u_k(t')dt', \quad 1 \leqslant i \leqslant n \tag{1.109}$$

which can be evaluated by analytic or numerical evaluation of the integrals. The resulting state responses are then obtained from the relation $x(t) = Tz(t)$.

Writing

$$T = [w_1, w_2, \ldots, w_n] \tag{1.110}$$

where w_k is the kth column of T, $1 \leqslant k \leqslant n$, then $|T| \neq 0$ requires that $w_k \neq 0$, $1 \leqslant k \leqslant n$, and the matrix identity

$$AT = [Aw_1, Aw_2, \ldots, Aw_n] = TA^* \tag{1.111}$$

$$= [w_1, w_2, \ldots, w_n] \begin{bmatrix} \lambda_1 & 0 & \cdots & 0 \\ 0 & \lambda_2 & & \vdots \\ \vdots & & \ddots & 0 \\ 0 & \cdots & 0 & \lambda_n \end{bmatrix}$$

$$= [\lambda_1 w_1, \lambda_2 w_2, \ldots, \lambda_n w_n] \tag{1.112}$$

indicates (by equating columns) that

$$Aw_j = \lambda_j w_j, \quad 1 \leqslant j \leqslant n \tag{1.113}$$

and hence that λ_j is an *eigenvalue* of A and w_j is a non-zero *eigenvector* of A corresponding to that eigenvalue. The resulting matrix T is a (so-called) *eigenvector matrix* of A.

The above analysis can be summarized in terms of the following computational procedure.

Step 1: Compute the eigenvalues $\lambda_1, \lambda_1, \ldots, \lambda_n$ of A by evaluating the roots of the *characteristic polynomial*

$$\rho(s) = |sI_n - A| \tag{1.114}$$

by analytical means or by the use of commonly available eigenvalue evaluation subroutines on a digital computer.
Step 2: Solve equation (1.113) for any non-zero eigenvector w_j, $1 \leqslant j \leqslant n$, (by hand where possible but, if not, using available computer programmes) and construct T.
Step 3: If T is singular, the method fails (see Section 1.5.3). If, however, T is nonsingular, compute $B^* = T^{-1}B$ and $z(0) = T^{-1}x(0)$.
Step 4: Evaluate the expressions in (1.109) by analytical or numerical evaluation of the integrals and construct the state response from $x(t) = Tz(t)$.
(Note: although A is real, its eigenvalues may be real or they may be complex numbers occurring in complex conjugate pairs. In such a case the matrices T and B^* will have complex entries and the transformed state response $z(t)$ will be complex. The resulting state response $x(t) = Tz(t)$ will, however, always be real).

EXAMPLE 1.5.2. Consider the linearized liquid level system of (1.71) with $a_1 = a_2 = \beta = 1$,

$$\frac{dx}{dt} = \begin{bmatrix} -1 & 1 \\ 1 & -1 \end{bmatrix} x + \begin{bmatrix} 1 & 0 \\ 0 & 1 \end{bmatrix} u \tag{1.115}$$

initially in the equilibrium position $x(0) = (0, 0)^T$. Consider the problem of the evaluation of the responses of the liquid levels if a volume V_0 of liquid is quickly poured into vessel one at time $t = 0$. This can be modelled by considering the input vector

$$u(t) = \begin{bmatrix} V_0 \delta(t) \\ 0 \end{bmatrix} \tag{1.116}$$

where $\delta(t)$ is the unit Dirac delta/impulse function. The characteristic polynomial

$$\rho(s) = |sI - A| = \begin{bmatrix} s+1 & -1 \\ -1 & s+1 \end{bmatrix} = s(s+2) \tag{1.117}$$

indicating that A has eigenvalues $\lambda_1 = 0$ and $\lambda_2 = -2$. It is left as an exercise for the reader to consider the eigenvalue equations $Aw_j = \lambda_j w_j$, $j = 1, 2$, and show that $w_1 = \begin{bmatrix} 1 \\ 1 \end{bmatrix}$ and $w_2 \begin{bmatrix} -1 \\ 1 \end{bmatrix}$ are suitable eigenvectors (Note: there is a degree of non-uniqueness here as eigenvectors can be scaled by any non-zero scalar and still remain eigenvectors). The corresponding eigenvector matrix is

$$T = \begin{bmatrix} 1 & -1 \\ 1 & 1 \end{bmatrix} \tag{1.118}$$

(see equation (1.110)), and hence

$$A^* = T^{-1}AT = \begin{bmatrix} 0 & 0 \\ 0 & -2 \end{bmatrix}, \qquad B^* = T^{-1}B = \begin{bmatrix} \frac{1}{2} & \frac{1}{2} \\ -\frac{1}{2} & \frac{1}{2} \end{bmatrix} \tag{1.119}$$

The resulting transformed model takes the form

$$\frac{dz(t)}{dt} = \begin{bmatrix} 0 & 0 \\ 0 & -2 \end{bmatrix} z(t) + \begin{bmatrix} \frac{1}{2} & \frac{1}{2} \\ -\frac{1}{2} & \frac{1}{2} \end{bmatrix} \begin{bmatrix} V_0\delta(t) \\ 0 \end{bmatrix} \tag{1.120}$$

with initial condition $z(0) = \begin{bmatrix} 0 \\ 0 \end{bmatrix}$. Writing these equations down in element form yields

$$dz_1/dt = \tfrac{1}{2}V_0\delta(t), z_1(0) = 0,$$

$$dz_2/dt = -2z_2 - \tfrac{1}{2}V_0\delta(t), z_2(0) = 0,$$

which have the solutions $z_1(t) = \frac{1}{2}V_0$ and $z_2(t) = -\frac{1}{2}V_0e^{-2t}$ in the

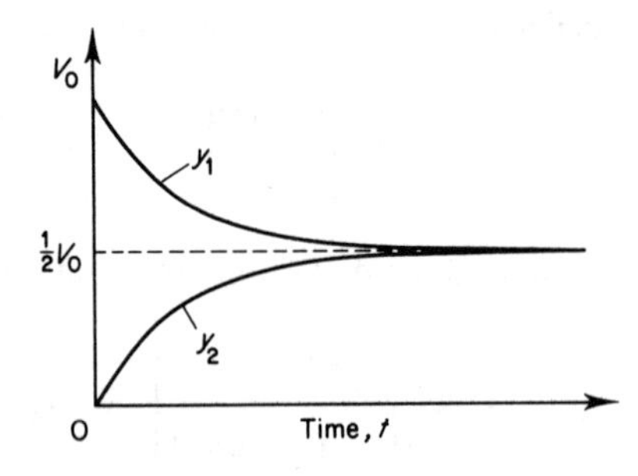

Fig. 14. Response of liquid level system to impulse in vessel one.

interval $t > 0$. The corresponding state solution is obtained as

$$\begin{bmatrix} x_1(t) \\ x_2(t) \end{bmatrix} = Tz(t) = \begin{bmatrix} 1 & -1 \\ 1 & 1 \end{bmatrix} \frac{V_0}{2} \begin{bmatrix} 1 \\ -e^{-2t} \end{bmatrix} = \frac{V_0}{2} \begin{bmatrix} 1+e^{-2t} \\ 1-e^{-2t} \end{bmatrix} \tag{1.121}$$

which is sketched in Fig. 14. As expected physically, the first vessel empties itself into the second vessel until the levels settle out.

EXAMPLE 1.5.3. Consider the classical problem of finding the solution of the second order scalar differential equation

$$\frac{d^2y(t)}{dt^2} + 3\frac{dy(t)}{dt} + 2y(t) = 2u(t) \tag{1.122}$$

with a unit step input and zero initial conditions. The reader should have no problem solving this problem directly using Laplace transform techniques. Alternatively the following state-variable approach can be used. Define the state variables $x_1(t) = y(t)$ and $x_2(t) = dy(t)/dt$ then it is seen that $dx_1/dt = x_2$ and that (1.122) reduces to $dx_2/dt = -2x_1 - 3x_2 + 2u$. The initial conditions are $x_1(0) = x_2(0) = 0$. In matrix form,

$$\frac{dx(t)}{dt} = \begin{bmatrix} 0 & 1 \\ -2 & -3 \end{bmatrix} x(t) + \begin{bmatrix} 0 \\ 2 \end{bmatrix} u(t), \qquad x(0) = \begin{bmatrix} 0 \\ 0 \end{bmatrix} \tag{1.123}$$

with the output equation $y(t) = (1 \quad 0)x(t)$. The characteristic polynomial $\rho(s) = |sI - A| = (s+1)(s+2)$ indicates that A has eigenvalues $\lambda_1 = -1$ and $\lambda_2 = -2$. The reader should verify that a suitable choice of eigenvector matrix is

$$T = \begin{bmatrix} 1 & 1 \\ -1 & -2 \end{bmatrix} \tag{1.124}$$

and hence that

$$A^* = \begin{bmatrix} -1 & 0 \\ 0 & -2 \end{bmatrix}, \qquad B^* = \begin{bmatrix} 2 \\ -2 \end{bmatrix}, \qquad C^* = [1 \quad 1] \tag{1.125}$$

The corresponding transformed equations are simply $dz_1(t)/dt = -z_1 + 2u$, $z_1(0) = 0$, $dz_2/dt = -2z_2 - 2u$, $z_2(0) = 0$ which have the solutions $z_1(t) = 2(1 - e^{-t})$ and $z_2(t) = -(1 - e^{-2t})$. The output response follows directly from $y(t) = C^*z(t) = z_1(t) + z_2(t) = 1 - 2e^{-t} + e^{-2t}$.

1.5.3 Multiple eigenvalues and the Jordan form

The transformation T producing the diagonal form of (1.107) is quite obviously a powerful theoretical and computational tool. The existence of such a T is not guaranteed, however, as evidenced by a consideration of the non-zero 2×2 matrix

$$A = \begin{bmatrix} 0 & 1 \\ 0 & 0 \end{bmatrix} \tag{1.126}$$

which has multiple eigenvalues $\lambda_1 = \lambda_2 = 0$. Suppose that a matrix T exists satisfying (1.107) in this case, i.e. $T^{-1}AT = 0$ (2×2 matrix of zeros) or $A = T(T^{-1}AT)T^{-1} = 0$ which is evidently nonsense. The assumption that T exists is hence incorrect in this case. Fortunately this problem does not occur often in practice as, in almost all situations of practical relevance, the matrix A has distinct eigenvalues (see following exercise). For this reason (and to retain a fairly simple treatment) the "pathological" cases described in this section will be considered only occasionally in this text.

EXERCISE 1.5.1. Suppose that the $n \times n$ real matrix A has distinct eigenvalues $\lambda_1, \lambda_2, \ldots, \lambda_n$ with corresponding non-zero eigenvectors $w_1, w_2, \ldots, w_n$ satisfying $Aw_k = \lambda_k w_k$, $1 \leq k \leq n$. Let $v_1^T, v_2^T, \ldots, v_n^T$ be non-zero *dual eigenvectors* satisfying $v_k^T A = \lambda_k v_k^T$, $1 \leq k \leq n$. Prove that $v_k^T w_i = 0$ $(i \neq k)$ and hence, by scaling each v_k so that $v_k^T w_k = 1$, that $VT = I_n$ where T is defined by (1.110) and

$$V = \begin{bmatrix} v_1^T \\ v_2^T \\ \vdots \\ v_n^T \end{bmatrix}$$

Deduce that the eigenvector matrix T is nonsingular and hence that $T^{-1}AT$ takes the diagonal form of (1.107).

The exercise indicates that the inability to transform A to diagonal form is connected with the existence of multiple/repeated eigenvalues. Although some matrices possessing multiple eigenvalues can be transformed to diagonal form, it is only possible in many cases to obtain the so-called "Jordan Form",

$$T^{-1}AT = \begin{bmatrix} J_1 & 0 & \cdots & 0 \\ 0 & J_2 & & \vdots \\ \vdots & & \ddots & 0 \\ 0 & \cdots & 0 & J_q \end{bmatrix} \tag{1.127}$$

where each $n_k \times n_k$ "Jordan Block" J_k has the structure $J = \eta_k$ if $n_k = 1$, or, for $n_k > 1$, for non-diagonal structure

$$J_k = \begin{bmatrix} \eta_k & 1 & 0 & \cdot & \cdot & 0 \\ 0 & \eta_k & 1 & & & \cdot \\ \cdot & 0 & \eta_k & & & \cdot \\ \cdot & & & & & \cdot \\ \cdot & & & & & 0 \\ \cdot & & & & & 1 \\ 0 & \cdot & \cdot & \cdot & 0 & \eta_k \end{bmatrix} \tag{1.128}$$

where η_k is an eigenvalue of A. Quite obviously $n = n_1 + n_2 + \ldots + n_q$.

The matrix of equation (1.126) is clearly already in its Jordan form, as is

$$A = \begin{bmatrix} 1 & 1 & 0 \\ 0 & 1 & 0 \\ 0 & 0 & -1 \end{bmatrix} \tag{1.129}$$

with $n_1 = 2, n_2 = 1, \eta_1 = 1, \eta_2 = -1$. In contrast, the matrix

$$A = \begin{bmatrix} 1 & 1 & -1 \\ 0 & 2 & -1 \\ 0 & 1 & 0 \end{bmatrix} \tag{1.130}$$

is not in Jordan form but can be transformed to the Jordan form

$$T^{-1}AT = \begin{bmatrix} 1 & 1 & 0 \\ 0 & 1 & 0 \\ 0 & 0 & 1 \end{bmatrix} \tag{1.131}$$

using the transformation

$$T = \begin{bmatrix} 1 & -1 & 0 \\ 1 & 1 & 1 \\ 1 & 0 & 1 \end{bmatrix} \tag{1.132}$$

EXERCISE 1.5.2. Given the linear system (1.89), suppose that the transformation T transforms A to the Jordan form of (1.127). Using the change of variable $z(t) = T^{-1}x(t)$ show that

$$\frac{dz_k(t)}{dt} = J_k z_k(t) + B_k^* u(t), \qquad 1 \leqslant k \leqslant q \tag{1.133}$$

where

$$B^* = \begin{bmatrix} B_1^* \\ B_2^* \\ \cdot \\ \cdot \\ B_q^* \end{bmatrix}, B_k^* \text{ has dimension } n_k \times l$$

$$\text{and } z(t) = \begin{bmatrix} z_1(t) \\ z_2(t) \\ \cdot \\ \cdot \\ z_q(t) \end{bmatrix}$$

where $z_k(t)$ has dimension $n_k \times 1$. Note that the upper triangular structure of J_k can make possible the analytic solution of (1.133) as described in Example 1.5.1.

1.6 Stability of Linear Time-invariant Systems

The reader will be familiar with the notion of the asymptotic stability of the single-input/single-output system (1.6), i.e. the requirement that, in the absence of inputs, the system output $\to 0$ as $t \to +\infty$ independent of the system initial conditions. He will also recall that the system is stable if, and only if, the solutions of the equation $P(\lambda) = 0$ (i.e. the poles of the system transfer function (1.7)) have strictly negative real parts. The approach taken to the definition and characterization of the stability of the l-input/m-output linear time-invariant system

$$\frac{dx(t)}{dt} = Ax(t) + Bu(t), \qquad x(0) = x_0$$

$$y(t) = Cx(t) + Du(t) \tag{1.134}$$

is somewhat different however, stability being defined in terms of asymptotic properties of the *state* rather than the output. Later chapters will prove, however, that the approaches are equivalent in the case of $m = l = 1$.

The natural intuitive idea of linear system stability is independent of the choice of outputs, the particular input or the initial state of the system.

Definition 1.6.1. *The system* (*1.134*) *is said to be* asymptotically stable *if, and only if, all solutions of the* homogenous *system*

$$\frac{dx(t)}{dt} = Ax(t), \qquad x(0) = x_0 \tag{1.135}$$

satisfy the relation, as $t \to +\infty$, $\lim x(t) = 0$ *independent of the choice of initial condition* $x(0)$ (*the limit being interpreted as stating that each element of* $x(t) \to 0$ *as* $t \to +\infty$).

A convenient test for asymptotic stability can be obtained in terms of the eigenvalues of A. Taking, for simplicity, the situation when A has distinct eigenvalues then, by Exercise 1.5.1., there exists a nonsingular eigenvector matrix T such that equation (1.107) holds. Defining $z(t) = T^{-1}x(t)$, then the techniques of Section 1.5.2 indicate that $z_i(t) = e^{\lambda_i t} z_i(0)$, $1 \leqslant i \leqslant n$, and hence that

$$\begin{aligned} x(t) = Tz(t) &= T \begin{bmatrix} e^{\lambda_1 t} & \cdots & 0 \\ \vdots & e^{\lambda_2 t} \ddots & \vdots \\ 0 & \cdots & e^{\lambda_n t} \end{bmatrix} z(0) \\ &= T \begin{bmatrix} e^{\lambda_1 t} & \cdots & 0 \\ \vdots & e^{\lambda_2 t} \ddots & \vdots \\ 0 & \cdots & e^{\lambda_n t} \end{bmatrix} T^{-1}x(0) \end{aligned} \tag{1.136}$$

The following result is hence easily proved in this case by application of the definition.

Theorem 1.6.1. *The linear time-invariant system* (*1.134*) *is asymptotically stable if, and only if, the system characteristic polynomial*

$$\rho(s) = |sI_n - A| = (s - \lambda_1)(s - \lambda_2) \ldots (s - \lambda_n) \tag{1.137}$$

has roots $\lambda_1, \lambda_2, \ldots, \lambda_n$ *with strictly negative real parts*, $\mathrm{Re}\,\lambda_k < 0$, $1 \leqslant k \leqslant n$ (*i.e. the eigenvalues of* A *all lie in the familiar left-half complex plane*).

(Note: the reader will see that the system characteristic polynomial is just the characteristic polynomial of A.)

EXERCISE 1.6.1. Verify theorem 1.6.1 if A only has a Jordan form. (Hint: use the methods of Exercise 1.5.2.)

As in classical control theory, the following terminology is used

(i) if Re $\lambda_k < 0, 1 \leqslant k \leqslant n$, the system is stable.

(ii) If Re $\lambda_k > 0$ for some k, the system is unstable.

(iii) If Re $\lambda_k \leqslant 0$, $1 \leqslant k \leqslant n$, and Re $\lambda_{k_1} = 0$ for some k_1, the system is critically or marginally stable.

Finally, note that stability is independent of the magnitude of the imaginary parts of the eigenvalues of A. If λ_k is real then the relation $z_k(t) = e^{\lambda_k t} z_k(0)$ indicates a simple non-oscillatory type of dynamics. The existence of eigenvalues with non-zero imaginary parts does, in fact, correspond to the presence of an oscillatory mode in the system. To illustrate this, consider the two-dimensional system

$$\frac{dx(t)}{dt} = \begin{bmatrix} \sigma & -\omega \\ \omega & \sigma \end{bmatrix} x(t) \tag{1.138}$$

where $\omega \neq 0$. This system has eigenvalues $\sigma \pm j\omega$ with eigenvectors $\begin{bmatrix} \pm j \\ 1 \end{bmatrix}$. Applying the technique of Section 1.5.2 the reader can verify that

$$x(t) = \tfrac{1}{2}\begin{bmatrix} j & -j \\ 1 & 1 \end{bmatrix} \begin{bmatrix} e^{(\sigma+j\omega)t} & 0 \\ 0 & e^{(\sigma-j\omega)t} \end{bmatrix} \begin{bmatrix} -j & 1 \\ j & 1 \end{bmatrix} x(0)$$

$$= e^{\sigma t} \begin{bmatrix} \cos \omega t & -\sin \omega t \\ \sin \omega t & \cos \omega t \end{bmatrix} x(0) \tag{1.139}$$

indicating an oscillatory response of frequency ω equal to the imaginary parts of the eigenvalues and envelope described by the real part of the eigenvalue σ.

1.7 Use of the Matrix Exponential

If (1.134) were a scalar rather than a matrix equation, the reader could easily apply the "integrating factor method" to obtain the general solution

$$x(t) = e^{At}x_0 + \int_0^t e^{A(t-t^1)} Bu(t^1)\, dt^1 \tag{1.140}$$

The purpose of this section is to show how this formula has a natural formal counterpart in the matrix case.

1.7.1 Properties of e^M

The first step in the generalization of (1.140) to the matrix case is the extension of the notion of exponential to include matrix exponents. Let M be any $n \times n$ matrix. The *exponential matrix* e^M is defined by the infinite series

$$e^M = I_n + M + \frac{M^2}{2!} + \frac{M^3}{3!} + \ldots \tag{1.141}$$

This series converges and is the natural generalization of the notion of scalar exponential.

The matrix exponential has a number of important properties:

(a) If M is diagonal with diagonal elements $\lambda_1, \lambda_1, \ldots, \lambda_n$ then M^k is diagonal

$$M^k = \begin{bmatrix} \lambda_1^k & & & \\ & \lambda_2^k & & \\ & & \ddots & \\ & & & \lambda_n^k \end{bmatrix} \quad k \geqslant 1 \tag{1.142}$$

Substituting into (1.141) and summing the series, it is easily verified that e^M is diagonal of the form

$$e^M = \begin{bmatrix} e^{\lambda_1} & & & \\ & e^{\lambda_2} & & \\ & & \ddots & \\ & & & e^{\lambda_n} \end{bmatrix} \tag{1.143}$$

(b) Let T be any nonsingular $n \times n$ matrix then the reader should verify that

$$\begin{aligned} (T^{-1}MT)^k &= T^{-1}MTT^{-1}MTT^{-1} \ldots TT^{-1}MT \\ &= T^{-1}M^kT, \qquad k \geqslant 1 \end{aligned} \tag{1.144}$$

and hence that

$$\begin{aligned} e^{T^{-1}MT} &= I + T^{-1}MT + \frac{(T^{-1}MT)^2}{2!} + \ldots \\ &= I + T^{-1}MT + \frac{T^{-1}M^2T}{2!} + \ldots \\ &= T^{-1}\left(I + M + \frac{M^2}{2!} + \ldots\right)T \\ &= T^{-1}e^MT \end{aligned} \tag{1.145}$$

In particular, if M has eigenvalues $\lambda_1, \lambda_2, \ldots, \lambda_n$ and nonsingular eigenvector matrix T, then

$$T^{-1}MT = \begin{bmatrix} \lambda_1 & & \\ & \ddots & \\ & & \lambda_n \end{bmatrix} \tag{1.146}$$

and it follows directly from (a) and (1.145) that

$$e^M = T e^{T^{-1}MT} T^{-1} = T \begin{bmatrix} e^{\lambda_1} & \cdots & & 0 \\ 0 & e^{\lambda_2} & & \vdots \\ \vdots & & \ddots & \\ 0 & \cdots & & e^{\lambda_n} \end{bmatrix} T^{-1} \tag{1.147}$$

which is a convenient way of calculating e^M if $\lambda_1, \ldots, \lambda_n$ and T are known.

(c) If M_1 and M_2 are $n \times n$ matrices then the reader should be careful as

$$e^{M_1 + M_2} \neq e^{M_1} e^{M_2} \tag{1.148}$$

unless M_1 and M_2 commute, i.e. $M_1M_2 = M_2M_1$. Particularly useful examples of commuting pairs are $M_1 = M$, $M_2 = \lambda I$ (λ a real or complex scalar) when (using (a))

$$e^{\lambda I + M} = e^{\lambda I} e^M = e^{\lambda} e^M \tag{1.149}$$

or $M_1 = \alpha M, M_2 = \beta M$ (α and β scalars) when

$$e^{\alpha M + \beta M} = e^{\alpha M} e^{\beta M} \tag{1.150}$$

(d) It is easily verified from the series expansion that $e^0 = I$ and hence, from (1.150) with $\alpha = -\beta = 1$

$$(e^M)^{-1} = e^{-M} \tag{1.151}$$

(e)

$$e^M M^k = M^k e^M \tag{1.152}$$

EXAMPLE 1.7.1. Given the matrix $A = \begin{bmatrix} 0 & 1 \\ 0 & 0 \end{bmatrix}$ consider the problem of calculating e^{At} where t is a scalar. It is easily verified that $A^2 = 0$ and hence $A^k = 0$ for $k \geqslant 2$. Substituting this into the series expansion yields

$$e^{At} = I + At = \begin{bmatrix} 1 & t \\ 0 & 1 \end{bmatrix} \tag{1.153}$$

EXAMPLE 1.7.2. Consider the problem of calculating e^{At} for the liquid level system of Example 1.5.2. Using the specified values of λ_1, λ_2 and T in equation (1.147) yields

$$
\begin{aligned}
e^{At} &= T\, e^{T^{-1}ATt} T^{-1} \\
&= \begin{bmatrix} 1 & -1 \\ 1 & 1 \end{bmatrix} \begin{bmatrix} 1 & 0 \\ 0 & e^{-2t} \end{bmatrix} \tfrac{1}{2} \begin{bmatrix} 1 & 1 \\ -1 & 1 \end{bmatrix} \\
&= \tfrac{1}{2} \begin{bmatrix} 1 + e^{-2t} & 1 - e^{-2t} \\ 1 - e^{-2t} & 1 + e^{-2t} \end{bmatrix} \qquad (1.154)
\end{aligned}
$$

1.7.2 Formal solution of the state equations

We now verify that (1.140) is the solution of (1.134). Note that $x(0) = x_0$ as required, i.e. it satisfies the initial conditions. We must now show that it satisfies the state equation. To do this we need the following formulae

(a) if $M_1(t)$ and $M_2(t)$ are time-varying matrices, then

$$
\frac{d}{dt}(M_1(t)M_2(t)) = M_1(t)\left(\frac{dM_2(t)}{dt}\right) + \left(\frac{dM_1(t)}{dt}\right) M_2(t) \qquad (1.155)
$$

This follows from the formula

$$
\frac{d}{dt} \sum_r (M_1(t))_{ir}(M_2(t))_{rk} = \sum_r (M_1(t))_{ir} \left(\frac{dM_2(t)}{dt}\right)_{rk} + \sum_r \left(\frac{dM_1(t)}{dt}\right)_{ir} (M_2(t))_{rk} \qquad (1.156)
$$

expressed in matrix form. Note that the ordering of the matrices in (1.55) is important.

(b) Using the series expansion and (1.152), we have

$$
\begin{aligned}
\frac{d}{dt} e^{At} &= \frac{d}{dt}\left(I + At + \frac{A^2t^2}{2!} + \frac{A^3t^3}{3!} + \ldots\right) \\
&= A + A^2t + \frac{A^3t^2}{2!} + \ldots \\
&= A\left(I + At + \frac{A^2t^2}{2!} + \ldots\right) \\
&= A\, e^{At} = e^{At} A \qquad (1.157)
\end{aligned}
$$

indicating that the matrix exponential behaves exactly as the scalar exponential when differentiated.

Consider now the transformation $\psi(t) = e^{-At}x(t)$, then $\psi(0) = x(0)$ and

$$\frac{d}{dt}\psi(t) = \left(\frac{d}{dt}e^{-At}\right)x(t) + e^{-At}\frac{dx(t)}{dt} \qquad \text{(by (1.155))}$$

$$= -e^{-At}Ax(t) + e^{-At}(Ax(t) + Bu(t))$$

$$= e^{-At}Bu(t) \tag{1.158}$$

Integrating this equation yields

$$\psi(t) = e^{-At}x(t) = x_0 + \int_0^t e^{-At'}Bu(t')\,dt' \tag{1.159}$$

Equation (1.140) follows directly by multiplying both sides of the equation by e^{At} and using (1.150) and (1.151).

EXAMPLE 1.7.3. Consider the solution of

$$\frac{dx(t)}{dt} = \begin{bmatrix} 0 & 1 \\ 0 & 0 \end{bmatrix} x(t) + \begin{bmatrix} 0 \\ 1 \end{bmatrix} u(t), \qquad x(0) = \begin{bmatrix} 0 \\ 1 \end{bmatrix} \tag{1.160}$$

with the step input $u(t) \equiv 1$, $t \geqslant 0$. Using (1.153) directly in (1.140) indicates that

$$x(t) = \begin{bmatrix} 1 & t \\ 0 & 1 \end{bmatrix} \begin{bmatrix} 0 \\ 1 \end{bmatrix} + \int_0^t \begin{bmatrix} 1 & t-t' \\ 0 & 1 \end{bmatrix} \begin{bmatrix} 0 \\ 1 \end{bmatrix} dt'$$

$$= \begin{bmatrix} t \\ 1 \end{bmatrix} + \int_0^t \begin{bmatrix} (t-t') \\ 1 \end{bmatrix} dt' = \begin{bmatrix} t + \frac{1}{2}t^2 \\ 1 + t \end{bmatrix} \tag{1.161}$$

Finally, note that the formal solution (1.140) can be obtained directly from the analysis of Section 1.5.2 when A can be diagonalized. More precisely, from (1.109)

$$z(t) = e^{A^*t}z(0) + \int_0^t e^{A^*(t-t')}B^*u(t')\,dt' \tag{1.162}$$

and hence

$$\begin{aligned} x(t) &= Tz(t) \\ &= T\mathrm{e}^{A^*t}T^{-1}Tz(0) + \int_0^t T\mathrm{e}^{A^*(t-t')}T^{-1}TB^*u(t')\,\mathrm{d}t' \\ &= \mathrm{e}^{TA^*T^{-1}t}x(0) + \int_0^t \mathrm{e}^{TA^*T^{-1}(t-t')}Bu(t')\,\mathrm{d}t' \\ &= \mathrm{e}^{At}x(0) + \int_0^t \mathrm{e}^{A(t-t')}Bu(t')\,\mathrm{d}t' \end{aligned} \tag{1.163}$$

which is just (1.140). The formal solution is, however, quite general, holding for any choice of matrices A and B.

EXERCISE 1.7.1. Show that the system (1.134) is asymptotically stable if, and only if,

$$\lim_{t\to+\infty} \mathrm{e}^{At} = 0.$$

1.7.3 Impulse, step and sinusoidal responses

Using (1.140) the output response of the system (1.134) takes the form, $t \geqslant 0$,

$$y(t) = C\mathrm{e}^{At}x_0 + \int_0^t C\mathrm{e}^{A(t-t')}Bu(t')\,\mathrm{d}t' + Du(t) \tag{1.164}$$

or, writing $Du(t) = \int_0^t D\delta(t-t')u(t')\,\mathrm{d}t'$ (where δ is the unit Dirac delta function) and defining the $m \times l$ *weighting matrix*,

$$H(t) = C\mathrm{e}^{At}B + D\delta(t) \tag{1.165}$$

the solution takes the form

$$y(t) = C\mathrm{e}^{At}x_0 + \int_0^t H(t-t')u(t')\,\mathrm{d}t' \tag{1.166}$$

The first term represents the effect of initial conditions. The second term is a matrix *convolution integral* describing the contribution of the input to the response. For the case of zero initial conditions $x_0 = 0$,

$$y(t) = \int_0^t H(t-t')u(t')\,\mathrm{d}t' \tag{1.167}$$

which is a direct generalization of (1.9), suggesting that $H(t)$ is the generalization of the notion of impulse function.

(a) Impulse Responses. For an l-input system, the general form of impulsive input at time $t = 0$ takes the form of l impulsive inputs $u_k(t) = \alpha_k \delta(t)$, $1 \leqslant k \leqslant l$, applied simultaneously in the form of the $l \times 1$ input vector

$$u(t) = \begin{bmatrix} \alpha_1 \delta(t) \\ \alpha_2 \delta(t) \\ \vdots \\ \alpha_l \delta(t) \end{bmatrix} = \alpha \delta(t) \tag{1.168}$$

where $\alpha = [\alpha_1, \alpha_2, \ldots, \alpha_l]^T$ is the $l \times 1$ vector of impulse amplitudes. Taking the case of zero initial conditions, the output response is

$$y(t) = \int_0^t H(t-t')\alpha\delta(t')\,\mathrm{d}t' = H(t)\alpha \tag{1.169}$$

which is simply the product of the weighting matrix $H(t)$ and the impulse amplitude vector α. For this reason $H(t)$ is frequently called the *impulse response matrix*.

(b) Step Responses. For an l-input system, the general form of step input at $t = 0$ takes the form of step inputs $u_k(t) = \alpha_k, t > 0, u_k(t) = 0, t \leqslant 0, 1 \leqslant k \leqslant l$, applied simultaneously. In matrix form (with notation as above)

$$u(t) = \begin{cases} \alpha, & t > 0 \\ 0, & t \leqslant 0 \end{cases} \tag{1.170}$$

With zèro initial conditions the system response takes the form

$$y(t) = \int_0^t H(t-t')\alpha\,\mathrm{d}t' = \left\{\int_0^t H(t-t')\,\mathrm{d}t'\right\}\alpha \tag{1.171}$$

which is simply the product of $\int_0^t H(t-t')\mathrm{d}t'$ and the step amplitude vector α.

(c) Sinusoidal Responses. For an l-input system, a general form of sinusoidal input can be represented by the input vector

$$u(t) = \begin{bmatrix} \alpha_1 \cos(\omega t + \beta_1) \\ \alpha_2 \cos(\omega t + \beta_2) \\ \vdots \\ \alpha_l \cos(\omega t + \beta_l) \end{bmatrix} \tag{1.172}$$

consisting of l scalar inputs of equal frequency ω but different amplitudes and phases. Defining

$$\alpha = \begin{bmatrix} \alpha_1 e^{j\beta_1} \\ \vdots \\ \alpha_l e^{j\beta_l} \end{bmatrix}, \qquad \hat{u}(t) = \alpha e^{j\omega t} \tag{1.173}$$

the reader will easily verify that $u(t)$ is simply the real part of $\hat{u}(t)$, i.e.

$$u(t) = \operatorname{Re} \hat{u}(t) \tag{1.174}$$

Taking the case of zero initial conditions, the response $\hat{y}(t)$ to the complex input $\hat{u}(t)$ takes the form

$$\hat{y}(t) = \int_0^t H(t-t')\hat{u}(t')\,dt' = \left\{ \int_0^t H(t-t')e^{j\omega t'}\,dt' \right\}\alpha \tag{1.175}$$

The real system response $y(t)$ to the real input $u(t)$ is just

$$\begin{aligned} y(t) &= \int_0^t H(t-t')u(t')\,dt' = \int_0^t H(t-t') \operatorname{Re} \hat{u}(t')\,dt' \\ &= \operatorname{Re} \int_0^t H(t-t')\hat{u}(t')\,dt' = \operatorname{Re} \hat{y}(t) \end{aligned} \tag{1.176}$$

The form of these solutions can be obtained by evaluating

$$\begin{aligned} \int_0^t H(t-t')e^{j\omega t'}\,dt' &= \int_0^t \{D\delta(t-t') + Ce^{A(t-t')}B\}e^{j\omega t'}\,dt' \\ &= De^{j\omega t} + Ce^{At}\left\{ \int_0^t e^{(j\omega I - A)t'}\,dt' \right\} B \end{aligned} \tag{1.177}$$

Noting that integration of the relation $de^{Mt}/dt = e^{Mt}M$ yields the identity

$$e^{Mt} - I = \left\{ \int_0^t e^{Mt'}\,dt' \right\} M, \tag{1.178}$$

we see that

$$\int_0^t e^{(j\omega I - A)t'}\,dt' = (e^{(j\omega I - A)t} - I)(j\omega I - A)^{-1} \tag{1.179}$$

provided that $j\omega I - A$ is nonsingular, i.e. provided that A has no eigenvalue equal to $j\omega$. With this proviso, it follows that

$$\begin{aligned} \hat{y}(t) &= (D + C(j\omega I - A)^{-1}B)\hat{u}(t) - Ce^{At}(j\omega I - A)^{-1}B\alpha \\ &= G(j\omega)\hat{u}(t) - Ce^{At}(j\omega I - A)^{-1}B\alpha \end{aligned} \tag{1.180}$$

where $G(j\omega)=D+C(j\omega I-A)^{-1}B$ is the $m\times l$ system frequency response matrix (a matrix of frequency dependent but time independent complex gains).

It is the oscillating first term of (1.180) that is of greatest significance. More precisely, if the system is asymptotically stable then (Exercise 1.7.1) the second term tends to zero as $t\to+\infty$, leading to the *asymptotic approximation*

$$\hat{y}(t)\simeq G(j\omega)\hat{u}(t),\qquad t\geqslant 0 \tag{1.181}$$

The term $G(j\omega)\hat{u}(t)$ is hence the generalization of the idea of the *steady state* sinusoidal response to a sinusoidal input. In the case of a single-input/single-output system ($m=l=1$), $G(j\omega)$ is just the well known frequency response function that has proved to be so valuable in stability studies using the Nyquist diagram or (factoring $G(j\omega)$ into gain and phase form) the Bode diagram. In the multi-input/multi-output case the system is represented by the ml Nyquist or Bode plots generated by the elements of $G(j\omega)$. We will see more of these important considerations later.

EXERCISE 1.7.2. Expressing each element of $G(j\omega)$ in the gain/phase form $G_{ik}(j\omega)=a_{ik}(\omega)\,\mathrm{e}^{j\phi_{ik}(\omega)}$, $1\leqslant i\leqslant m$, $1\leqslant k\leqslant l$, show that the steady state sinusoidal response takes the form of the vector with elements, $1\leqslant i\leqslant m$, expressed by the summation

$$\begin{aligned} y_i(t) &= a_{i1}(\omega)\alpha_1\cos(\omega t+\beta_1+\phi_{i1}(\omega))+\ldots \\ &\quad + a_{il}(\omega)\alpha_l\cos(\omega t+\beta_l+\phi_{il}(\omega)) \end{aligned} \tag{1.182}$$

or, equivalently, if we define $G_{i1}(j\omega)\alpha_1\mathrm{e}^{j\beta_1}+G_{i2}(j\omega)\alpha_2\mathrm{e}^{j\beta_2}+\ldots+G_{il}(j\omega)\alpha_l\mathrm{e}^{j\beta_l}=\hat{a}_i\mathrm{e}^{j\hat{\phi}_i}$, then $y_i(t)=\hat{a}_i\cos(\omega t+\hat{\phi}_i)$, $1\leqslant i\leqslant m$.

EXERCISE 1.7.3. Show that the steady state constant response from a stable system to the step input (1.170) is obtained from the relation $y_\infty=\lim_{t\to+\infty}y(t)=G(0)\alpha$.

EXAMPLE 1.7.4. Verify that the system

$$\frac{\mathrm{d}x(t)}{\mathrm{d}t}=\begin{bmatrix}0&1&0\\0&1&0\\0&0&-1\end{bmatrix}x(t)+\begin{bmatrix}0&0\\1&0\\0&1\end{bmatrix}u(t)$$

$$y(t)=\begin{bmatrix}1&0&1\\2&0&1\end{bmatrix}x(t) \tag{1.183}$$

has the weighting function

$$H(t) = \begin{bmatrix} e^t - 1 & e^{-t} \\ 2(e^t - 1) & e^{-t} \end{bmatrix} \quad (1.184)$$

Hence show that the response from zero initial conditions to simultaneous application of unit steps in both u_1 and u_2 at $t = 0$ is

$$y_1(t) = e^t - t - e^{-t}, \qquad y_2(t) = 2e^t - 1 - 2t - e^{-t} \quad (1.185)$$

(Hint: use (1.147) to calculate e^{At} and evaluate (1.167) noting that $u(t) = (1, 1)^T, t > 0$.)

EXERCISE 1.7.4. Consider the intuitive argument that, if $G(j\omega)$ is the generalization of the idea of frequency response, then the matrix of transfer functions $G(s) = C(sI - A)^{-1}B + D$ is likely to be the generalization of the notion of transfer function. More will be seen of this in Section (3.2)

1.8 Discrete State Variable Models

All of the previous sections have considered continuous multi-input/multi-output systems in which the elements of the input vector can vary in an arbitrary way in time producing arbitrary continuous variations in the elements of the output vector. Considerable simplification in the representation of process dynamics can be achieved, however, by the use of approximate (and in some cases exact) *difference equation* models relating the values of states, inputs and outputs at discrete points in time. This is of particular relevance in the development of control systems including mini- or micro-computer control elements where digital data collection and digital input actuation lead naturally to such considerations.

1.8.1 Sampling and discrete systems models

The operation of sampling a scalar function of time $f(t)$ at discrete intervals of time represented by the sampling interval h should be well known to the reader. In its simplest form the operation consists of replacing the function $f(t)$ by the sequence of sampled data points $f(0), f(h), f(2h), f(3h), \ldots$. In physical terms, the sampling operation can be illustrated as in Fig. 15, i.e. as the dynamic operation of replacing the continuous function by a pulse train of frequency h^{-1}. The implicit assumption that the first sample is taken at $t = 0$ is easily assumed by choice of origin in time.

The operation of sampling several functions of time $f_1(t), f_2(t), \ldots, f_p(t)$

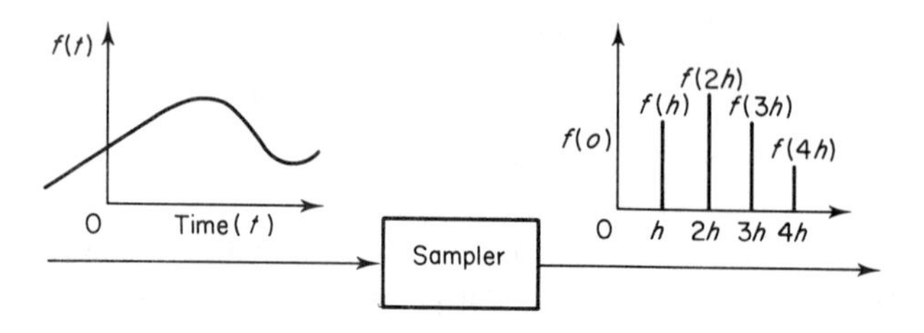

Fig. 15. Sampler frequency h.

consists of sampling each function as described above. More precisely, the kth function $f_k(t)$ is replaced by the sequence of data points $f_k(\alpha_k)$, $f_k(\alpha_k + h_k)$, $f_k(\alpha_k + 2h_k), \ldots$ separated by the sampling interval h_k and initiated at some time $\alpha_k \geqslant 0$. In general terms, there is no justification for assuming that the functions are sampled with the same frequency, nor that the sampling starts at the same point in time. Mathematical analysis of this general case is (to say the least) difficult. For this reason we will restrict our attention to the case of *synchronized sampling*, i.e. the situation when $f_1, f_2, \ldots, f_p$ are all sampled at the same frequency ($h_1 = h_2 = \ldots = h_p = h$) and are all first sampled at time $t = 0$. In this situation we can represent the sampling operation in matrix form. Defining the $p \times 1$ vector function of time $f(t) = (f_1(t), f_2(t), \ldots, f_p(t))^T$, the operation of synchronized sampling of $f_1, \ldots, f_p$ can be regarded as replacing the vector function $f(t)$ by the sequence of sampled data vectors $f(0)$, $f(h)$, $f(2h)$, $f(3h), \ldots$. On paper, therefore, a synchronous sampling of several functions takes the same form as sampling of a scalar function. A nice example of the simplifying effect of matrix notation!

EXERCISE 1.8.1. Show that synchronous sampling of the two functions $f_1(t) = t$, $f_2(t) = t^2$ at intervals repeated by $h = 1$ s generates the data sequence

$$\begin{bmatrix} 0 \\ 0 \end{bmatrix}, \begin{bmatrix} 1 \\ 1 \end{bmatrix}, \begin{bmatrix} 2 \\ 4 \end{bmatrix}, \begin{bmatrix} 3 \\ 9 \end{bmatrix}, \begin{bmatrix} 4 \\ 16 \end{bmatrix}, \ldots$$

Consider now the general nonlinear continuous system of (1.26)–(1.28) and suppose that we synchronously sample the states, inputs and outputs at intervals h. The problem of the construction of a *discrete model* of the system dynamics is that of developing exact recursion relations between the samples of the form (c.f. (1.26)–(1.28))

$$\begin{aligned} x_{k+1} &= \phi(x_k, u_k, k), \qquad (x_0 \text{ specified}) \\ y_k &= \Psi(x_k, u_k, k), \qquad k \geqslant 0 \end{aligned} \tag{1.186}$$

where, for notational convenience, we define $x_k = x(kh)$, $u_k = u(kh)$ and $y_k = y(kh)$, $k \geqslant 0$. If this is possible then, given the initial condition x_0, and the sampled input, the values of the states and outputs at the sample points can be obtained by repeated application of (1.186) for $k = 0, 1, 2, \ldots$, i.e.

$$\begin{aligned} &x_0 \text{ specified}, && y_0 = \Psi(x_0, u_0, 0) \\ &x_1 = \phi(x_0, u_0, 0), && y_1 = \Psi(x_1, u_1, 1) \\ &x_2 = \phi(x_1, u_1, 1), && y_2 = \Psi(x_2, u_2, 2) \\ &\quad\vdots \end{aligned} \tag{1.187}$$

It is important to emphasize, however, that except in very special cases (Section 1.8.3), it is impossible to find an exact discrete model of the process. This is in many ways self evident as the sampling process ignores the input dynamics between samples yet these can have a significant effect on the values of the sampled state. It is, however, very often possible to obtain useful approximate models. For example, the simple recursion relations (1.84) in Section 1.4 can be regarded as an approximate discrete model of process dynamics with $x_{k+1} = \phi(x_k, u_k, k) = x_k + hf(x_k, u_k, kh)$ and $y_k = \Psi(x_k, u_k, k) = g(x_k, u_k, kh)$. The approximation is good if the sampling interval h is small enough.

By far the most useful class of discrete models is the linear time-invariant discrete state-variable model of the general structure (c.f. (1.88))

$$\begin{aligned} x_{k+1} &= \Phi x_k + \Delta u_k && (x_0 \text{ specified}) \\ y_k &= Cx_k + Du_k, && k \geqslant 0 \end{aligned} \tag{1.188}$$

where Φ, Δ, C, D are constant real $n \times n$, $n \times l$, $m \times n$ and $m \times l$ matrices respectively. Conceptually, it can be regarded as a discrete form of the linear time-invariant continuous state-variable model (1.134). (Hence the retention of the notation C, D as the output equation can be sampled without error!) The precise relationship between the two models is explored in detail in Section 1.8.3.

1.8.2 Stability and solution of the discrete equations

In this section attention is focussed on the linear discrete model (1.188) and the development of results and techniques paralleling those of Sections 1.5.2, 1.6 and 1.7.2 for continuous systems.

The general solution of (1.188) is found by iteration as follows

$$\begin{aligned} k &= 0 : x_1 = \Phi x_0 + \Delta u_0 \\ k &= 1 : x_2 = \Phi x_1 + \Delta u_1 = \Phi^2 x_0 + \Phi \Delta u_0 + \Delta u_1 \end{aligned}$$

$$k = 2: x_3 = \Phi x_2 + \Delta u_2 = \Phi^2 x_1 + \Phi\Delta u_1 + \Delta u_2$$
$$= \Phi^3 x_0 + \Phi^2 \Delta u_0 + \Phi\Delta u_1 + \Delta u_2 \tag{1.189}$$
$$\vdots$$

to be of the form (remembering that $\Phi^0 = I_n$)

$$x_k = \Phi^k x_0 + \sum_{i=0}^{k-1} \Phi^{k-i-1} \Delta u_i, \qquad k \geqslant 1 \tag{1.190}$$

The corresponding output sequence is

$$y_k = \begin{cases} Cx_0 + Du_0, & k = 0 \\ C\Phi^k x_0 + \sum_{i=0}^{k-1} C\Phi^{k-i-1}\Delta u_i + Du_k, & k \geqslant 1 \end{cases} \tag{1.191}$$

Equations (1.190) and (1.191) are the equivalent of (1.140) and (1.164), expressing the state and output at a given point in time as an explicit function of the initial state and the (sampled) system input.

The above relationships are rather cumbersome and provide little insight into system behaviour. There are considerable benefits to be obtained by transformation of the state variable as in Sections (1.5.1) and (1.5.2). Let T be an arbitrary $n \times n$ nonsingular matrix and write (c.f. (1.97))

$$z_k = T^{-1} x_k, \qquad k \geqslant 0 \tag{1.192}$$

The reader should easily verify that the transformed state satisfies the discrete state equation

$$z_{k+1} = \Phi^* z_k + \Delta^* u_k, \qquad z_0 = T^{-1} x_0$$
$$\Phi^* = T^{-1}\Phi T, \qquad \Delta^* = T^{-1}\Delta \tag{1.193}$$

which should be compared with (1.98). The output equation becomes (c.f. (1.101), (1.102))

$$y_k = C^* z_k + D^* u_k, \qquad k \geqslant 0$$
$$C^* = CT, \qquad D^* = D \tag{1.194}$$

A particular advantage is obtained if we can choose T so that (see Section 1.5.2)

$$\Phi^* = T^{-1}\Phi T = \begin{bmatrix} \eta_1 & 0 & \cdots & 0 \\ 0 & \eta_2 & & \vdots \\ \vdots & & \ddots & \\ 0 & \cdots & & \eta_n \end{bmatrix} \tag{1.195}$$

i.e. Φ has eigenvalues $\eta_1, \eta_2, \ldots, \eta_n$ with nonsingular eigenvector matrix T. In this case (1.193) becomes

$$(z_{k+1})_i = \eta_i (z_k)_i + \Delta^*_{i1}(u_k)_1 + \ldots + \Delta^*_{il}(u_k)_l \qquad 1 \leqslant i \leqslant n \qquad (1.196)$$

which has explicit solution

$$(z_k)_i = \eta_i^k (z_0)_i + \sum_{r=0}^{k-1} \eta_i^{k-r-1} \sum_{p=1}^{l} \Delta^*_{ip}(u_r)_p \qquad 1 \leqslant i \leqslant n \qquad (1.197)$$

Having computed the sequence $z_0, z_1, z_2, \ldots$, the states $x_0, x_1, x_2, \ldots$ are obtained from the relation $x_k = Tz_k, k \geqslant 0$.

Finally the notions of stability discussed in Section (1.6) can be carried over to discrete systems.

Definition 1.8.1. *The system (1.188) is said to be* asymptotically stable *if, and only if, all solutions of the homogenous systems*

$$x_{k+1} = \Phi x_k \qquad (1.198)$$

tend to zero as $k \to +\infty$ *independent of the choice of initial condition* x_0.

Conditions for stability are devised as follows: assuming for simplicity that Φ can be diagonalized as in (1.95) and setting $z_k = T^{-1}x_k (k \geqslant 0)$, the solution (1.197) for z_k indicates that $(z_k)_i = \eta_i^k (z_0)_i$, $1 \leqslant i \leqslant n$, $k \geqslant 0$ and hence

$$x_k = Tz_k = T \begin{bmatrix} \eta_1^k & \cdot & \cdot & \cdot & 0 \\ & \eta_2^k & & & \vdots \\ & & & & \cdot \\ & & & & \\ 0 & \cdot & \cdot & \cdot & \eta_n^k \end{bmatrix} z_0$$

$$= T \begin{bmatrix} \eta_1^k & & \\ & \ddots & \\ & & \eta_n^k \end{bmatrix} T^{-1} x_0 \qquad (1.199)$$

The following result now follows directly from the definitions (c.f. Theorem 1.6.1 in Section 1.6).

Theorem 1.8.1. *The linear discrete system (1.188) is asymptotically stable if, and only if, the system characteristic polynomial*

$$\rho(z) = |zI - \Phi| = (z - \eta_1)(z - \eta_2) \ldots (z - \eta_n) \qquad (1.200)$$

has roots $\eta_1, \eta_2, \ldots, \eta_n$ with modulus strictly less than unity, i.e. $|\eta_k| < 1$, $1 \leqslant k \leqslant n$. Equivalently, all eigenvalues of Φ must lie inside the familiar unit circle in the complex plane.

1.8.3 Discrete models from continuous models

In certain circumstances, for certain types of input, it is possible to obtain an exact linear discrete model of the form (1.188) relating states, inputs and outputs at the sampling instants. These circumstances are precisely when the system input can be regarded as being generated by a sample-hold device. The reader will recall the notion of a sample-hold, i.e. a device that transforms a continuous signal $f(t)$ into the staircase function $\tilde{f}(t) = f(kh)$, $kh \leqslant t < (k+1)h$ as illustrated in Fig. 16. The idea of a synchronized sample-hold device is a natural extension of the discussion of Section (1.8.1) and, on paper, is defined as above replacing the scalar function $f(t)$ by a vector function.

If the system input vector $u(t)$ can be regarded as the output of a sample-hold unit (as in Fig. 17), it takes the (so-called) *piece-wise constant* form

$$u(t) = u_k, \qquad kh \leqslant t < (k+1)h \tag{1.201}$$

The system input commonly takes this form in computer control systems. Assuming that the system continuous model is linear and time-invariant

$$\frac{\mathrm{d}x(t)}{\mathrm{d}t} = Ax(t) + Bu(t), \qquad x(0) = x_0$$
$$y(t) = Cx(t) + Du(t) \tag{1.202}$$

the general solution (1.140) indicates that

$$x(h) = \mathrm{e}^{Ah}x_0 + \int_0^h \mathrm{e}^{A(h-t)}Bu(t)\,\mathrm{d}t$$
$$= \mathrm{e}^{Ah}x_0 + \int_0^h \mathrm{e}^{A(h-t)}Bu_0\,\mathrm{d}t$$

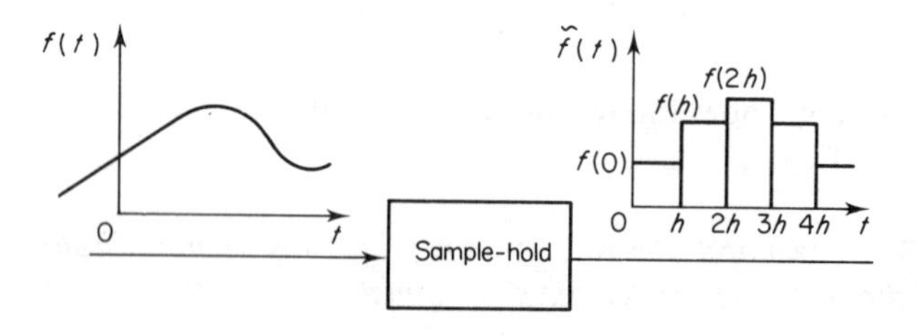

Fig. 16. Sampler and zero-order hold.

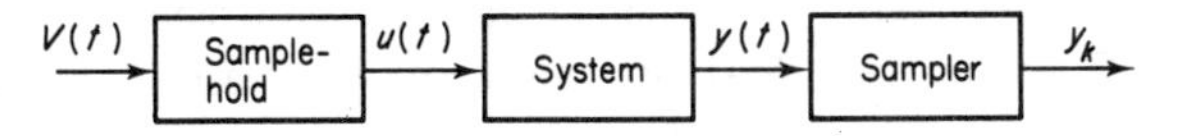

Fig. 17. Continuous system with sampled output and piece-wise constant input.

or, remembering that u_0 is a constant vector, it can be taken out of the integral to yield

$$x_1 = \Phi x_0 + \Delta u_0 \tag{1.203}$$

where Φ and Δ are given explicitly by the formulae

$$\Phi = e^{Ah}, \qquad \Delta = \int_0^h e^{A(h-t)} B \,dt \tag{1.204}$$

It is important to note that relation (1.203) is exact! It is also crucial to observe that, as the underlying continuous system is time-invariant, the relationship between x_2, x_1 and u_1 must be identical to (1.203), namely $x_2 = \Phi x_1 + \Delta u_1$. More generally we can deduce from the time-invariance property that we must have

$$x_{k+1} = \Phi x_k + \Delta u_k, \qquad k \geqslant 0 \tag{1.205}$$

and, trivially, $y_k = Cx_k + Du_k$, $k \geqslant 0$. Again, we emphasize that this discrete model is exact!

Calculation of Φ:

The construction of the discrete model requires the evaluation of Φ and Δ in (1.204). Φ can be approximated by truncation of the exponential series after a suitably large number of terms. The truncations after 2, 3 and 4 terms, for example, take the form

$$\Phi_1(h) = I + Ah$$

$$\Phi_2(h) = I + Ah + \frac{A^2h^2}{2!}$$

$$\Phi_3(h) = I + Ah + \frac{A^2h^2}{2!} + \frac{A^3h^3}{3!} \tag{1.206}$$

More generally, the truncation obtained from the first $M + 1$ terms indicates terms up to the power h^M and takes the form

$$\Phi_M(h) = I + Ah + \frac{A^2h^2}{2!} + \ldots + \frac{A^Mh^M}{M!} \tag{1.207}$$

The matrix Φ is approximated by $\Phi_M(h)$. It is clear that the accuracy of the approximation Φ_M to Φ depends upon the number of terms $M + 1$. The number of terms required for a given accuracy depends upon the step length h. If h is "small" then M can be small but if h is "large" then it is almost inevitable that a large number of terms must be calculated. The exception to this rule is when $A^{M+1}=0$ for some $M \geqslant 0$. In this case it is easily verfied that $\Phi = \Phi_{M+k}(h)$ for all $k \geqslant 0$.

An alternative approach if A has eigenvalues $\eta_1, \eta_2, \ldots, \eta_n$ and nonsingular eigenvector matrix T is obtained by the formula

$$\Phi = e^{Ah} = T e^{T^{-1}ATh} T^{-1}$$
$$= T \begin{bmatrix} e^{\eta_1 h} & & & \\ & e^{\eta_2 h} & & \\ & & \ddots & \\ & & & e^{\eta_n h} \end{bmatrix} T^{-1} \tag{1.208}$$

This expression is exact but requires calculation of $\eta_1, \eta_2, \ldots, \eta_n$ and T.

Calculation of Δ:

The calculation of Δ can proceed in a similar manner to the above. In particular, an approximation to Δ can be obtained from (1.204) by use of a truncated representation of $e^{A(h-t)}$. More precisely, we can use the approximation

$$\Delta_K(h) = \int_0^h \Phi_{K-1}(h-t)B\,dt \tag{1.209}$$

for some $K \geqslant 1$. This can be evaluated precisely by substitution from (1.207) and integration, i.e.

$$\Delta_K(h) = \int_0^h \left(I + A(h-t) + \frac{A^2(h-t)^2}{2!} + \ldots + \frac{A^{K-1}(h-t)^{K-1}}{(K-1)!} \right) B\,dt$$
$$= \left(Ih + \frac{Ah^2}{2!} + \frac{A^2h^3}{3!} + \ldots + \frac{A^{K-1}h^K}{K!} \right) B \tag{1.210}$$

which is easily evaluated on a digital computer. A particularly useful formula can be used if A is nonsingular, namely

$$\Delta_K(h) = \left(Ih + \frac{Ah^2}{2!} + \ldots + \frac{A^{K-1}h^K}{K!} \right) B$$
$$= \left(Ah + \frac{A^2h^2}{2!} + \ldots + \frac{A^K h^K}{K!} \right) A^{-1}B$$
$$= (\Phi_K(h) - I)A^{-1}B \tag{1.211}$$

In particular, letting $K \to +\infty$ yields the exact formula

$$\Delta = (\Phi - I)A^{-1}B \tag{1.212}$$

valid whenever A is nonsingular.

Finally, if the eigenvalues and eigenvector matrix of A are known, Δ could be obtained from the following exact formula,

$$\Delta = \int_0^h e^{A(h-t)} B \, dt = T \left(\int_0^h e^{T^{-1}AT(h-t)} dt \right) T^{-1} B$$

$$= T \begin{bmatrix} \int_0^h e^{\eta_1 (h-t)} dt & \cdots & 0 \\ \vdots & & \vdots \\ 0 & \cdots & \int_0^h e^{\eta_n (h-t)} dt \end{bmatrix} T^{-1} B \tag{1.213}$$

EXERCISE 1.8.2. Use (1.208) to show that the discrete system (1.205) is stable if, and only if, the underlying continuous system is stable. (Hint: look at theorems 1.6.1. and 1.8.1 and note that

$$\eta_k = e^{\lambda_k h}, \qquad 1 \leqslant k \leqslant n \tag{1.214}$$

Write $\lambda_k = \sigma_k + j\omega_k (\sigma_k, \omega_k$ real) and note that $|\eta_k| = e^{\sigma_k h}$ and $h > 0$.)

EXERCISE 1.8.3. Use (1.207) and (1.210) to verify that the discrete model (1.88) obtained from the Euler method is equivalent to (1.205) with Φ and Δ approximated by $\Phi_1(h)$ and $\Delta_1(h)$ respectively.

1.9 A Note on Controllability and Observability Problems

It is evident from all the preceding sections that the stability and dynamic responses of systems described by (continuous or discrete) state-variable models depend crucially upon the nature of the interconnections between the individual system states and the interconnections between these states and the system inputs and outputs. Consideration of the nature of these interconnections is of great significance if, say, a system is unstable; when it is necessary to know whether or not the system is stabilizable by the use of feedback. Alternatively, if a system is unstable, it is necessary to know whether unstable behaviour will be detected by monitoring the available measured outputs.

EXAMPLE 1.9.1. Consider the single-input/single-output continuous systems illustrated in Fig. 18. The system of Fig. 18(a) is unstable but the unstable behaviour of x_1 can be deduced by monitoring the output y. Unfortunately, the dynamics of x_1 are unaffected by the input u and

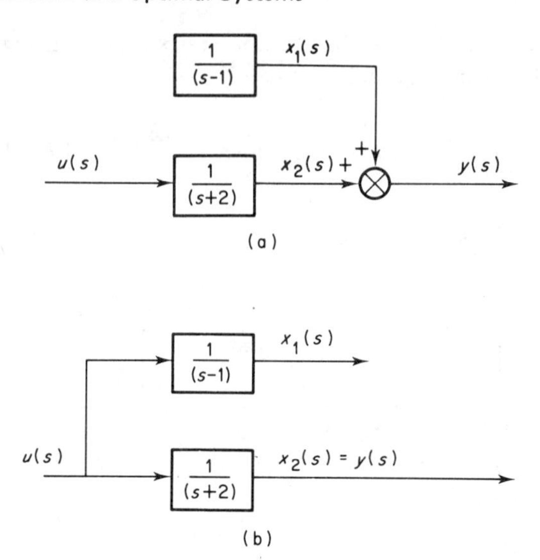

Fig. 18. Uncontrollable and unobservable systems.

hence no form of feedback control will remove this instability. The system of Fig. 18(b) is again unstable but here the instability cannot be detected by monitoring the output. Thus, although the unstable variable x_1 is affected by the input u, feedback control will not remove the instability.

In this section we sketch out the formal representation of these ideas in the form of the ideas of controllability and observability. In rough terms, the property of *controllability* will imply that the inputs can influence all state variables independently and the property of *observability* will imply that behaviour of every state variable can be detected by monitoring the system outputs over a period of time. The following definitions make these ideas more precise for the case of continuous systems. The definitions for discrete systems take the obvious form.

Definition 1.9.1. *A state variable model of a dynamic system is said to be* completely state-controllable (*or, for simplicity*, controllable) *if, given any states* $\mathbf{x}_0$ *and* $\mathbf{x}_f$ *and any initial time* t_0, *there exists a time* $t_1 > t_0$ *and an input* $\mathbf{u}(t)$ *such that the system state is transferred from the initial state* $\mathbf{x}(t_0) = \mathbf{x}_0$ *to the final state* $\mathbf{x}(t_1) = \mathbf{x}_f$.

In essence, the system is controllable if we can transfer any given initial state to any final state in a finite time by suitable choice of input, i.e. we have complete control of the dynamics of the state.

Definition 1.9.2. *A state variable model of a dynamic system is said to be* completely state-observable (*or, for simplicity,* observable) *if, for any time t_0, there exists a time $t_1 > t_0$ such that a knowledge of the output vector $y(t)$ and input vector $u(t)$ in the time interval $t_0 \leqslant t \leqslant t_1$ is sufficient to determine the initial state $x(t_0)$ uniquely.*

Conditions for the controllability and observability of a linear continuous time-invariant system are stated without proof below.

Theorem 1.9.1. *A necessary and sufficient condition for a linear continuous time-invariant system to be completely state controllable is that*

$$\text{rank}\ [B, AB, A^2B, \ldots, A^{n-1}B] = n \tag{1.215}$$

(Note: controllability depends only on the state matrix A and the input matrix B.)

Theorem 1.9.2. *A necessary and sufficient condition for a linear continuous time-invariant system to be completely state observable is that*

$$\text{rank}\ [C^T, A^TC^T, \ldots, (A^T)^{n-1}C^T] = n \tag{1.216}$$

(Note: controllability depends only on the state matrix A and the measurement matrix C.)

Before concluding this section note that

(a) The absence of controllability and/or observability is a pathological situation. In general a well designed plant will be controllable and observable. In this sense the above results are primarily of theoretical value.

(b) For those readers not familiar with the idea of the rank of a matrix, condition (1.215) can be replaced by the equivalent requirement that we can construct a $n \times n$ nonsingular matrix with columns selected from the columns of the $n \times ln$ matrix $[B, AB, \ldots, A^{n-1}B]$. Similar comments hold for (1.216).

EXAMPLE 1.9.2. The liquid level system is controllable as $n = 2$ and

$$[B, AB] = \begin{bmatrix} 1/a_1 & 0 & -\beta/a_1^2 & \beta/a_1a_2 \\ 0 & 1/a_2 & \beta/a_1a_2 & -\beta/a_2^2 \end{bmatrix} \tag{1.217}$$

has rank $= 2$ (the 2×2 matrix with columns equal to the first two columns of $[B, AB]$ is nonsingular). It is also observable as

$$[C^T, A^TC^T] = \begin{bmatrix} 1 & 0 & -\beta/a_1 & \beta/a_2 \\ 0 & 1 & \beta/a_1 & -\beta/a_2 \end{bmatrix} \tag{1.218}$$

has rank $= 2$ (take the first two columns again).

EXERCISE 1.9.1. Show that the system of Example (1.2.2) is both controllable and observable. Use the result of Exercise (1.2.3).

Finally conditions for the controllability and observability of a linear discrete time-invariant system are stated below.

Theorem 1.9.3. *A necessary and sufficient condition for the linear discrete, time-invariant system (1.188) to be controllable (resp. observable) is that*

$$\text{rank } [\Delta, \Phi\Delta, \ldots, \Phi^{n-1}\Delta] = n$$

$$(\textit{resp.}\ \text{rank } [C^T, \Phi^T C^T, \ldots, (\Phi^T)^{n-1} C^T] = n) \tag{1.219}$$

Problems

(1) Consider the RCL circuit of Fig. 19 with voltage source input $u(t)$ and output equal to the capacitor current $i_2(t)$. Verify that the circuit equations

$$L\frac{di_1(t)}{dt} = u(t) - v_2(t)$$

$$C\frac{dv_2(t)}{dt} = i_1(t) - \frac{1}{R}v_2(t)$$

give rise to the linear, time-invariant state variable model

$$\frac{dx(t)}{dt} = \begin{bmatrix} 0 & -\frac{1}{L} \\ \frac{1}{C} & -\frac{1}{RC} \end{bmatrix} x(t) + \begin{bmatrix} \frac{1}{L} \\ 0 \end{bmatrix} u(t)$$

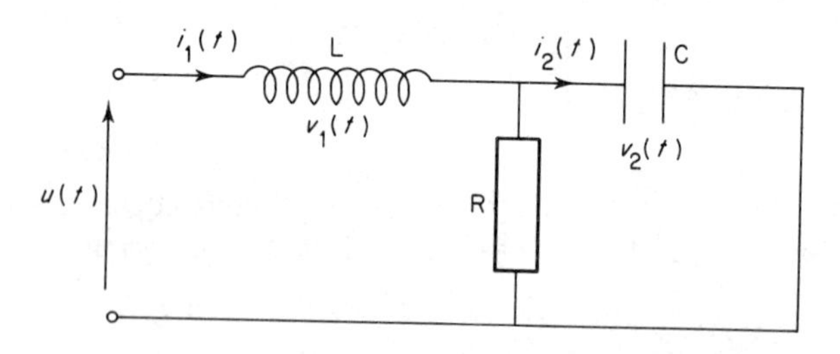

Fig. 19. An RCL network.

$$y(t) = \begin{bmatrix} 1 & -\frac{1}{R} \end{bmatrix} x(t)$$

(Hint: use Kirchoff's voltage and current laws and $x_1 = i_1, x_2 = v_2$.)

(2) Show that the network of Fig. 20 with voltage source input $u(t)$ and output equal to the voltage across the capacitor is represented by the linear time-invariant model

$$\frac{dx(t)}{dt} = \begin{bmatrix} -\frac{1}{C}\left\{\frac{1}{R_1} + \frac{1}{R_2}\right\} & -\frac{1}{C} \\ \frac{1}{L} & 0 \end{bmatrix} x(t) + \begin{bmatrix} \frac{1}{R_1 C} \\ 0 \end{bmatrix} u(t)$$

$$y(t) = [1 \quad 0]\, x(t)$$

where the state variables are $x_1 = V, x_2 = i$.

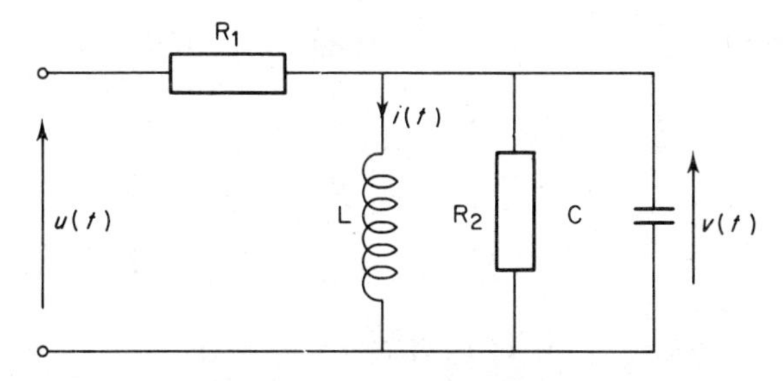

Fig. 20. Another RCL circuit.

(3) Verify that a single-input/single-output system with standard lead-lag, proportional plus integral transfer function

$$g(s) = k\frac{(1 + sT_1)}{(1 + sT_2)}\left\{1 + \frac{1}{sT_3}\right\}$$

can be represented in the form shown in Fig. 21. Hence show that the system has a linear, time-invariant model of the form

$$\frac{dx(t)}{dt} = \begin{bmatrix} 0 & \frac{1}{T_3} \\ 0 & -\frac{1}{T_2} \end{bmatrix} x(t) + \begin{bmatrix} \frac{kT_1}{T_2 T_3} \\ \frac{k(T_2 - T_1)}{T_2^2} \end{bmatrix} u(t)$$

$$y(t) = (1 \quad 1)x(t) + \frac{kT_1}{T_2} u(t)$$

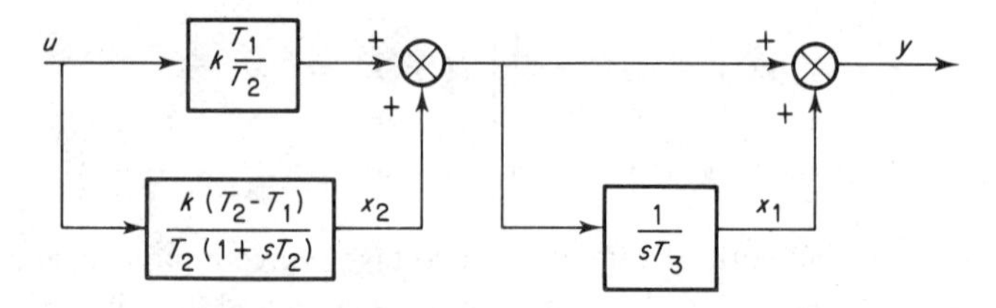

Fig. 21. Block diagram for lead-lag, proportional plus integral system.

(4) A single-input/single-output system of the form of equation (1.3) has transfer function (1.7) with *partial fraction* expansion of the form

$$g(s) = \frac{b_0}{a_0} + \frac{\alpha_1}{s-\lambda_1} + \frac{\alpha_2}{s-\lambda_2} + \ldots + \frac{\alpha_n}{s-\lambda_n}$$

if the n solutions $\lambda_1, \lambda_2, \ldots, \lambda_n$ of the relation $P(\lambda) = 0$ are distinct. Verify that the system can be represented in the block diagram form of Fig. 22. Hence deduce that the system has a state variable model of the form

$$\frac{\mathrm{d}x(t)}{\mathrm{d}t} = \begin{bmatrix} \lambda_1 & 0 & \ldots & 0 \\ 0 & \lambda_2 & & \vdots \\ \vdots & & & 0 \\ 0 & \ldots & 0 & \lambda_n \end{bmatrix} x(t) + \begin{bmatrix} \alpha_1 \\ \alpha_2 \\ \vdots \\ \alpha_n \end{bmatrix} u(t)$$

$$y(t) = [1 \quad 1 \ldots 1]\, x(t) + \frac{b_0}{a_0}\, u(t)$$

where the A matrix has a diagonal form (Hint: write down the differential equations for each block in Fig. 22.)

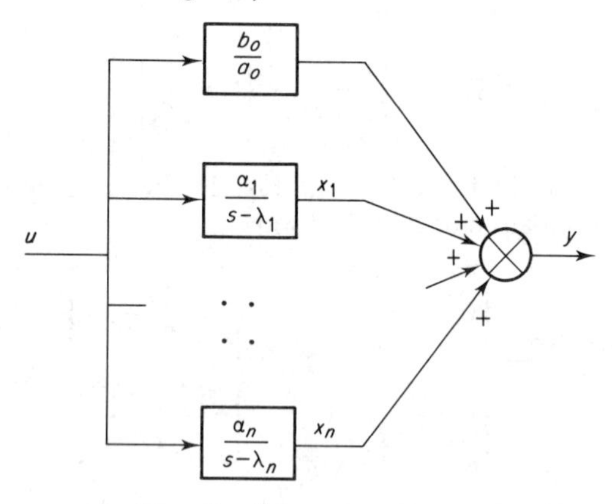

Fig. 22. Parallel system representation.

(5) A single-input/single-output system has transfer function $g(s) = \alpha/(s+\eta)^q$. By representing the system as in Fig. 23, deduce that the system has a state space representation of the form

$$\frac{dx(t)}{dt} = \begin{bmatrix} \eta & 1 & 0 & \cdot & \cdot & 0 \\ 0 & \eta & 1 & 0 & & \vdots \\ \vdots & & & & & 0 \\ \vdots & & & & & 1 \\ 0 & \cdot & \cdot & \cdot & 0 & \eta \end{bmatrix} x(t) + \begin{bmatrix} 0 \\ 0 \\ \vdots \\ 0 \\ \alpha \end{bmatrix} u(t)$$

$$y(t) = [1 \quad 0 \quad 0]\, x(t)$$

Consider how one might obtain a state-space representation of the system

$$g(s) = \frac{\alpha_1}{(s+\eta_1)^{n_1}} + \frac{\alpha_2}{(s+\eta_2)^{n_2}} + \ldots + \frac{\alpha_q}{(s+\eta_q)^{n_q}}$$

where the A matrix is in the Jordan form defined by (1.127) and (1.128). (Hint: use a system decomposition analogous to Fig. 22 consisting of q parallel branches with transfer functions $\alpha_j/(s+\eta_j)^{n_j}$, $1 \leqslant j \leqslant m$.)

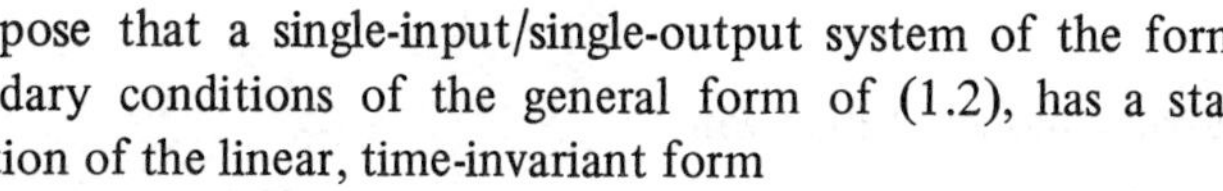

Fig. 23. Series system representation.

(6) (a) Suppose that a single-input/single-output system of the form of (1.3), with boundary conditions of the general form of (1.2), has a state variable representation of the linear, time-invariant form

$$\frac{dx(t)}{dt} = Ax(t) + Bu(t), \qquad y(t) = Cx(t) + Du(t)$$

Show that

$$y(t) = Cx(t) + Du(t)$$

$$\frac{dy(t)}{dt} = CAx(t) + CBu(t) + D\frac{du(t)}{dt}$$

$$\vdots$$

and hence that the matrix equation

$$\begin{bmatrix} d_0 \\ d_1 \\ d_2 \\ \vdots \\ d_{n-1} \end{bmatrix} = \begin{bmatrix} C \\ CA \\ CA^2 \\ \vdots \\ CA^{n-1} \end{bmatrix} x(0)$$

$$+ \begin{bmatrix} Du(0) \\ CBu(0) + D\left.\dfrac{du(t)}{dt}\right|_{t=0} \\ \vdots \\ CA^{n-2}Bu(0) + CA^{n-3}B\left.\dfrac{du(t)}{dt}\right|_{t=0} + \ldots + D\left.\dfrac{d^{n-1}u(t)}{dt^{n-1}}\right|_{t=0} \end{bmatrix}$$

enables the calculation of the initial state $x(0)$ from the initial output and input data if the system state-variable model is observable.

(b) If the state-variable model is observable show that the condition of zero output and input initial conditions is equivalent to the state variable initial condition $x(0) = 0$.

(7) Use the results of problems (4) and (6) to deduce that a system with transfer function

$$g(s) = \frac{2}{s(s+1)(s+2)}$$

has a state variable model of the form

$$\frac{dx(t)}{dt} = \begin{bmatrix} 0 & 0 & 0 \\ 0 & -1 & 0 \\ 0 & 0 & -2 \end{bmatrix} x(t) + \begin{bmatrix} 1 \\ -2 \\ 1 \end{bmatrix} u(t)$$

$$y(t) = (1 \quad 1 \quad 1)x(t)$$

Deduce that the system response to a unit step demand from zero initial conditions is $y(t) = t - 2(1 - e^{-t}) + \frac{1}{2}(1 - e^{-2t})$. Check this result by direct calculation of the inverse Laplace transform of $s^{-1}g(s)$.

(8) For the RCL circuit of problem (1), show that the system matrix

$$A = \begin{bmatrix} 0 & -\dfrac{1}{L} \\ \dfrac{1}{C} & -\dfrac{1}{RC} \end{bmatrix}$$

has eigenvalues

$$\lambda_1 = -1/2RC + \sqrt{((1/2RC)^2 - (1/LC))}$$
$$\lambda_2 = -1/2RC - \sqrt{((1/2RC)^2 - (1/LC))}$$

and an eigenvector matrix

$$T = \begin{bmatrix} 1 & 1 \\ -\lambda_1 L & -\lambda_2 L \end{bmatrix}$$

which is nonsingular if, and only if, $\lambda_1 \neq \lambda_2$. Show also that the system is asymptotically stable for all choice of parameters $R > 0$, $C > 0$, $L > 0$ but that exponentially decaying oscillations are present if $L < 4R^2C$.

(9) Show that the single-input/single-output second order lag

$$(D^2 + 2\xi\omega_0 D + \omega_0^2)y(t) = \omega_0^2 u(t), \qquad \omega_0^2 > 0, \qquad \xi > 0$$

has the state variable description

$$\frac{dx(t)}{dt} = \begin{bmatrix} 0 & 1 \\ -\omega_0^2 & -2\xi\omega_0 \end{bmatrix} x(t) + \begin{bmatrix} 0 \\ \omega_0^2 \end{bmatrix} u(t)$$

$$y(t) = [1 \quad 0]\, x(t)$$

if we choose states $x_1 = y, x_2 = dy/dt$. Show that the system has the characteristic polynomial $\rho(s) = |sI - A| = s^2 + 2\xi\omega_0 s + \omega_0^2$ and hence eigenvalues $\lambda_1 = \omega_0(-\xi + \sqrt{(\xi^2 - 1)})$ and $\lambda_2 = \omega_0(-\xi - \sqrt{(\xi^2 - 1)})$ with eigenvector matrix $T = \begin{bmatrix} 1 & 1 \\ \lambda_1 & \lambda_2 \end{bmatrix}$ that is nonsingular if $\lambda_1 \neq \lambda_2$. (Equivalently if the damping ratio $\xi \neq 1$ and hence the system is either underdamped or overdamped.) Defining $z(t) = T^{-1}x(t)$, show in this case that

$$\frac{dz(t)}{dt} = \begin{bmatrix} \lambda_1 & 0 \\ 0 & \lambda_2 \end{bmatrix} z(t) + \frac{\omega_0^2}{(\lambda_2 - \lambda_1)} \begin{bmatrix} -1 \\ 1 \end{bmatrix} u(t)$$

$$y(t) = [1 \quad 1]\, z(t)$$

is the diagonal form of state equations. Deduce that the response from zero initial conditions to a unit step input is simply

$$y(t) = \frac{\omega_0^2}{(\lambda_2 - \lambda_1)} \{\lambda_1^{-1}(1 - e^{-\lambda_1 t}) - \lambda_2^{-1}(1 - e^{-\lambda_2 t})\}.$$

In the case of equal eigenvalues $\lambda_1 = \lambda_2 = -\omega_0$, use the transformation $T = \begin{bmatrix} 1 & 0 \\ -\omega_0 & 1 \end{bmatrix}$ to transform the system to the Jordan form

$$\frac{dz(t)}{dt} = \begin{bmatrix} -\omega_0 & 1 \\ 0 & -\omega_0 \end{bmatrix} z(t) + \begin{bmatrix} 0 \\ \omega_0^2 \end{bmatrix} u(t)$$

$$y(t) = [1 \quad 0]\, z(t)$$

Noting that $\lambda_1 = \lambda_2$ if, and only if, the damping ratio $\xi = 1$ we see that although the system cannot be transformed to diagonal form if it is critically damped, it can be transformed to Jordan form.

(10) A two-input/two-output system has the structure defined by Fig. 24. Using the defined states x_1, x_3 and $x_2 = dx_1/dt$, show that the system has a state variable representation of the form

$$\frac{dx(t)}{dt} = \begin{bmatrix} 0 & 1 & 0 \\ 0 & -1 & 1 \\ K & 0 & -2 \end{bmatrix} x(t) + \begin{bmatrix} 0 & 0 \\ 1 & 0 \\ 0 & 1 \end{bmatrix} u(t)$$

$$y(t) = \begin{bmatrix} 1 & 0 & 0 \\ 0 & 0 & 1 \end{bmatrix} x(t)$$

Verify that the system has the characteristic polynomial

$$\rho(s) = s(s+1)(s+2) - K = s^3 + 3s^2 + 2s - K$$

and use the Routh stability criterion (see, for example, Raven (1978)) to show that the system is stable if K lies in the range $-6 < K < 0$. In the case of $K =$

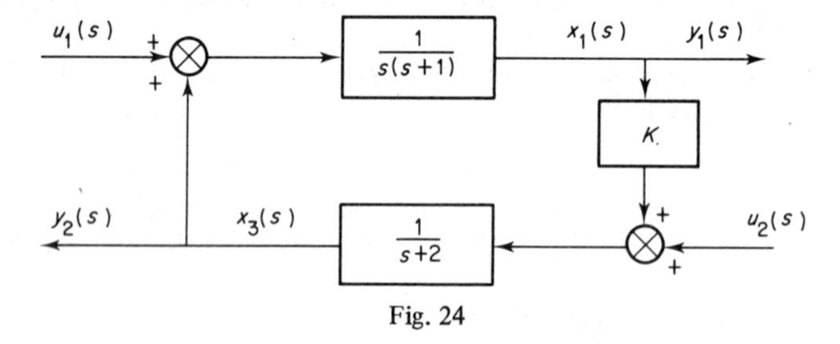

Fig. 24

-6, verify that the system is critically stable and capable of sustained self-oscillation at a frequency of $\sqrt{2}\,\mathrm{rad\,s^{-1}}$ (show that the system has eigenvalues $\lambda_1 = j\sqrt{2}, \lambda_2 = -j\sqrt{2}, \lambda_3 = -3$).

Finally, using the state transformation $z(t) = T^{-1}x(t)$ with

$$T = \begin{bmatrix} 1 & 1 & 1 \\ j\sqrt{2} & -j\sqrt{2} & -3 \\ j\sqrt{2}-2 & -j\sqrt{2}-2 & 6 \end{bmatrix}$$

and considering the case of zero systems inputs (i.e. $u_1(t) \equiv u_2(t) \equiv 0$), verify that

$$\frac{dz(t)}{dt} = \begin{bmatrix} j\sqrt{2} & 0 & 0 \\ 0 & -j\sqrt{2} & 0 \\ 0 & 0 & -3 \end{bmatrix} z(t)$$

$$y(t) = \begin{bmatrix} 1 & 1 & 1 \\ j\sqrt{2}-2 & -j\sqrt{2}-2 & 6 \end{bmatrix} z(t)$$

and hence find the response of the output from the initial condition $x_1(0) = 1$, $x_2(0) = x_3(0) = 0$.

(11) Verify for the matrices

(a) $A = \begin{bmatrix} \alpha & \beta \\ \beta & \alpha \end{bmatrix}$, (b) $A = \begin{bmatrix} 1 & -2 \\ 1 & -2 \end{bmatrix}$, (c) $A = \begin{bmatrix} 0 & 1 \\ -1 & 0 \end{bmatrix}$

that the matrix exponential e^{At} takes the form

(a) $$e^{At} = \tfrac{1}{2}\begin{bmatrix} e^{(\alpha+\beta)t} + e^{(\alpha-\beta)t} & e^{(\alpha+\beta)t} - e^{(\alpha-\beta)t} \\ e^{(\alpha+\beta)t} - e^{(\alpha-\beta)t} & e^{(\alpha+\beta)t} + e^{(\alpha-\beta)t} \end{bmatrix}$$

(b) $$e^{At} = \begin{bmatrix} 2-e^{-t}, & -2+2e^{-t} \\ 1-e^{-t}, & -1+2e^{-t} \end{bmatrix}$$

(c) $$e^{At} = \begin{bmatrix} \cos t & \sin t \\ -\sin t & \cos t \end{bmatrix}$$

(12) For the homogenous system $dx(t)/dt = Ax(t)$, Section 1.7.2 has indicated that $x(t) = e^{At}x(0)$. An alternative approach is to take Laplace transforms

to obtain $sx(s) - x(0) = Ax(s)$ and hence $x(t) = \mathscr{L}^{-1}((sI - A)^{-1}x(0)) = (\mathscr{L}^{-1}(sI - A)^{-1})x(0)$ (where the inverse Laplace transform of a matrix is just the matrix of inverse Laplace transforms). In particular, it follows that $e^{At} = \mathscr{L}^{-1}((sI - A)^{-1})$ which provides a simple technique for calculating e^{At}. Taking, for example, the matrix of problem 11(b) we deduce that

$$e^{At} = \mathscr{L}^{-1}(sI - A)^{-1} = \mathscr{L}^{-1}\begin{bmatrix} \dfrac{s+2}{s(s+1)} & \dfrac{-2}{s(s+1)} \\ \dfrac{1}{s(s+1)} & \dfrac{s-1}{s(s+1)} \end{bmatrix}$$

$$= \begin{bmatrix} 2 - e^{-t} & -2 + 2e^{-t} \\ 1 - e^{-t} & -1 + 2e^{-t} \end{bmatrix}$$

in agreement with previous results. Repeat the calculations for the other matrices in problem (11).

(13) Show that the nonlinear system

$$\frac{dx_1}{dt}(t) = (x_1(t))^2 + x_2(t)(x_2(t) - 1) + u_1(t)$$

$$\frac{dx_2}{dt}(t) = x_1(t)(x_1(t) - 2) - x_2(t)(x_2(t) - 1) + u_2(t)$$

has two *real* equilibrium states

$$\text{(a)} \quad x_s = \begin{bmatrix} 0 \\ 0 \end{bmatrix}, \qquad \text{(b)} \quad x_s = \begin{bmatrix} 0 \\ 1 \end{bmatrix}$$

corresponding to the case of zero inputs. Hence show that the corresponding linearized models describing dynamics about the equilibrium points are

(a) $$\frac{dx(t)}{dt} = \begin{bmatrix} 0 & -1 \\ -2 & 1 \end{bmatrix} x(t) + \begin{bmatrix} 1 & 0 \\ 0 & 1 \end{bmatrix} u(t)$$

(b) $$\frac{dx(t)}{dt} = \begin{bmatrix} 0 & 1 \\ -2 & -1 \end{bmatrix} x(t) + \begin{bmatrix} 1 & 0 \\ 0 & 1 \end{bmatrix} u(t)$$

respectively. Investigate the stability of the models in each case.

(14) Show that the state-variable model

$$\frac{dx(t)}{dt} = \begin{bmatrix} 0 & 1 & 0 \\ 0 & 0 & 1 \\ -1 & -2 & -3 \end{bmatrix} x(t) + \begin{bmatrix} 0 \\ 0 \\ 1 \end{bmatrix} u(t)$$

$$y(t) = (1 \quad 1 \quad 1)x(t)$$

is a representation of the differential system

$$(D^3 + 3D^2 + 2D + 1)y(t) = (D^2 + D + 1)u(t)$$

More generally show that the state-variable model

$$\frac{dx(t)}{dt} = \begin{bmatrix} 0 & 1 & 0 & \cdots & 0 \\ 0 & 0 & 1 & & \vdots \\ \vdots & & & & 0 \\ 0 & \cdots & \cdots & 0 & 1 \\ -a_n & -a_{n-1} & & \cdots & -a_1 \end{bmatrix} x(t) + \begin{bmatrix} 0 \\ 0 \\ \vdots \\ 0 \\ 1 \end{bmatrix} u(t)$$

$$y(t) = [b_n \quad b_{n-1} \ \cdots \ b_1]\, x(t)$$

is a representation of the differential system

$$(D^n + a_1 D^{n-1} + \ldots + a_{n-1} D + a_n)y(t) = (b_1 D^{n-1} + \ldots + b_n)u(t)$$

Hence find a state variable model of each of the systems

(a) $$(D^4 + 2D^2 + 3)y(t) = (D - 1)u(t)$$

(b) $$D^2 y(t) + 2Dy(t) + y(t) = u(t)$$

(c) $$(D^2 - 1)y(t) = (D - 1)u(t)$$

and check your results by recalculation of the differential equation. Verify in each case that the system is controllable and that, with the exception of (c), each system is also observable.

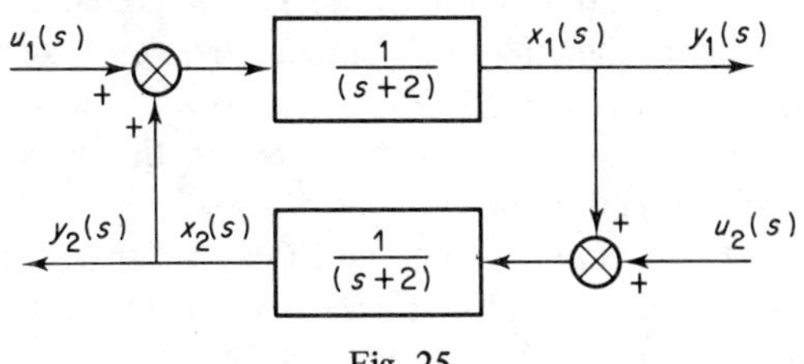

Fig. 25

(b) Verify that the system of Fig. 25 has the state-variable model

$$\frac{dx(t)}{dt} = \begin{bmatrix} -2 & 1 \\ 1 & -2 \end{bmatrix} x(t) + \begin{bmatrix} 1 & 0 \\ 0 & 1 \end{bmatrix} u(t)$$

$$y(t) = \begin{bmatrix} 1 & 0 \\ 0 & 1 \end{bmatrix} x(t)$$

If the inputs u_1 and u_2 are generated from synchronized sample-hold units of period $h = 1$ s, verify that the system has a discrete model of the form

$$x_{k+1} = \begin{bmatrix} 0.209 & 0.159 \\ 0.159 & 0.209 \end{bmatrix} x_k + \begin{bmatrix} 0.47 & 0.16 \\ 0.16 & 0.47 \end{bmatrix} u_k$$

$$y_k = \begin{bmatrix} 1 & 0 \\ 0 & 1 \end{bmatrix} x_k$$

Check the stability of the two models.

Use your results to find the response of the system at the sample instants to a unit step input in $u_1(t)$. Assume zero initial conditions.

Remarks and Further Reading

This chapter has provided the minimum foundation required for further study in control theory. The reader should not be alarmed by this as, although minimal, the material will take him a long way before his mathematics fails him. It is emphasized that the reader should be familiar with classical control theory (see, for example, Raven, 1978; Shinners, 1978; Marshall, 1978; Power and Simpson, 1978; Takahashi *et al.*, 1972) and should brush up on his matrix theory with particular emphasis on eigenvalues and eigenvectors (see, for example, Gantmacher, 1959; Mirsky, 1963; Bellman, 1970; Tropper, 1969). More detailed comments on individual topics follow.

(1) The construction of state-variable models is described in many textbooks on control science and engineering (see, for example, Power and Simpson, 1978; Chen, 1970; Ogata, 1967; Layton, 1976; Owens, 1978; Takahashi *et al.*, 1972; Porter, 1966). In most cases emphasis is placed on the construction of linear time-invariant models. In fact, in many cases, the texts are solely concerned with linear systems (see, for example, Zadeh and Desoer, 1963; Rugh, 1975; Chen, 1970; Barnett, 1975; Kailath, 1980). For a text including a large number of examples see Swisher (1976) or Takahashi *et al.* (1972). For an analysis of more

general classes of model see Rosenbrock (1970), Wolovich (1974) or Kailath (1980).

(2) The idea of a system equilibrium state is fundamental to the analysis and design of any system designed for "base-load" operation. In such cases a lot of thought has to go into the design of the system to produce acceptable steady state values of the state variables (e.g. in the design of nuclear reactors for steady state operation, the average temperature of fuel elements in the reactor core may be specified to produce the required power output whilst the peak fuel element temperature is constrained for safety reasons). In most cases the design of the equilibrium state requires the solution of (1.43) for a number of trial inputs u_s to produce the required form of x_s. Some indication of the techniques that can be used can be obtained from Ortega and Rheinboldt (1970).

(3) Linearization is a widely used technique for obtaining approximate linear models of process dynamics valid in the vicinity of an equilibrium point. For more details on the mathematical background see Brockett (1970) or Holtzmann (1970). For an alternative simple approach, see Power and Simpson (1978), Raven (1978) or Rugh (1975).

(4) It is a useful fact that every single-input/single-output system described by the ordinary differential equation (1.3) can be represented in state variable form. This has been illustrated in the preceding problems (see, for example, problems (4)–(7), (14)) and is treated in more depth by Ogata (1967), Takahashi *et al.* (1972) and Owens (1978), for example.

(5) Numerical simulation of sets of first order ordinary differential equations is a standard problem. Library subroutines to perform this operation are normally available on a scientific computer. For examples of programs, see Goult *et al.* (1974), Melsa (1970) or Dorf (1974).

(6) For a rigorous development and use of the Jordan form, see Chen (1970) or Ogata (1967). For geometric interpretations of eigenvalues and eigenvectors, see Takahashi *et al.* (1972) or Porter (1966).

(7) The concepts of stability used here are the natural extensions of the idea of pole used in classical theory. More general approaches using the nice notions of Liapunov function can be found in Willems (1970), Takahashi *et al.* (1972), Ogata (1967), Layton (1978) or Munro (1979a).

(8) For a rigorous development of the notion of matrix exponential and its generalization used in the solution of linear, time-varying systems, see, for example, Zadeh and Desoer (1963) or Brockett (1970).

(9) The notion of sampling is a prerequisite for this text and can be found in many textbooks, e.g. Power and Simpson (1978), Raven (1978), Cappellini *et al.* (1978), Cadzow (1973) or Jury (1964). For an insight into nonlinear discrete system stability theory, see Vidal (1969).

(10) Controllability and observability are fundamentally important in general systems theory via the mathematical ideas of minimality (see, for example, Brockett, 1970). They also play a central role in the theory of stabilization by pole allocation (see, for example, Wonham, 1974; Porter and Crossley, 1972). For these (and other) reasons a large research effort has been invested in their characterization. The interested reader should refer to Wonham (1974) for a geometric treatment or to Wolovich (1974) or Rosenbrock (1970) for an algebraic treatment. Standard matrix treatments can also be found in textbooks such as Brockett (1970), Ogata (1967), Barnett (1975), Rugh (1975), Owens (1978) or Kailath (1980).

(11) It is worth the reader's while reassessing any prejudice that he may have held against the use of matrix methods. A quick review of the material of this chapter should underline the formal simplicity made possible by the use of matrix notation. The reader should also note that the solution of the state equations naturally gives rise to the notion of eigenvector transformations. It is also clear that the analysis of stability requires the use of eigenvalues and that the general solution described in Section 1.7 and its application in Section 1.8 is not possible without the introduction of matrix exponentials. I rest my case!

2. State Feedback

In this chapter we explore the basic ideas and matrix representations of feedback systems for the control of dynamic systems described by state-variable models of the form introduced in Chapter one. There will be no attempt at an exhaustive treatment, emphasis being placed on concepts and methodology. Interested readers are referred to the reading list for details of theoretical development and applications.

2.1 Feedback

In general terms the objective of control action on an l-input/m-output dynamic system is to manipulate all the system inputs $u_1(t), u_2(t), \ldots, u_l(t)$ in such a way that all of the outputs $y_1(t), y_2(t), \ldots, y_m(t)$ behave in an acceptable manner in situations of interest. In matrix terms, the system input vector $u(t)$ is manipulated to ensure that the output vector $y(t)$ behaves in an acceptable manner. The precise definition of "acceptable" is left for Chapter three. The purpose of this section is to identify general characteristics of feedback systems in this general multi-input/multi-output case.

(a) *The* m x *1 demand vector.* The reader should be familiar with the idea of demanded output signal in the context of the single-input/single-output feedback system of Fig. 4. More precisely, the demand $r(t)$ is a function of time t equal to the *ideal* closed-loop system response. It is the input to the closed-loop system.

In an l-input/m-output system there are, by definition, m outputs $y_1(t), \ldots, y_m(t)$ and hence, for control purposes, we must specify m demanded output signals $r_1(t), r_2(t), \ldots, r_m(t)$ where $r_k(t)$ is the ideal response from the output $y_k(t)$. These demand signals will be inputs to the closed-loop system or, in matrix form, the $m \times 1$ *demand* vector

$$r(t) = \begin{bmatrix} r_1(t) \\ r_2(t) \\ \vdots \\ r_m(t) \end{bmatrix} \tag{2.1}$$

will be a vector input to the closed-loop system.

(b) *The* r × *1 measurement vector.* For the purposes of this chapter we shall extend the state-variable models to include situations where measurements of the elements of the output vector $y(t)$ are not necessarily available but where measurements of r systems variables $z_1(t), z_2(t), \ldots, z_r(t)$ are available in the form of the *measurement vector*

$$z(t) = \begin{bmatrix} z_1(t) \\ z_2(t) \\ \vdots \\ z_r(t) \end{bmatrix} \tag{2.2}$$

Of course, if $m = r$ and $z_k(t) \equiv y_k(t)$, $1 \leqslant k \leqslant m$, we revert to the situation when measurements of outputs only are available. If $r = n$ and $z_k(t) = x_k(t)$, $1 \leqslant k \leqslant n$, then measurements of all the state variables are available.

If we include these measurements in the state variable model, the model of equations (1.26)–(1.28) generalizes to the form

$$\begin{aligned} \frac{\mathrm{d}x(t)}{\mathrm{d}t} &= f(x(t), u(t), t) && \text{(state equation)} \\ y(t) &= g(x(t), u(t), t) && \text{(output equation)} \\ z(t) &= h(x(t), u(t), t) && \text{(measurement equation)} \end{aligned} \tag{2.3}$$

where $h(x, u, t)$ is an $r \times 1$ vector function of state and input variables. The situation of primary interest here is the case of linear time-invariant systems when

$$\begin{aligned} \frac{\mathrm{d}x(t)}{\mathrm{d}t} &= Ax(t) + Bu(t) \\ y(t) &= Cx(t) + Du(t) \\ z(t) &= Ex(t) \end{aligned} \tag{2.4}$$

where E is a constant $r \times n$ matrix. That is, the measurements are assumed to be linear functions of the state variables only,

$$z_k(t) = E_{k1}x_1(t) + \ldots + E_{kn}x_n(t), \qquad 1 \leqslant k \leqslant r \tag{2.5}$$

(c) *The feedback system.* The general form of feedback system for the control of an l-input/m-output system is illustrated in Fig. 26, where the control element constructs the input vector from a knowledge of the demanded outputs $r_1(t), \ldots, r_m(t)$ and the available measurements $z_1(t), \ldots, z_r(t)$. The resulting closed-loop system must, of course, be stable and be such that the system outputs respond in a "satisfactory" manner to their demands.

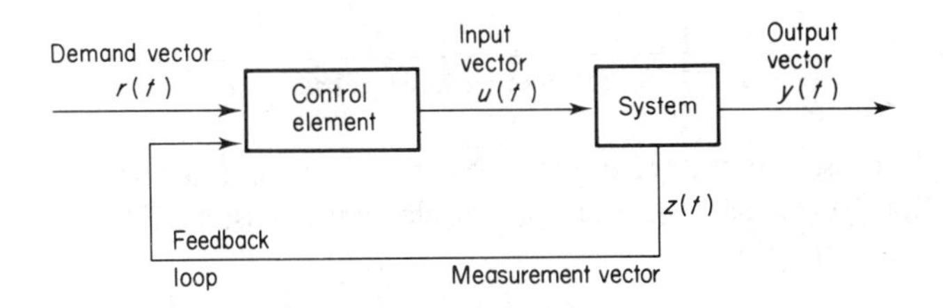

Fig. 26. The general structure of a feedback system.

Our main concern in this chapter is the use of feedback to stabilize the linear, time-invariant system of equation (2.4).

EXAMPLE 2.1.1. The reader will recall the following example from classical control theory. The second order lag $(D^2 + 2\xi\omega_0 D + \omega_0^2)y(t) = \omega_0^2 u(t)$ subjected to unit negative proportional feedback of the system output alone will tend to oscillate. Oscillations will set in at even low gains if the system damping ratio ξ is small. This situation can be improved if a measurement of the rate $dy(t)/dt$ is available, by use of the configuration shown in Fig. 27 and suitable choice of rate feedback constant k_2. We can put this problem into the state-variable format described above by noting that the system has a state-variable description (problem 9, Chapter one)

$$\frac{dx(t)}{dt} = \begin{bmatrix} 0 & 1 \\ -\omega_0^2 & -2\xi\omega_0 \end{bmatrix} x(t) + \begin{bmatrix} 0 \\ \omega_0^2 \end{bmatrix} u(t)$$

$$y(t) = [1 \quad 0]\, x(t) \tag{2.6}$$

where $x_1 = y$, $x_2 = dy/dt$. Assuming that measurements of output y and rate dy/dt are available and setting $z_1 = y$, $z_2 = dy/dt$, we see that

$$z(t) = \begin{bmatrix} y(t) \\ \dfrac{dy(t)}{dt} \end{bmatrix} = \begin{bmatrix} x_1(t) \\ x_2(t) \end{bmatrix} = x(t) \tag{2.7}$$

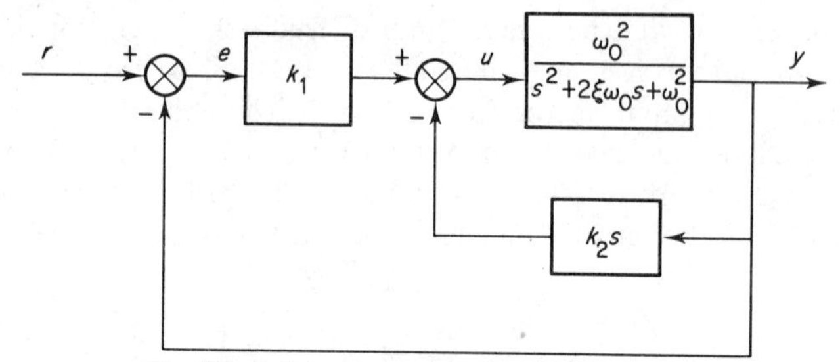

Fig. 27. Second order rate feedback system.

That is, the system model takes the form of (2.4) with $E = I_2$ (the 2×2 unit matrix). The control element is described by the feedback equation

$$\begin{aligned} u(t) &= k_1(r(t) - y(t)) - k_2 \frac{dy(t)}{dt} \\ &= k_1 r(t) - k_1 y(t) - k_2 \frac{dy(t)}{dt} \\ &= k_1 r(t) - (k_1 \quad k_2) z(t) \end{aligned} \tag{2.8}$$

and hence the closed-loop system has the structure shown in Fig. 26.

2.2 Linear Constant State Feedback

Consider a linear, time-invariant dynamical system of the form of (2.4). A general form of linear, constant feedback for this system takes the form

$$\begin{aligned} u_k(t) = {} & K_{k1} r_1(t) + K_{k2} r_2(t) + \ldots + K_{km} r_m(t) - F_{k1} z_1(t) - \ldots \\ & - F_{kr} z_r(t), \qquad 1 \leqslant k \leqslant l \end{aligned} \tag{2.9}$$

where each input to the system at time t is constructed as a linear combination of the demand signals and the available measurements at time t, the coefficients in the summation being constant. In a sense, it is a generalization of the idea of proportional feedback control.

Regarding K_{ki} and F_{ki} as elements of $l \times m$ and $l \times r$ constant matrices K and F respectively, the feedback equation (2.9) reduces to the matrix relation

$$u(t) = Kr(t) - Fz(t) \tag{2.10}$$

A familiar example of such a feedback law has been illustrated in Example 2.1.1. The block structure of the closed-loop system in the general case is illustrated in Fig. 28.

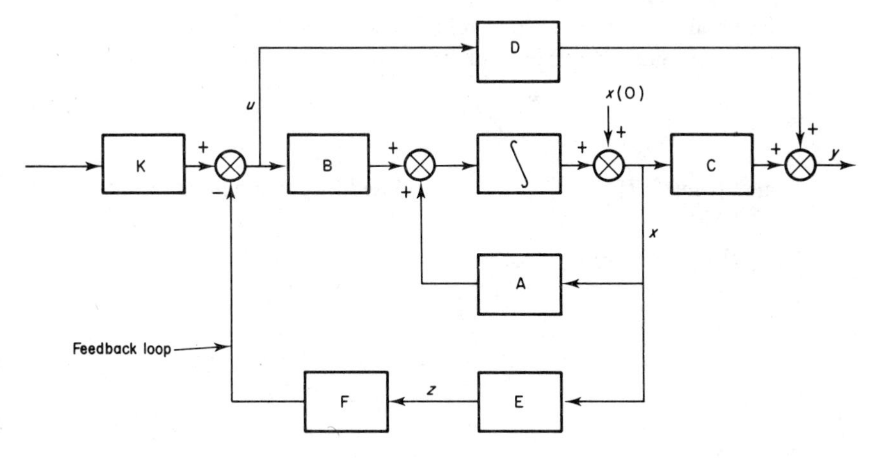

Fig. 28. Linear system with linear constant feedback.

The mathematical form of the closed-loop system is obtained by substituting (2.10) into the system equations, $\mathrm{d}x(t)/\mathrm{d}t = Ax(t) + Bu(t) = Ax(t) + B(Kr(t) - Fz(t))$ or, writing $z(t) = Ex(t)$

$$\frac{\mathrm{d}x(t)}{\mathrm{d}t} = (A - BFE)x(t) + BKr(t) \tag{2.11}$$

and, in a similar manner,

$$y(t) = \{C - DFE\}x(t) + DKr(t) \tag{2.12}$$

The closed-loop system is hence described by a linear, time-invariant model with the A, B, C, D matrices replaced by $A - BFE$, BK, $C - DFE$ and DK respectively. In fact, the action of feedback has changed the system dynamics considerably. In particular, although the stability of the open loop system (i.e. the system (2.4) without control) is described by the *open-loop characteristic polynomial*

$$\rho_0(s) = |sI - A| \tag{2.13}$$

the stability of the closed-loop system (2.11)–(2.12) is described by the *closed-loop characteristic polynomial*

$$\rho_c(s) = |sI - A + BFE| \tag{2.14}$$

More precisely, the closed-loop system is asymptotically stable if, and only if, all solutions of the equation $\rho_c(s) = 0$ (i.e. eigenvalues of $A - BFE$) have strictly negative real parts.

EXAMPLE 2.2.1. Considering the second-order plant $(\mathrm{D}^2 + 2\xi\omega_0\mathrm{D} + \omega_0^2)y(t) = \omega_0^2 u(t)$ in the control set-up illustrated in Fig. 27, it has been shown in

Example 2.1.1 that the control law takes the form of equation (2.8). In the above notation, this is equivalent to

$$K = k_1, \qquad F = (k_1 \quad k_2) \tag{2.15}$$

and, from (2.7), $E = \begin{bmatrix} 1 & 0 \\ 0 & 1 \end{bmatrix}$. The closed-loop characteristic polynomial $\rho_c(s)$ takes the form

$$\rho_c(s) = \begin{vmatrix} s & -1 \\ \omega_0^2(1+k_1) & s + 2\xi\omega_0 + \omega_0^2 k_2 \end{vmatrix}$$

$$= s^2 + (2\xi\omega_0 + \omega_0^2 k_2)s + \omega_0^2(1+k_1)$$

The closed-loop system is hence stable if, and only if, $2\xi + \omega_0 k_2 > 0$ and $1 + k_1 > 0$. This result is easily checked by transfer function methods noting from Fig. 27 that the closed-loop system has the transfer function $k_1 \omega_0^2 / \rho_c(s)$.

Finally, in the particular case when measurements of all the state variables are available, we have $r = n$ and, without loss of generality, can take $z_k(t) = x_k(t)$, $1 \leqslant k \leqslant n$. Equivalently $z(t) = x(t)$ and $E = I_n$, the feedback law of (2.10) reducing to

$$u(t) = Kr(t) - Fx(t) \tag{2.16}$$

In this special case the control law is termed a linear, constant *state feedback* control law. In many ways this is an abuse of terminology as, in the more general case of (2.10), substitution for $z(t)$ yields $u(t) = Kr(t) - FEx(t)$ which is a state feedback control law with F replaced by $F' = FE$. It is customary, however, to reserve the term "state feedback" for cases when all states are measured.

EXAMPLE 2.2.2. Using problem (14) of Chapter one, the third order single-input/single-output system $(D^3 + 3D^2 + 3D + 1)y(t) = (D + 2)u(t)$ has a state-variable representation of the form

$$\frac{dx(t)}{dt} = \begin{bmatrix} 0 & 1 & 0 \\ 0 & 0 & 1 \\ -1 & -3 & -3 \end{bmatrix} x(t) + \begin{bmatrix} 0 \\ 0 \\ 1 \end{bmatrix} u(t)$$

$$y(t) = [2 \quad 1 \quad 0]\, x(t) \tag{2.17}$$

Suppose that measurements $z_1(t) = y(t)$ and $z_2(t) = \mathrm{d}y(t)/\mathrm{d}t$ are available, then

$$z_1(t) = [2 \quad 1 \quad 0]\, x(t)$$

$$z_2(t) = [2 \quad 1 \quad 0]\, \frac{\mathrm{d}x(t)}{\mathrm{d}t} = [0 \quad 2 \quad 1]\, x(t) \qquad (2.18)$$

Combining these equations in matrix form yields

$$z(t) = \begin{bmatrix} 2 & 1 & 0 \\ 0 & 2 & 1 \end{bmatrix} x(t) \qquad (2.19)$$

and hence E by inspection.

EXERCISE 2.2.1. Suppose that the single-input/single-output system (1.3) has rank $k \geq 1$ (i.e. $b_i = 0$, $0 \leq i \leq k-1$) and that measurements $z_1(t) = y(t)$, $z_2(t) = \mathrm{d}y(t)/\mathrm{d}t, \ldots, z_k(t) = \mathrm{d}^{k-1}y(t)/\mathrm{d}t^{k-1}$ are available. If $\mathrm{d}x(t)/\mathrm{d}t = Ax(t) + Bu(t)$, $y(t) = Cx(t)$ is a state-variable model of the process (see, for example, problem (14) of Chapter one), show that $z(t) = Ex(t)$ where

$$E = \begin{bmatrix} C \\ CA \\ \vdots \\ CA^{k-1} \end{bmatrix} \qquad (2.20)$$

Use your result to verify (2.19) in Example 2.2.2.

2.3 Pole Allocation

It is clear from the closed-loop equations (2.11) and (2.12) that the linear, constant feedback law (2.10) can be used to change the dynamic behaviour of the open-loop system, e.g. to achieve the stabilization of a previously unstable system or to speed up the system response. In this section, we illustrate the sort of techniques that can be used and the theoretical possibilities by consideration of the problem of stabilization. More precisely we consider the following problem.

The pole-allocation problem. Given a specified $r \times 1$ measurement vector $z(t)$ and a specified set of complex numbers $\mu_1, \mu_2, \ldots, \mu_n$, find a $l \times r$ feedback matrix F such that

$$\rho_c(s) = |sI - A + BFE| = (s - \mu_1)(s - \mu_2) \ldots (s - \mu_n) \qquad (2.21)$$

If it is remembered that the eigenvalues of $A - BFE$ describe the stability and degree of oscillation in the closed-loop system, the requirement of closed-loop stability reduces to specifying $\mu_1, \ldots, \mu_n$ so that $\operatorname{Re} \mu_k < 0$, $1 \leqslant k \leqslant n$. If we also choose $|\operatorname{Im} \mu_k| \leqslant \alpha$, $1 \leqslant k \leqslant n$, for some $\alpha \geqslant 0$, then the resulting closed-loop system is stable with a limited frequency of oscillation.

The existence of a suitable matrix F depends upon the measurement set $z(t)$ and the choice of closed-loop eigenvalues $\mu_1, \ldots, \mu_n$. If, however, we demand that a suitable F should exist, independent of how we choose $\mu_1, \ldots, \mu_n$, the conditions for existence are stated, without proof, below.

Theorem 2.3.1. *The pole allocation problem is solvable for all choices of μ_1, $\mu_2, \ldots, \mu_n$ if, and only if, the open-loop system (2.4) is completely controllable and measurements of all state variables are available.*

In other words if we want to allocate the closed-loop eigenvalues arbitrarily we must use a linear, constant state feedback controller and the system under consideration must be controllable. In the following sections, some computational methods are described. More sophisticated results can be found in the reading list.

2.3.1 *Single-input systems*

Suppose initially that the open-loop system has only a single input (i.e. $l = 1$) and hence that B is an $n \times 1$ matrix. Suppose also that the A and B matrices take the "controllable canonical form"

$$A = \begin{bmatrix} 0 & 1 & 0 & & \\ 0 & 0 & 1 & & \\ & & & & 0 \\ 0 & & & 0 & 1 \\ -a_n & -a_{n-1} & & & -a_1 \end{bmatrix}, \quad B = \begin{bmatrix} 0 \\ 0 \\ \\ 0 \\ 1 \end{bmatrix} \tag{2.22}$$

then it is easily verified that $[B, AB, \ldots, A^{n-1}B]$ is nonsingular and hence the system is controllable. It can also be verified (try a few examples!) that the open-loop characteristic polynomial is

$$\rho_0(s) = |sI_n - A| = s^n + a_1 s^{n-1} + \ldots + a_{n-1}s + a_n \tag{2.23}$$

which provides an alternative route to calculating $a_1, \ldots, a_n$.

Suppose, by direct calculation, that

$$(s - \mu_1)(s - \mu_2) \ldots (s - \mu_n) = s^n + b_1 s^{n-1} + \ldots + b_n \tag{2.24}$$

where $\mu_1, \mu_2, \ldots, \mu_n$ are the desired closed-loop eigenvalues and consider the state feedback law

$$u(t) = Kr(t) - [b_n - a_n, b_{n-1} - a_{n-1}, \ldots, b_1 - a_1]x(t) \qquad (2.25)$$

where K (which plays no role in the pole-allocation problem) is arbitrary. It is easily verified that

$$A - BF = \begin{bmatrix} 0 & 1 & 0 \ldots 0 \\ 0 & 0 & 1 \\ & & & 1 \\ -b_n & -b_{n-1} & \ldots & -b_1 \end{bmatrix} \qquad (2.26)$$

and hence that the closed-loop characteristic polynomial is

$$\rho_c(s) = |sI - A + BF| = s^n + b_1 s^{n-1} + \ldots + b_{n-1}s + b_n$$
$$= (s - \mu_1)(s - \mu_2) \ldots (s - \mu_n) \qquad (2.27)$$

The state feedback control law (2.25) hence solves the pole-allocation problem in this special case.

EXAMPLE 2.3.1. Consider the single-input/single-output system described by the equation $(D^3 + 3D^2 + 3D + 1)y(t) = u(t)$ with state-variable model

$$\frac{dx(t)}{dt} = \begin{bmatrix} 0 & 1 & 0 \\ 0 & 0 & 1 \\ -1 & -3 & -3 \end{bmatrix} x(t) + \begin{bmatrix} 0 \\ 0 \\ 1 \end{bmatrix} u(t)$$

$$y(t) = [1 \quad 0 \quad 0]\, x(t) \qquad (2.28)$$

with A and B in the form of (2.22) with $a_3 = 1, a_2 = a_1 = 3$. Suppose that it is desired to use the state feedback law

$$u(\mathrm{t}) = k_0 r(t) - (k_1 \quad k_2 \quad k_3)x(t) \qquad (2.29)$$

to allocate the closed-loop eigenvalues to the positions $\mu_1 = -2$, $\mu_2 = -1 + j, \mu_3 = -1 - j$. The first step is to evaluate

$$(s - \mu_1)(s - \mu_2)(s - \mu_3) = s^3 + 4s^2 + 6s + 4 \qquad (2.30)$$

indicating that $b_1 = 4$, $b_2 = 6$ and $b_3 = 4$. Comparing (2.29) with (2.25) it is clear that we need $k_1 = b_3 - a_3 = 3$, $k_2 = b_2 - a_2 = 3$, $k_3 = b_1 - a_1 = 1$, i.e. $u(t) = k_0 r(t) - (3 \quad 3 \quad 1)x(t)$ will do the trick! It is left as an exercise for the reader to show that the closed-loop system has the structure shown in Fig. 29 if $k_0 = 3$.

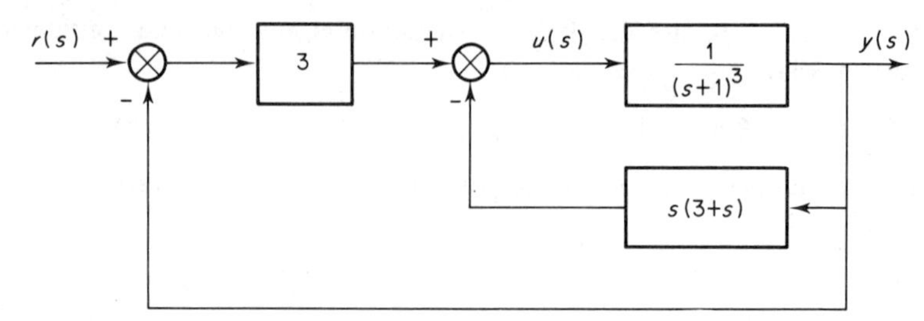

Fig. 29.

To deal with the general case of a single-input system we use transformation techniques similar to those used in Section 1.5.1 to transform A and B to the form of (2.22). More precisely, suppose there exists an $n \times n$ nonsingular matrix T such that

$$A^* = T^{-1}AT = \begin{bmatrix} 0 & 1 & 0 \dots & 0 \\ \cdot & & 1 & \cdot \\ \cdot & & & 0 \\ 0 & & & 1 \\ -a_n & \dots & \dots & -a_1 \end{bmatrix}, \quad B^* = T^{-1}B = \begin{bmatrix} 0 \\ 0 \\ \vdots \\ 0 \\ 1 \end{bmatrix} \tag{2.31}$$

then the identity

$$\begin{aligned} \rho_0(s) &= |sI - A| = |T^{-1}(sI - A)T| \\ &= |sI - A^*| \equiv s^n + a_1 s^{n-1} + \dots + a_{n-1}s + a_n \end{aligned} \tag{2.32}$$

indicates that $a_1, a_2, \dots, a_n$ are simply the coefficients in the characteristic polynomial of A.

Choosing now the state feedback law $u(t) = Kr(t) - Fx(t)$ with

$$F = [b_n - a_n, b_{n-1} - a_{n-1}, \dots, b_1 - a_1]\, T^{-1} \tag{2.33}$$

the identity

$$\begin{aligned} |sI - A + BF| &= |T^{-1}(sI - A + BF)T| \\ &= |sI - A^* + B^* FT| \\ &= \begin{vmatrix} s & -1 & 0 \dots\dots & 0 \\ 0 & s & -1 & \cdot \\ \cdot & & & 0 \\ 0 \dots\dots & & s & -1 \\ b_n & b_{n-1} \dots\dots & & s + b_1 \end{vmatrix} \end{aligned}$$

$$= s_n + b_1 s^{n-1} + \ldots + b_n$$
$$= (s - \mu_1)(s - \mu_2)\ldots(s - \mu_n) \quad (2.34)$$

indicates that this control law solves our problem. It remains to find the transformation matrix T.

Let t_k be the kth column of T, i.e.

$$T = [t_1, t_2, \ldots, t_n] \quad (2.35)$$

and write (2.31) in the form $TA^* = AT$, $B = TB^*$. It can then be verified (by equating columns on the left- and right-hand sides of these equations) that $t_1, t_2, \ldots, t_n$ are the solutions of the recursion relations

$$\begin{aligned} t_n &= B \\ At_n &= t_{n-1} - a_1 t_n \\ At_{n-1} &= t_{n-2} - a_2 t_n \\ &\vdots \\ At_2 &= t_1 - a_{n-1} t_n \quad (2.36) \\ At_1 &= -a_n t_n \quad (2.37) \end{aligned}$$

More precisely, (2.36) is solved recursively for t_n, then t_{n-1}, then $t_{n-2}, \ldots,$ and finally t_1. Equation (2.37) is then automatically satisfied (write $At_1 + a_n t_n = A^2 t_2 + a_{n-1} A t_n + a_n t_n = A^3 t_3 + a_{n-2} A^2 t_n + a_{n-1} A t_n + a_n t_n = \ldots = (A^n + a_1 A^{n-1} + \ldots + a_{n-1} A + a_n I) t_n = \rho_0(A) t_n = \underset{\sim}{0}$ as (using the Cayley-Hamilton theorem) every matrix A satisfies its characteristic equation $\rho_0(A) = 0$). The mathematically inclined reader should have no problem proving the following result concerning the nonsingularity of T.

EXERCISE 2.3.1. Verify from (2.36) that T takes the form

$$T = [B, AB, \ldots, A^{n-1}B] \begin{bmatrix} X & X \ldots X & 1 \\ \vdots & & 1 & 0 \\ X & & & \\ X & 1 & & \\ 1 & 0 \ldots & \ldots & 0 \end{bmatrix} \quad (2.38)$$

where the Xs denote unspecified scalars. Hence deduce that T is nonsingular if, and only if, the system is controllable, i.e. T is nonsingular whenever (theorem 2.3.1) the pole-allocation problem is solvable.

EXAMPLE 2.3.2. Consider the single-input mechanical system of Exercise 1.2.3 with $k_1 = k_2 = m_1 = m_2 = 1$ for simplicity, i.e.

$$\frac{dx(t)}{dt} = \begin{bmatrix} 0 & 0 & 1 & 0 \\ 0 & 0 & 0 & 1 \\ -2 & 1 & 0 & 0 \\ 1 & -1 & 0 & 0 \end{bmatrix} x(t) + \begin{bmatrix} 0 \\ 0 \\ 1 \\ 0 \end{bmatrix} u(t) \qquad (2.39)$$

Assume that all states of the system are measured (i.e. measurements of the position and velocities of both masses are made). Noting that

$$[B, AB, A^2B, A^3B] = \begin{bmatrix} 0 & 1 & 0 & -1 \\ 0 & 0 & 0 & 1 \\ 1 & 0 & -1 & 0 \\ 0 & 0 & 1 & 0 \end{bmatrix}$$

is nonsingular, it is seen that the system is controllable and hence that the pole allocation problem for this system is solvable (Theorem 2.3.1).

The first step is the calculation of the system characteristic polynomial

$$\rho_0(s) = |sI - A| = s^4 + 3s^2 + 1 \qquad (2.40)$$

i.e. the open-loop system has eigenvalues $\pm j\,0.62$ and $\pm j\,1.62$ and hence (as expected physically) is marginally stable and highly oscillatory. Note also that $a_1 = a_3 = 0, a_2 = 3$ and $a_4 = 1$.

The next step in the synthesis of a stabilizing controller is the choice of the desired closed-loop eigenvalues μ_1, μ_2, μ_3 and μ_4. It is at this stage that we hit the first problem with pole allocation, namely specification of the closed-loop performance in terms of numerical choice of eigenvalues. For stability it is natural to require that $\mathrm{Re}\,\mu_j \leqslant -\alpha$, $1 \leqslant j \leqslant 4$, for some real number $\alpha > 0$ representing the "degree of stability". The degree of oscillation present in the system can also be limited by imposing the requirement that $|\mathrm{Im}\,\mu_j/\mathrm{Re}\,\mu_j| \leqslant \beta$, $1 \leqslant j \leqslant 4$. In diagramatic terms, the eigenvalues of the closed-loop system are constrained to lie in the region of the complex plane shown in Fig. 30. This constraint still leaves an infinite number of possibilities.

For numerical simplicity, consider the choice of $\mu_1 = \mu_2 = -1 + j$, $\mu_3 = \mu_4 = -1 - j$ leading to

$$\begin{aligned}(s-\mu_1)(s-\mu_2)(s-\mu_3)(s-\mu_4) &= (s^2 + 2s + 2)^2 \\ &= s + 4s^3 + 8s^2 + 8s + 4 \qquad (2.41)\end{aligned}$$

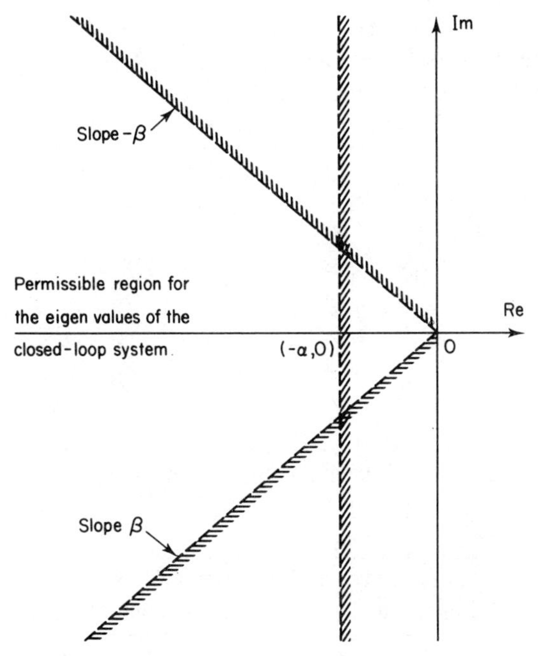

Fig. 30. System poles and performance specifications

and hence $b_1 = 4$, $b_2 = b_3 = 8$ and $b_4 = 4$. We must now find the matrix T using (2.35) and (2.36), i.e.

$$t_4 = B = \begin{bmatrix} 0 \\ 0 \\ 1 \\ 0 \end{bmatrix}, \qquad t_3 = At_4 + a_1 t_4 = \begin{bmatrix} 1 \\ 0 \\ 0 \\ 0 \end{bmatrix}$$

$$t_2 = At_3 + a_2 t_4 = \begin{bmatrix} 0 \\ 0 \\ 1 \\ 1 \end{bmatrix}, \qquad t_1 = At_2 + a_3 t_4 = \begin{bmatrix} 1 \\ 1 \\ 0 \\ 0 \end{bmatrix} \tag{2.42}$$

and hence

$$T = \begin{bmatrix} 1 & 0 & 1 & 0 \\ 1 & 0 & 0 & 0 \\ 0 & 1 & 0 & 1 \\ 0 & 1 & 0 & 0 \end{bmatrix}, \qquad T^{-1} = \begin{bmatrix} 0 & 1 & 0 & 0 \\ 0 & 0 & 0 & 1 \\ 1 & -1 & 0 & 0 \\ 0 & 0 & 1 & -1 \end{bmatrix} \tag{2.43}$$

(Note: we see that $At_1 = -a_4 t_4$ verifying (2.37) in this case.) The resulting state feedback control law $u(t) = Kr(t) - Fx(t)$ is then specified by choosing F according to (2.33)

$$F = [3 \quad 8 \quad 5 \quad 4] \begin{bmatrix} 0 & 1 & 0 & 0 \\ 0 & 0 & 0 & 1 \\ 1 & -1 & 0 & 0 \\ 0 & 0 & 1 & -1 \end{bmatrix}$$

$$= [5 \quad -2 \quad 4 \quad 4] \tag{2.44}$$

and choosing K arbitrarily. (Note: the choice of K will affect closed-loop *transient* performance but has no influence on the solution of the pole-allocation problem.)

EXERCISE 2.3.2. Verify that the closed-loop system in the above example does indeed have the required eigenvalues. (Hint: calculate $\rho_c(s)$ directly.)

2.3.2 Multi-input systems

The case of a general linear, time-invariant system with more than one input ($l > 1$) is a subject of research interest (see reading list at end of chapter) and well beyond the scope of this text. We shall, however, illustrate the foundations of the methodology by showing how the multi-input case can be reduced to the single-input case by suitable choice of controller. Once this has been done, of course, the methods of Section 2.3.1 can be used to find the appropriate state feedback controller.

Suppose that we are to use the state feedback controller $u(t) = Kr(t) - Fx(t)$ and set

$$F = pF_0 \tag{2.45}$$

where $p = (p_1, p_2, \ldots, p_l)^T$ is a $l \times 1$ vector of real numbers and F_0 is a $1 \times n$ matrix of real numbers. The important point to note is that the resulting closed-loop characteristic polynomial

$$\rho_c(s) = |sI_n - A + BpF_0| \tag{2.46}$$

is identical to the closed-loop characteristic polynomial of the single-input system

$$\frac{dx(t)}{dt} = Ax(t) + Bp\hat{u}(t) \tag{2.47}$$

with state feedback controller $\hat{u}(t) = K_0 r(t) - F_0 x(t)$. It follows directly that the pole-allocation problem for our original multi-input system reduces to the choice of F_0 for the single-input system (2.47) using the techniques of Section 2.3.1.

The only problem in the above argument is that the pole-allocation problem for (2.47) must be solvable, i.e. the single-input system (2.47) must be controllable in the sense that $|[Bp, ABp, \ldots, A^{n-1}Bp]| \neq 0$. This requirement depends on the choice of p and also on the original multi-input system, as illustrated by the two-input examples

$$A = \begin{bmatrix} -2 & 1 \\ 1 & 1 \end{bmatrix}, \qquad B = \begin{bmatrix} 1 & 1 \\ 0 & 1 \end{bmatrix} \tag{2.48}$$

when

$$|Bp, ABp| = \begin{bmatrix} p_1 + p_2 & -p_2 - 2p_1 \\ p_2 & p_1 + 2p_2 \end{bmatrix} \tag{2.49}$$

and

$$A = \begin{bmatrix} -1 & 0 \\ 0 & -1 \end{bmatrix}, \qquad B = \begin{bmatrix} 1 & 1 \\ 0 & 1 \end{bmatrix} \tag{2.50}$$

when

$$|Bp, ABp| = \begin{bmatrix} p_1 + p_2 & -p_1 - p_2 \\ p_2 & -p_2 \end{bmatrix} \tag{2.51}$$

In the first case, the resulting single-input system is controllable for all choice of p satisfying $p_1^2 + 5p_1 p_2 + 3p_2^2 \neq 0$ whereas, in the second case, no choice of p will ensure the controllability of the single-input system. Note in both cases that the original multi-input systems were controllable.

It is fortunate that, for most systems, a suitable choice of p will guarantee the controllability of (2.47). Such systems are said to be *cyclic.*

EXAMPLE 2.3.3. Consider the linearized model 1.71 of the liquid level system introduced in Example 1.2.1 and illustrated in Fig. 8. Assume the data $a_1 = a_2 = \beta = 1$ and hence

$$A = \begin{bmatrix} -1 & 1 \\ 1 & -1 \end{bmatrix}, \qquad B = \begin{bmatrix} 1 & 0 \\ 0 & 1 \end{bmatrix} \tag{2.52}$$

and $\rho_0(s) = |sI - A| = s^2 + 2s$ indicating that $a_1 = 2$, $a_2 = 0$. The system is hence marginally stable.

Consider the synthesis of a state feedback law to stabilize the system and allocate the closed-loop eigenvalues to the positions $\mu_1 = \mu_2 = -4$, i.e. $\rho_c(s) = (s+4)^2 = s^2 + 8s + 16$ and $b_1 = 8$, $b_2 = 16$. As we are only interested in stabilization, the controller will only be operating in a regulatory mode. Set $K = 0$, therefore, and consider the state feedback $u(t) = -Fx(t)$ with $F = pF_0$. It is easily verified that

$$[Bp, ABp] = \begin{bmatrix} p_1 & -p_1 + p_2 \\ p_2 & p_1 - p_2 \end{bmatrix} \tag{2.53}$$

and hence that the single-input system (2.47) will be controllable for any choice of p satisfying $p_1 + p_2 \neq 0$ and $p_1 - p_2 \neq 0$. But what values of p_1 and p_2 do we choose? A single calculation indicates that

$$u_1(t) = p_1 F_0 x(t), \qquad u_2(t) = p_2 F_0 x(t) \tag{2.54}$$

and hence, if we choose $p_1 = 0$, the closed-loop system will be controlling using the second input $u_2(t)$ only (i.e. the flow into vessel two only). Choose therefore $p_1 = 0$, $p_2 = 1$ when the single input system (2.47) is defined by the data

$$A = \begin{bmatrix} -1 & 1 \\ 1 & -1 \end{bmatrix}, \qquad Bp = \begin{bmatrix} 0 \\ 1 \end{bmatrix} \tag{2.55}$$

Following the techniques of Section 2.3.1, $F_0 = (b_2 - a_2, b_1 - a_1)T^{-1}$ where T is derived from (2.35) and (2.36) with B replaced by Bp and $n = 2$, i.e.

$$t_2 = Bp = \begin{bmatrix} 0 \\ 1 \end{bmatrix}, \qquad t_1 = At_2 + a_1 t_2 = \begin{bmatrix} 1 \\ 1 \end{bmatrix}, \qquad T = \begin{bmatrix} 1 & 0 \\ 1 & 1 \end{bmatrix} \tag{2.56}$$

and hence $F_0 = [16,6]\,T^{-1} = [10 \quad 6]$. The final control system takes the explicit form

$$u_1(t) \equiv 0, \qquad u_2(t) = -10x_1(t) - 6x_2(t) \tag{2.57}$$

and is illustrated in Fig. 31.

Finally, we observe that, for a given choice of $\mu_1, \mu_2, \ldots, \mu_n$, it may be possible to convert the state feedback law $u(t) = Kr(t) - Fx(t)$ into a law of the form $u(t) = Kr(t) - F_1 z(t)$ involving constant feedback of the measurements $z(t) = Ex(t)$ only. This is obviously only possible if $F_1 Ex = Fx$ is independent of the vector x, i.e. if

$$F = F_1 E \tag{2.58}$$

for some $l \times r$ matrix F_1.

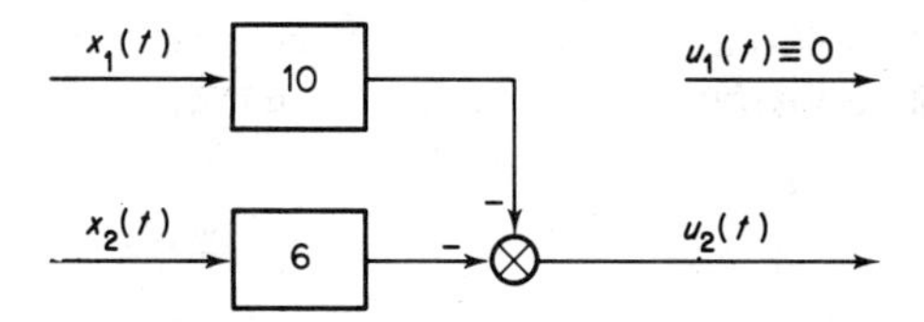

Fig. 31. Block representation of a state feedback control law.

Equation (2.58) is rarely satisfied indicating that full state feedback is necessary in general for pole allocation. There are ways around this problem however involving the use of dynamic control elements that construct estimates of the state. These elements are called *state observers* or, more simply, *observers*.

2.4 Observers

Suppose that measurements of all state variables $x_1(t), \ldots, x_n(t)$ are not available but that the measurements $z_1(t), \ldots, z_r(t)$ specified by the matrix equation $z(t) = Ex(t)$ are made. Given the system matrices A and B it is still possible to compute a state feedback law $u(t) = Kr(t) - Fx(t)$ to allocate the closed-loop system eigenvalues to desired positions $\mu_1, \ldots, \mu_n$ (if the system is controllable, of course). This controller cannot be implemented, however, using information concerning $z(t)$ alone unless (2.58) holds for some F_1. Consider therefore the possibility of constructing a dynamic control element (termed an observer) that takes information on the system inputs $u(t')$ and measurements $z(t')$ for $t' \leqslant t$ and uses it to construct an estimate $\hat{x}(t)$ of the system state $x(t)$(see Fig. 32(a)). If this is possible and the estimate is asymptotically correct in the sense that

$$\lim_{t \to +\infty} (x(t) - \hat{x}(t)) = 0 \tag{2.59}$$

(i.e. the error in the estimate approaches zero after a long enough period of time) then it is natural to envisage the approximate implementation of the feedback $u(t) = Kr(t) - Fx(t)$ by

$$u(t) = Kr(t) - F\hat{x}(t) \tag{2.60}$$

as shown in Fig. 32(b). These ideas are formalized below.

2.4.1 Observers and open-loop dynamics

Consider initially the open-loop system (2.4)

$$\begin{aligned} \frac{\mathrm{d}x(t)}{\mathrm{d}t} &= Ax(t) + Bu(t) \\ y(t) &= Cx(t) + Du(t) \\ z(t) &= Ex(t) \end{aligned} \tag{2.61}$$

Being linear and time-invariant, it is natural to consider the possibility of using linear, time-invariant observers described by the state-variable model,

$$\frac{d\hat{x}(t)}{dt} = M_1\hat{x}(t) + M_2u(t) + M_3z(t) \tag{2.62}$$

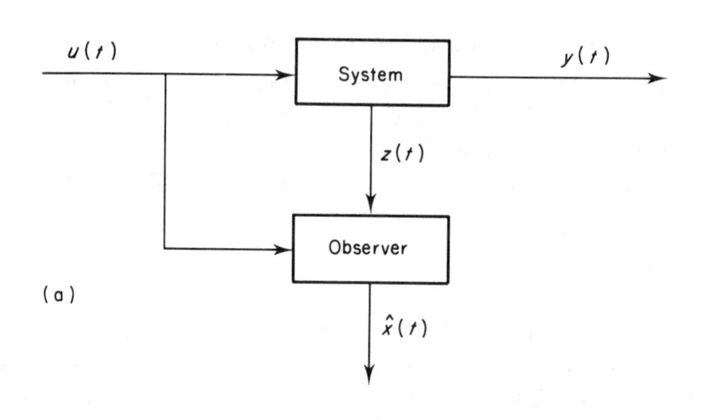

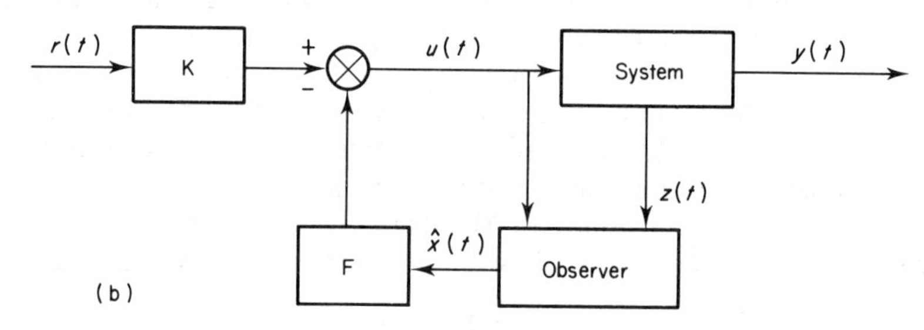

Fig. 32. (a) Open-loop system with observer. (b) Implementation of $u = Kr - Fx$ using an observer and measurements z.

where M_1, M_2, M_3 are (as yet unspecified) constant matrices of dimension $n \times n$, $n \times l$ and $n \times r$ respectively. Note that the observer is driven by the present inputs and measurements made on the system. Defining the error vector

$$e(t) = \hat{x}(t) - x(t) \tag{2.63}$$

and subtracting (2.61) from (2.62) and using the relation $z = Ex$ yields

$$\begin{aligned}\frac{de(t)}{dt} &= M_1\hat{x}(t) + M_2u(t) + M_3Ex(t) - Ax(t) - Bu(t)\\ &= M_1e(t) + (M_1 - A + M_3E)x(t) + (M_2 - B)u(t)\end{aligned} \tag{2.64}$$

In practice, we would like the error dynamics to be independent of the input $u(t)$ and the present value of $x(t)$. To eliminate the dependence on $u(t)$, set

$$M_2 = B \tag{2.65}$$

and to remove the explicit dependence on $x(t)$, set

$$M_1 = A - M_3 E \tag{2.66}$$

Equation (2.64) now reduces to

$$\frac{de(t)}{dt} = (A - M_3 E)e(t) \tag{2.67}$$

The solution of this equation is clearly independent of input $u(t)$ and has no explicit dependence on the state $x(t)$. It takes the simple form

$$e(t) = e^{(A-M_3E)t}e(0) \tag{2.68}$$

where $e(0)$ is the (normally unknown) value of the error vector at the initial time $t = 0$.

If we choose M_3 such that the eigenvalues of $A - M_3 E$ all have strictly negative real parts, then (Exercise 1.7.1)

$$\lim_{t \to +\infty} e(t) = 0 \tag{2.69}$$

and (see equation (2.59)) the estimate produced by the observer is asymptotically correct.

In summary, a linear time-invariant observer for the linear time-invariant system (2.61) takes the form

$$\frac{d\hat{x}(t)}{dt} = (A - M_3 E)\hat{x}(t) + Bu(t) + M_3 z(t) \tag{2.70}$$

where M_3 is specified to ensure that the eigenvalues of $A - M_3 E$ have strictly negative real parts. The resulting estimation error $e(t)$ asymptotically approaches zero as $t \to +\infty$ and the observer is said to be asymptotically stable.

EXAMPLE 2.4.1. Consider the single-input/single-output system $(D^2 + D + 1)y(t) = u(t)$ with state variable model

$$\frac{dx(t)}{dt} = \begin{bmatrix} 0 & 1 \\ -1 & -1 \end{bmatrix} x(t) + \begin{bmatrix} 0 \\ 1 \end{bmatrix} u(t)$$

$$y(t) = (1 \quad 0)x(t) \tag{2.71}$$

Suppose that the only measurement available is that of the output $y(t)$, i.e. $z(t) = y(t)$ and hence $E = (1 \quad 0)$. The observer for the system takes the form of (2.70), i.e.

$$\frac{\mathrm{d}\hat{x}(t)}{\mathrm{d}t} = \begin{bmatrix} -\alpha & 1 \\ -(1+\beta) & -1 \end{bmatrix} \hat{x}(t) + \begin{bmatrix} 0 \\ 1 \end{bmatrix} u(t) + \begin{bmatrix} \alpha \\ \beta \end{bmatrix} y(t) \quad (2.72)$$

together with an initial condition $\hat{x}(0)$ equal to an initial guess at the (unknown) value of $x(t)$ at time $t = 0$ (typically we take $\hat{x}(0) = 0$ unless there are strong reasons to suggest a non-zero value). A block diagram of this observer is given in Fig. 33.

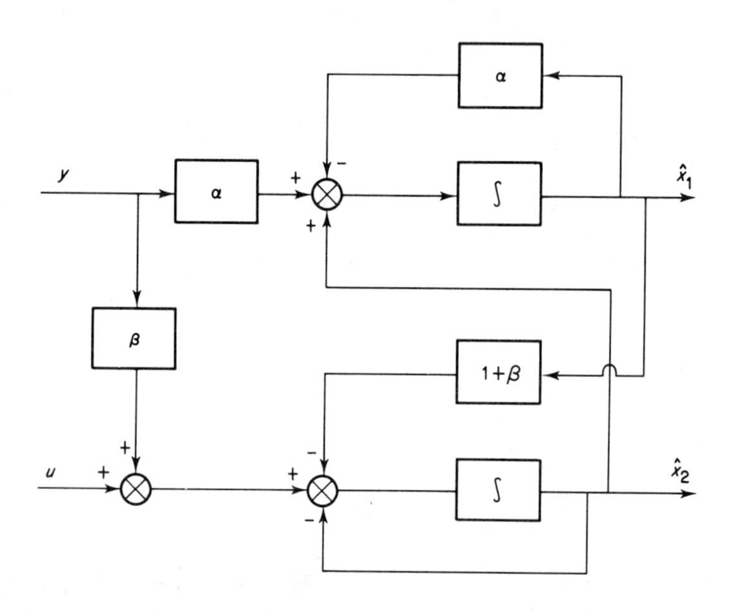

Fig. 33.

2.4.2 *Observers and closed-loop dynamics*

Consider now the feedback system of Fig. 32(b) consisting of the linear system (2.61) with asymptotically stable observer (2.70) and the feedback law (2.60). The control law (2.60) takes the form

$$u(t) = Kr(t) - F\hat{x}(t) = \{Kr(t) - Fx(t)\} - Fe(t) \quad (2.73)$$

and, using (2.69), asymptotically approaches the correct control $u(t) = Kr(t) - Fx(t)$ as $t \to +\infty$. It is intuitively obvious, therefore, that the closed-loop system is stable if the state feedback law $u(t) = Kr(t) - Fx(t)$ stabilizes (2.61). To prove this rigorously, substitute (2.73) into (2.61) to give

$$\frac{dx(t)}{dt} = (A - BF)x(t) + BKr(t) - BFe(t)$$

$$y(t) = (C - DF)x(t) - DFe(t) + DKr(t) \tag{2.74}$$

Noting that the state of the feedback system is specified uniquely by $x(t)$ and $e(t)$, a state-variable model is obtained by combining (2.67) with (2.74) in the form

$$\frac{d\xi(t)}{dt} = \begin{bmatrix} A - BF & -BF \\ 0 & A - M_3E \end{bmatrix} \xi(t) + \begin{bmatrix} BK \\ 0 \end{bmatrix} r(t)$$

$$y(t) = [C - DF, -DF]\, \xi(t) + DKr(t) \tag{2.75}$$

where the $2n \times 1$ vector $\xi(t) = \begin{bmatrix} x(t) \\ e(t) \end{bmatrix}$. The characteristic polynomial of the closed-loop system is hence

$$\rho(s) = \begin{vmatrix} sI - A + BF & BF \\ 0 & sI - A + M_3E \end{vmatrix}$$

$$\equiv |sI - A + BF| \cdot |sI - A + M_3E| \tag{2.76}$$

which is simply the product of the closed-loop characteristic polynomial $|sI - A + BF|$ of the system (2.61) with the "ideal" state feedback law $u(t) = Kr(t) - Fx(t)$ and the characteristic polynomial $|sI - A + M_3E|$ of the observer (2.70). The eigenvalues of the closed-loop system are hence simply the eigenvalues $\mu_1, \mu_2, \ldots, \mu_n$ specified for the calculation of the state feedback matrix F together with the extra eigenvalues of $A - M_3E$ introduced by the observer. In practice all these eigenvalues have strictly negative real parts and hence the closed-loop system is stable.

2.4.3 *Specifying the observer eigenvalues*

The state feedback matrix F is specified to allocate the eigenvalues of $A - BF$. The only remaining design parameters are K and M_3. There is no space in this text for a detailed consideration of the choice of K but it is possible to give some indication of techniques for choosing M_3. The key to the procedure is to note from (2.68) that the estimation error $e(t)$ decays at a rate dependent on the eigenvalues of $A - M_3E$ (and hence dependent on M_3 as A and E are fixed by the system). It is natural, therefore, to choose M_3 in such a way that the eigenvalues of $A - M_3E$ have specified desirable values $\eta_1, \eta_2, \ldots, \eta_n$, i.e. M_3 is specified by solving a pole allocation problem!

We can, in fact, apply the methods of Section 2.3. to solve this problem by writing the observer characteristic polynomial in the form

$$\rho_{\text{obs}}(s) = |sI - A + M_3E| = |sI - A^T + E^T M_3^T| \tag{2.77}$$

and noting that M_3 can hence be computed by finding a state feedback control law $u(t) = K_0 r(t) - F_0 x(t)$ for the linear time-invariant system

$$\frac{dx(t)}{dt} = A^T x(t) + E^T u(t) \tag{2.78}$$

that allocates the closed-loop eigenvalues to positions $\eta_1, \eta_2, \ldots, \eta_n$ and setting $M_3 = F_0^T$.

EXERCISE 2.4.1. Choosing $z(t) = y(t)$, prove that the pole-allocation problem for the observer is solvable independent of the choice of $\eta_1, \ldots, \eta_n$ if, and only if, the original system (2.61) is observable.

Consider the mechanical system of Exercise 1.2.3 with $k_1 = k_2 = m_1 = m_2 = 1$, i.e.

$$\frac{dx(t)}{dt} = \begin{bmatrix} 0 & 0 & 1 & 0 \\ 0 & 0 & 0 & 1 \\ -2 & 1 & 0 & 0 \\ 1 & -1 & 0 & 0 \end{bmatrix} x(t) + \begin{bmatrix} 0 \\ 0 \\ 1 \\ 0 \end{bmatrix} u(t) \tag{2.79}$$

If all the states are measured then it has been shown in Example 2.3.2 that the state feedback controller $u(t) = Kr(t) - Fx(t)$ with $F = [5 \quad -2 \quad 4 \quad 4]$ generates a closed-loop system with eigenvalues $\mu_1 = \mu_2 = -1 + j$, $\mu_3 = \mu_4 = -1 - j$. Consider now the problem of implementing this controller if measurements of the outputs only are available, i.e.

$$z(t) = y(t) = \begin{bmatrix} 1 & 0 & 0 & 0 \\ 0 & 1 & 0 & 0 \end{bmatrix} x(t) \tag{2.80}$$

from which E is obtained by inspection. It is clear that the state feedback controller cannot be implemented as a constant feedback of the output only as the presence of non-zero terms in the third and fourth elements of F requires the presence of the third and fourth states in the measurement set. More precisely, if we explicitly evaluate the control law

$$u(t) = Kr(t) - Fx(t) = Kr(t) - 5y_1(t) + 2y_2(t) - 4(x_3(t) + x_4(t)) \tag{2.81}$$

it is clear that we need an additional measurement of $x_3(t) + x_4(t)$ for its exact implementation. Consider, therefore, the use of an observer of the form of equation (2.70). The observer is completely specified by $A, B, E(= C)$ and M_3. We will choose M_3 by specification of the eigenvalues of $A - M_3E$.

The poles specified in the pole-allocation problem generating F suggest that the closed-loop system will have time-constants of the order of 1 s. It is intuitively obvious, therefore, that if we require the error introduced by the observer to be small for $t > 1$, we should ensure that the error $e(t)$ decays with time constants much shorter than this. There is a bit of a trade-off here as small error time-constants tend to lead to high observer gains (represented by the elements of M_3) and high gains can lead to excessive amplification of measurement noise. We will compromise here by choosing the observer eigenvalues $\eta_1 = \eta_2 = \eta_3 = \eta_4 = -4$ and hence error time-constants of the order of $\tau = \frac{1}{4} = 0.25$ s.

Our observer design now reduces to the choice of M_3 such that

$$|sI_4 - A + M_3E| \equiv |sI_4 - A^T + E^TM_3^T| \equiv (s+4)^4$$
$$= s^4 + 16s^3 + 96s^2 + 256s + 256 \tag{2.82}$$

This is done using the pole-allocation procedures of Section 2.3 with A replaced by A^T and B replaced by E^T and F replaced by M_3^T. More precisely (2.82) yields the data $b_1 = 16$, $b_2 = 96$, $b_3 = b_4 = 256$. The characteristic polynomial of A^T is $|sI_4 - A^T|$ which is identical to the characteristic polynomial of A, i.e. $|sI - A|$ and hence from Example 2.3.2,

$$a_1 = 0, a_2 = 3, a_3 = 0, a_4 = 1.$$

There is one more design decision to make as E^T has dimension 4×2 and hence the pole-allocation problem is a "multi-input" problem. If we apply the procedure of Section 2.3.2, we must write $M_3^T = pM_4^T$ where $p = (p_1, p_2)^T$ is a 2×1 vector and M_4 is a 4×1 column vector of real numbers. It is the choice of p_1 and p_2 that represents design flexibility. We can obtain some insight into the physical significance of p by examining the contribution $M_3z(t)$ of the measurement to the observer equation (2.70). For this example $M_3z(t) = M_4p^Ty(t) = M_4(p_1y_1(t) + p_2y_2(t))$, i.e. p_1 and p_2 represent the weightings of the output measurements $y_1(t)$ and $y_2(t)$ respectively in the observer dynamics. A typical example is the choice of $p_1 = 1$ and $p_2 = 0$ which corresponds to the situation when the observer estimates the state based on measurement of the output $y_1(t)$ only. This is the case taken in this example.

The matrix M_4^T now takes the form $M_4^T = (b_4 - a_4, \ldots, b_1 - a_1)T^{-1}$ where $T = [t_1, t_2, t_3, t_4]$ is the matrix that transforms A^T and Ep^T to controllable canonical form. Applying the procedure of Section 2.3 it is clear that

$$t_4 = E^T p = \begin{bmatrix} 1 \\ 0 \\ 0 \\ 0 \end{bmatrix}, \quad t_3 = A^T t_4 + a_1 t_4 = \begin{bmatrix} 0 \\ 0 \\ 1 \\ 0 \end{bmatrix}$$

$$t_2 = A^T t_3 + a_2 t_4 = \begin{bmatrix} 1 \\ 1 \\ 0 \\ 0 \end{bmatrix}, \quad t_1 = A^T t_2 + a_3 t_4 = \begin{bmatrix} 0 \\ 0 \\ 1 \\ 1 \end{bmatrix} \tag{2.83}$$

leading to

$$T = \begin{bmatrix} 0 & 1 & 0 & 1 \\ 0 & 1 & 0 & 0 \\ 1 & 0 & 1 & 0 \\ 1 & 0 & 0 & 0 \end{bmatrix} \quad (|T| \neq 0) \tag{2.84}$$

and $M_4^T = (16 \quad 240 \quad 93 \quad 162)$. The required observer "gain matrix" M_3 takes the form

$$M_3 = M_4 p^T = \begin{bmatrix} 16 & 0 \\ 240 & 0 \\ 93 & 0 \\ 162 & 0 \end{bmatrix} \tag{2.85}$$

and gives rise to the final observer

$$\frac{d\hat{x}(t)}{dt} = \begin{bmatrix} -16 & 0 & 1 & 0 \\ -240 & 0 & 0 & 1 \\ -95 & 1 & 0 & 0 \\ -161 & -1 & 0 & 0 \end{bmatrix} \hat{x}(t) + \begin{bmatrix} 0 \\ 0 \\ 1 \\ 0 \end{bmatrix} u(t) + \begin{bmatrix} 16 & 0 \\ 240 & 0 \\ 93 & 0 \\ 162 & 0 \end{bmatrix} y(t) \tag{2.86}$$

It is left as an exercise for the reader to check that the observer does indeed have the required eigenvalues and to consider how it could be implemented using an analogue or microcomputer. (Hint: write down the observer equations element by element and construct an analogue simulation diagram.)

EXERCISE 2.4.2. The observer (2.70) is just a linear time-invariant system and, as such, all the ideas of Chapter one can be applied to it. In particular, it may be desirable to use the transformation $\xi(t) = T_0^{-1}\hat{x}(t)$ (where T_0 is a constant nonsingular matrix) to obtain the equivalent description

$$\begin{aligned}\frac{d\xi(t)}{dt} &= T_0^{-1}(A - M_3E)T_0\xi(t) + T_0^{-1}Bu(t) + T_0^{-1}M_3z(t)\\ \hat{x}(t) &= T_0\xi(t)\end{aligned} \tag{2.87}$$

T_0 could, for example, be chosen to make $T_0^{-1}(A - M_3E)T_0$ diagonal or of Jordan form, hence simplifying the problem of the realization of the observer.

2.5 State Feedback in Discrete Systems Problems

The development and definition of feedback concepts, pole-allocating controllers and observers for the linear, time-invariant discrete system

$$\begin{aligned}x_{k+1} &= \Phi x_k + \Delta u_k\\ y_k &= Cx_k + Du_k\end{aligned} \tag{2.88}$$

are a direct parallel of those of the previous sections. In fact all of the algorithms carry over with a simple change of notation replacing A by Φ and B by Δ. The following summarizes the required formulae:

(a) *The demand sequence.* The continuous demand vector (2.1) is replaced by the demand *sequence* $\{r_0, r_1, r_2, \ldots\}$ where r_k is the $m \times 1$ vector of demanded outputs at the kth sample instant.

(b) *The measurement equation.* The availability of measurements not necessarily equal to the outputs is represented by the matrix relation (c.f. (2.4))

$$z_k = Ex_k \tag{2.89}$$

where z_k is a $r \times 1$ vector of new measurements available at the kth sample instant.

(c) *Linear constant feedback.* The general form of linear, constant feedback takes the (matrix) form (c.f. (2.10))

$$u_k = Kr_k - Fz_k, \qquad k \geqslant 0 \tag{2.90}$$

where K and F are real, constant $l \times m$ and $l \times r$ matrices respectively. The form of the closed-loop system is obtained by substituting (2.90) and (2.89) into (2.88)

$$\begin{aligned} x_{k+1} &= (\Phi - \Delta FE)x_k + \Delta Kr_k \\ y_k &= (C - DFE)x_k + DKr_k, \qquad k \geqslant 0 \end{aligned} \tag{2.91}$$

and indicates that the dynamics of the open-loop system (2.88) are substantially modified by feedback. In particular, the "open-loop characteristic polynomial"

$$\rho_0(s) = |sI_n - \Phi| \tag{2.92}$$

describing the stability of the open-loop system (2.88) is transformed into the "closed-loop characteristic polynomial"

$$\rho_c(s) = |sI_n - \Phi + \Delta FE| \tag{2.93}$$

describing the stability of the closed-loop system (2.91). More precisely, the closed-loop system is asymptotically stable if, and only if, all solutions of the equation $\rho_c(s) = 0$ (i.e. all eigenvalues of $\Phi - \Delta FE$) lie inside the unit circle in the complex plane.

(d) *Linear constant state feedback.* In the case when measurements of all state variables are available, we have $r = n$ and $z_k = x_k$, $k \geqslant 0$, and $E = I_n$. The feedback law of (2.90) takes the form

$$u_k = Kr_k - Fx_k \tag{2.94}$$

and is termed a linear, constant, state-feedback control law.

(e) *Pole allocation.* The pole-allocation problem in discrete systems (c.f. Section 2.3) can be stated as follows: given a specified $r \times 1$ measurement vector z_k and a specified set of complex numbers $\mu_1, \mu_2, \ldots, \mu_n$, find an $l \times r$ feedback matrix F such that

$$\rho_c(s) = |sI - \Phi + \Delta FE| = \prod_{k=1}^{n} (s - \mu_k) \tag{2.95}$$

A comparison with (2.21) indicates that the pole-allocation problem in discrete systems theory is, on paper, simply the pole-allocation problem for continuous systems with Φ and Δ replacing A and B respectively. This is not really surprising as pole allocation, in mathematical terms, is a problem in the structure of polynomials rather than a problem of the structure of differential or difference equations. The only way that the system equation appears in the problem is in the choice of $\mu_1, \ldots, \mu_n$. In particular, the stable region $\text{Re}\,\mu_j < 0$, $1 \leqslant j \leqslant n$, for the continuous system is replaced by $|\mu_j| < 1$, $1 \leqslant j \leqslant n$ for the discrete system.

The following theorem follows directly from the above discussion and theorem 2.3.1, 1.9.1 and 1.9.3.

Theorem 2.5.1. *The pole-allocation problem is solvable for all choices of* $\mu_1, \ldots, \mu_n$ *if, and only if, the open-loop system (2.88) is controllable and measurements of all state variables are available.*

Equivalently our system must be controllable and constant state feedback must be used, as in the continuous case.

Finally, it is clear that all the techniques described in Sections 2.3.1 and 2.3.2 for computational solutions for continuous systems will carry over to discrete systems. All we have to do is replace A and B by Φ and Δ respectively!

(f) *Observers.* If measurements of all state variables are not available but measurements $z_k = Ex_k$, $k \geq 0$, are made, it is still possible to compute a state-feedback law $u_k = Kr_k - Fx_k$ to solve the pole-allocation problem for a specified set $\mu_1, \mu_2, \ldots, \mu_n$. This controller cannot be implemented directly, however, unless $F = F_1 E$ (for some F_1) when $u_k = Kr_k - F_1 z_k$ feeds back available measurements only. Just as in the continuous case of Section 2.4, we therefore consider the possibility of constructing a dynamic control element (the observer!) that takes information on system inputs and available measurements at sample instants $k^1 \leq k$ and uses it to construct an estimate $\hat{x}_k$ of the system state x_k. If this is possible and the estimate is asymptotically correct in the sense that

$$\lim_{k \to +\infty} (\hat{x}_k - x_k) = 0 \tag{2.96}$$

the state-feedback controller can be implemented in an approximate manner by

$$u_k = Kr_k - F\hat{x}_k, \qquad k \geq 0 \tag{2.97}$$

The methods of Section 2.4 can be carried through directly to discrete systems (this is left as an exercise for the reader!) to suggest the observer structure

$$\hat{x}_{k+1} = (\Phi - M_3 E)\hat{x}_k + \Delta u_k + M_3 z_k, \qquad k \geq 0 \tag{2.98}$$

The observer error $e_k = \hat{x}_k - x_k$ is then the solution of the discrete equation

$$e_{k+1} = (\Phi - M_3 E)e_k \tag{2.99}$$

indicating that the observer estimates are asymptotically correct if we choose M_3 such that $\Phi - M_3 E$ has eigenvalues $\eta_1, \eta_2, \ldots, \eta_n$ satisfying $|\eta_k| < 1$, $1 \leq k \leq n$. This is a pole-allocation problem that can be solved along the lines indicated by Section 2.4.3 with A replaced by Φ.

EXERCISE 2.5.1. Verify that system (2.88) with feedback (2.97) is described by the closed-loop system equations

$$\xi_{k+1} = \begin{bmatrix} \Phi - \Delta F & -\Delta F \\ 0 & \Phi - M_3 E \end{bmatrix} \xi_k + \begin{bmatrix} \Delta K \\ 0 \end{bmatrix} r_k$$

$$y_k = [C - DF, -DF]\,\xi_k + DKr_k \tag{2.100}$$

and that the closed-loop system has the characteristic polynomial

$$\rho(s) = |sI - \Phi + \Delta F| \cdot |sI - \Phi + M_3 E| \tag{2.101}$$

Hence deduce that the closed-loop system is stable if the state-feedback law $u_k = Kr_k - Fx_k$ stabilizes (2.88) and if $\Phi - M_3 E$ has eigenvalues $\eta_1, \ldots, \eta_n$ inside the unit circle in the complex plane.

Problems

(1) Show that the unstable single-input system

$$\frac{\mathrm{d}x(t)}{\mathrm{d}t} = \begin{bmatrix} 0 & 1 & 0 \\ 0 & 0 & 1 \\ 1 & 0 & -3 \end{bmatrix} x(t) + \begin{bmatrix} 0 \\ 0 \\ 1 \end{bmatrix} u(t)$$

with single output

$$y(t) = [1 \quad 0 \quad 0]\,x(t)$$

and two measurements described by the equation

$$z(t) = \begin{bmatrix} 1 & 0 & 0 \\ 0 & 1 & 0 \end{bmatrix} x(t)$$

is stabilized by the feedback law

$$u(t) = r(t) - [2 \quad 3]\,z(t)$$

(Hint: calculate the closed-loop characteristic polynomial and apply Routh's test.)

(2) Consider the simple mechanical system shown in Fig. 34(a) consisting of a friction-free vehicle moving on a horizontal surface and operated upon by an effective force input $u(t)$. The output from the system $y(t)$ is simply the distance of the vehicle from a specified point. For simplicity, take the case of unit mass $m = 1$.

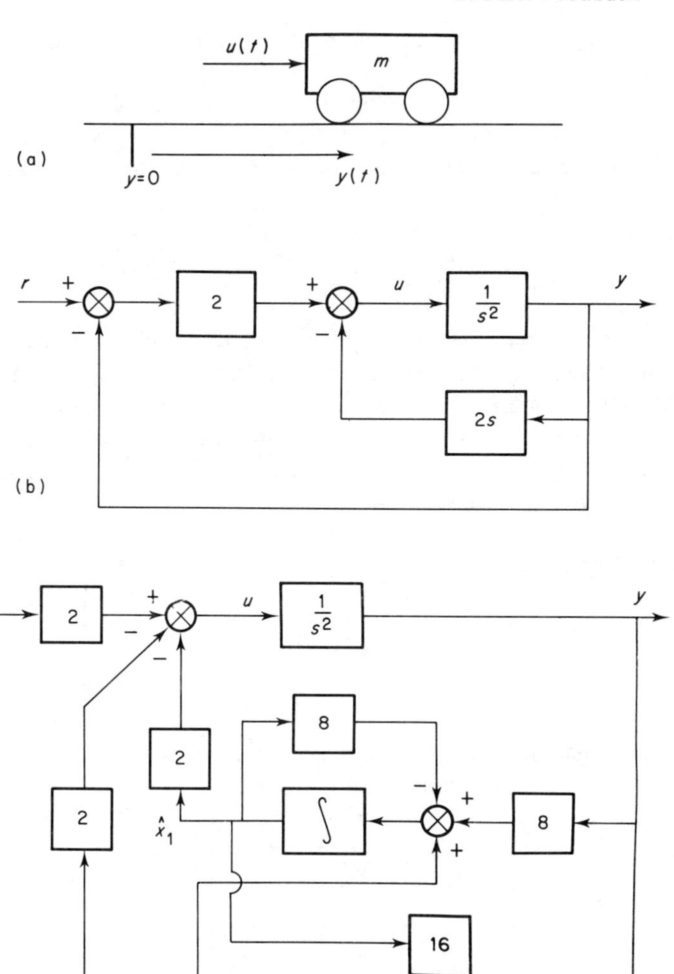

Fig. 34.

Verify that the system has a state-variable model

$$\frac{dx(t)}{dt} = \begin{bmatrix} 0 & 1 \\ 0 & 0 \end{bmatrix} x(t) + \begin{bmatrix} 0 \\ 1 \end{bmatrix} u(t), \; y(t) = \begin{bmatrix} 1 & 0 \end{bmatrix} x(t)$$

and that the state feedback law $u(t) = 2r(t) - [2 \quad 2]x(t)$ generates a closed-loop system with eigenvalues $\mu_1 = -1 + j$, $\mu_2 = -1 - j$. Show also that the closed-loop system can be represented by the classical configuration of Fig. 34(b).

The implementation of the state-feedback law requires the measurement of the rate $x_2(t) = dy(t)/dt$. Suppose that the designer only has access to measurement of the position $y(t)$. Verify that an observer for the system takes the general form

$$\frac{d\hat{x}(t)}{dt} = \begin{bmatrix} -\alpha & 1 \\ -\beta & 0 \end{bmatrix} \hat{x}(t) + \begin{bmatrix} 0 \\ 1 \end{bmatrix} u(t) + \begin{bmatrix} \alpha \\ \beta \end{bmatrix} y(t)$$

and generates the estimate $\hat{x}(t)$ of the state $x(t)$. Hence find the (stable) observer with eigenvalues $\eta_1 = \eta_2 = -4$ (Ans: choose $\alpha = 8$, $\beta = 16$) and verify that Fig. 34(c) represents an approximate implementation of the original state-feedback law. If you get the opportunity, simulate the responses of the systems in Fig. 34(b) and (c) from zero initial conditions to unit step inputs and assess the degradation in performance due to the use of the observer.

(3) Given the matrices

$$A = \begin{bmatrix} 1 & 0 & 1 \\ 2 & 1 & 0 \\ 0 & 1 & -1 \end{bmatrix}, \qquad b = \begin{bmatrix} 0 \\ 1 \\ 0 \end{bmatrix}$$

show that A has the characteristic polynomial $\rho_0(s) = s^3 - s^2 - s - 1$. Hence, derive the nonsingular transformation.

$$T = \begin{bmatrix} 1 & 0 & 0 \\ -1 & 0 & 1 \\ -1 & 1 & 0 \end{bmatrix}$$

that transforms A and b to controllable canonical form. Finally use your results to compute a state-feedback control law for the system

$$\frac{dx(t)}{dt} = \begin{bmatrix} 1 & 0 & 1 \\ 2 & 1 & 0 \\ 0 & 1 & -1 \end{bmatrix} x(t) + \begin{bmatrix} 1 & 1 \\ 2 & 1 \\ 1 & 1 \end{bmatrix} u(t)$$

that produces a closed-loop system with eigenvalues $\mu_1 = \mu_2 = \mu_3 = -1$. (Hint: find a 2×1 vector p such that $b = Bp$.) Check your result by direct evaluation of the closed-loop characteristic polynomial.

(4) A linear, time-invariant discrete system is said to be *dead-beat* if it is

represented by a state-variable model $x_{k+1} = \Phi x_k + \Delta u_k$, $k \geqslant 0$, where $\Phi^N = 0$ for some integer $N \geqslant 1$. Show that
(a) $\Phi^{N+k} = 0$ for all $k \geqslant 0$;
(b) $N = 1$ if, and only if, $\Phi = 0$;
(c) all eigenvalues of Φ are zero and Φ has characteristic polynomial $\rho(s) = s^n$;
(d) there is a nonsingular $n \times n$ matrix T such that

$$T^{-1}\Phi T = \begin{bmatrix} J_1 & 0 & \cdots & 0 \\ 0 & J_2 & & \vdots \\ \vdots & & & 0 \\ 0 & \cdots & 0 & J_q \end{bmatrix}$$

where $J_k = 0$ if $n_k = 1$. Otherwise it is an $n_k \times n_k$ Jordan block of the form

$$J_k = \begin{bmatrix} 0 & 1 & 0 & \cdots & 0 \\ 0 & 0 & 1 & & \vdots \\ \vdots & & & & 0 \\ \vdots & & & & 1 \\ 0 & \cdots & & \cdots & 0 \end{bmatrix}$$

(e) $N = \max_k n_k$;
(f) a necessary condition for a single-input discrete system to be controllable is that $N = n$.

(5) Show that a dead-beat discrete system $x_{k+1} = \Phi x_k + \Delta u_k$ has the property that, in the absence of any input, $x_k = \Phi^k x_0 = 0$ for $k \geqslant N$, i.e. the effect of the initial condition on transient response is eliminated after, at most, N samples. The system is an ideal regulator! This fact can make dead-beat action an attractive property in design, in the sense that, if the system is not dead-beat, it may be possible to choose a state-feedback law $u_k = Kr_k - Fx_k$ to make the closed-loop system $x_{k+1} = (\Phi - \Delta F)x_k + \Delta K r_k$ dead-beat. All one has to do is choose F to ensure that $\Phi - \Delta F$ has only zero eigenvalues.

Find a state-feedback law for the system

$$x_{k+1} = \begin{bmatrix} -2 & 1 \\ 1 & 1 \end{bmatrix} x_k + \begin{bmatrix} 1 \\ 1 \end{bmatrix} u_k$$

to produce a dead-beat closed-loop system. If access is only available to the single measurement $z_k = [1 \quad 1]x_k$, design a dead-beat observer to estimate the system state. Discuss the advantages of using a dead-beat observer. (Hint: show that the estimation error $e_k = 0$ for $k \geqslant 2$, independent of the initial error e_0.)

Remarks and Further Reading

The theory of control system *synthesis* using linear constant state feedback is extremely well developed (see, for example, Wonham, 1974; Wonham, 1978; Munro, 1979a; Porter and Crossley, 1972; Fossard, 1977) and has played a major role in the development of geometric (Wonham, 1974; Wonham, 1978) and algebraic (Rosenbrock, 1970) systems theory. In contrast the theoretical analysis of controller synthesis in the presence of only limited measurements has not reached the same degree of completeness, despite the fact that this is certainly the most common situation met in practice. This is probably due to the apparent difficulty in posing design problems in this case that have a unique mathematical solution. They either have no solution (as can occur if pole allocation is attempted without measurements of all states) or an infinite number of acceptable solutions as in classical design where the poles and zeros of compensation networks can very often vary over quite wide bands without unacceptable changes in closed-loop performance. This is not necessarily a bad thing as it involves the designer in the control synthesis making decisions and (quite probably) compromises, and introduces the possibility of using physical insight into process dynamics. Theoretical developments in this area have had a different character. They have tended to be frequency-domain based and aimed at generating a toolbag of techniques to aid the designer in decision-making. An introduction to these methods is described in the next chapter.

The following notes should help those readers who wish to follow up the large number of possibilities in state feedback control.

(1) The original proof of the pole-allocation theorem was provided by Wonham (1967) and alternative proofs (of varying complexity and generality) can be found in Wonham (1974), Barnett (1975) and Owens (1978).
(2) Pole allocation has been extensively studied with some attempts to assess the effect of limited state information (Munro, 1979b; Fallside, 1977) but has not been extensively applied in industry. The reader can, however, find applications described in Porter and Crossley (1972) and their chapter in the text edited by Munro (1979a). It has not been extensively applied primarily because of the measurement problem but also because no acceptable answer has yet been provided to the question as to where to put the closed-loop system eigenvalues.
(3) Pole allocation has obvious connections with the well known root-locus method (Shinners, 1978; Raven, 1978) and its multivariable generalization

(Owens, 1978; Owens, 1979b; Owens, 1981; MacFarlane, 1980; Postlethwaite and MacFarlane, 1979).

(4) Calculation of the coefficients $a_1, a_2, \ldots, a_n$ in (2.23) can be a numerical problem for high order systems. They can be computed, however, from knowledge of the eigenvalues of A or using the Faddeev algorithm (see Owens, 1978).

(5) Constant state feedback can be used to achieve other objectives such as non-interaction in the closed-loop system and disturbance isolation and rejection and even occurs naturally in other topics such as optimal control (see later chapters). The ideas can also be generalized to include dynamic elements in the feedback loop (so-called "dynamic state feedback"). Illustrative examples of these ideas can be found in Wonham (1974) and in the IEE paper by Munro (1979b).

(6) The design of observers is a subject in its own right. The ideas were introduced by Luenberger (1964, 1966) and have developed considerably since that time. More information can be found in the text edited by Munro (1979a), Layton (1976) and Kailath (1980).

3. Continuous Output Feedback

In this chapter we move back into more familiar territory and examine the natural generalizations of the familiar transfer function methods to the case of l-input/m-output linear, time-invariant systems. All of the generalizations will follow clearly in matrix notation and a number of new fundamentally multivariable phenomena will appear in our design vocabulary. Again there is no attempt at an exhaustive treatment, the emphasis being laid on illustrating new concepts and methodologies that form the foundations of more advanced study. The interested reader is referred to the reading list for further developments. All readers are urged to compare the material with the familiar methods for single-input/single-output systems.

In contrast to the last chapter where measurements of all state variables were assumed to be available, it is emphasized that the control designer can only use measurements of the system outputs $y_1(t), y_2(t), \ldots, y_m(t)$ in the control scheme. This is by far the most common situation met in practice as measurements mean instrumentation and instrumentation means money. The control benefits have got to be considerable, therefore, to merit investment in extensive measurement equipment.

3.1 Systems in Series and Feedback Configurations

In general terms, feedback control systems are built (a) by using the outputs from one subsystem to act as inputs to another and (b) by introducing unity feedback loops. In this section, we build up the basic relationships describing these operations in terms of the state space models of the subsystems.

Consider a l-input/m-output system (termed the *plant* to be controlled) described by the linear, time-invariant model.

$$\frac{dx_p(t)}{dt} = A_p x_p(t) + B_p u(t)$$

$$y(t) = C_p x_p(t) \tag{3.1}$$

with $l \times 1$ input vector $u(t)$, $n_1 \times 1$ state vector $x_p(t)$ and $m \times 1$ output vector $y(t)$. Note that we are assuming that there is no "D-term" in $y(t)$ as (a) this is the situation most commonly met in practice and (b) this assumption tends to simplify the analysis.

Suppose that the input $u(t)$ to the plant is generated as the output of the *forward path controller* with $m \times 1$ (note the dimension) "input" vector $e(t) = (e_1(t), \ldots, e_m(t))^T$ as illustrated in Fig. 35. There are two possibilities here. Either the control element contains dynamic elements or it does not. These possibilities are separated in the following sections.

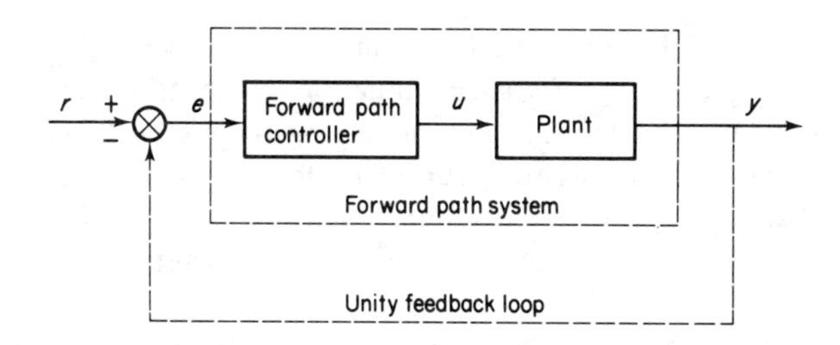

Fig. 35. Multivariable unity negative feedback system.

3.1.1 *Proportional control elements*

The simplest form of forward path controller simply sets each input to the plant equal to a linear combination $u_k(t) = K_{k1}e_1(t) + K_{k2}e_2(t) + \ldots + K_{km}e_m(t)$, $1 \leqslant k \leqslant l$, where the K_{ki} are real scalar "gains". Such a controller can always be represented in the matrix form

$$u(t) = Ke(t) \tag{3.2}$$

where K is the $l \times m$ matrix of constant gains K_{ki}. This form of controller is the generalization of the classical notion of *proportional control.*

The *forward path system* describing the relationship between $e(t)$ and $y(t)$ is obtained by substituting (3.2) into (3.1),

$$\frac{dx_p(t)}{dt} = Ax_p(t) + Be(t)$$

$$y(t) = Cx_p(t) \tag{3.3}$$

where

$$A = A_p, \qquad B = B_p K, \qquad C = C_p \tag{3.4}$$

The characteristic polynomial describing the stability of this system will be called the "open-loop characteristic polynomial"

$$\rho_0(s) = |sI_{n_1} - A| \tag{3.5}$$

and it is obviously identical to the characteristic polynomial of the plant.

Unity negative feedback is introduced into the picture by writing

$$e_k(t) = r_k(t) - y_k(t), \qquad 1 \leqslant k \leqslant m \tag{3.6}$$

where $r_k(t)$ is the ideal demanded response from $y_k(t)$. The function $e_k(t)$ is hence the error between the ideal and actual responses of $y_k(t)$. The vector $e(t)$ is termed the $m \times 1$ "error vector".

Collecting the equations in (3.6) together in matrix form yields the formally familiar equation

$$e(t) = r(t) - y(t) \tag{3.7}$$

where $r(t)$ is the $m \times 1$ demand vector (2.1). The final feedback configuration is illustrated in Fig. 35 (the differencing junction representing the vector difference operation defined by (3.7)) and, as in single-input/single-output control theory, is termed a "unity negative output feedback configuration". The term "output feedback" is used as the control element uses only information on output measurements and demands to construct the plant inputs $u_1, u_2, \ldots, u_l$.

The closed-loop system describing the output response $y(t)$ to a given demand vector $r(t)$ is obtained by substituting the equation $e = r - y = r - Cx_p$ into (3.3),

$$\frac{\mathrm{d}x_p(t)}{\mathrm{d}t} = (A - BC)x_p(t) + Br(t)$$

$$y(t) = Cx_p(t) \tag{3.8}$$

The characteristic polynomial describing the stability of this system will be called the "closed-loop characteristic polynomial" and takes the form

$$\rho_c(s) = |sI_{n_1} - A + BC| \tag{3.9}$$

It is clear that the inclusion of a forward path controller and a feedback loop has affected plant stability. More about this later!

3.1.2 Dynamic control elements

It is well known from classical control theory that the use of proportional control alone limits the attainable closed-loop performance and that a "high performance" controller tends to include dynamic terms such as integrators and phase compensation networks. In our multivariable notation, this possibility is represented by using a forward path controller described by a linear, time-invariant model of the form

$$\frac{dx_c(t)}{dt} = A_c x_c(t) + B_c e(t)$$

$$u(t) = C_c x_c(t) + D_c e(t) \tag{3.10}$$

where $x_c(t)$ is an $n_2 \times 1$ state vector describing the *dynamic* aspects of the control element. In contrast to the plant description (3.1) the controller does, in general, have a non-zero "D-term" in its state-variable description (see, for example, problem (3) in Chapter 1 where it is shown that the standard lead-lag, proportional plus integral controller for single-input/single-output system can be written in this form and also problem (3) at the end of this chapter).

The forward path system describing the dynamic relationship between $e(t)$ and $y(t)$ is obtained by introducing the $n \times 1$ state vector ($n = n_1 + n_2$)

$$x(t) = \begin{bmatrix} x_p(t) \\ x_c(t) \end{bmatrix} \tag{3.11}$$

and combining (3.1) and (3.10) in the form

$$\frac{dx(t)}{dt} = Ax(t) + Be(t), \qquad y(t) = Cx(t) \tag{3.12}$$

where

$$A = \begin{bmatrix} A_p & B_p C_c \\ 0 & A_c \end{bmatrix}, \qquad B = \begin{bmatrix} B_p D_c \\ B_c \end{bmatrix}, \qquad C = [C_p, 0] \tag{3.13}$$

The characteristic polynomial of this system is the open-loop characteristic polynomial

$$\rho_0(s) = |sI_n - A|$$

$$\equiv |sI_{n_1} - A_p| \cdot |sI_{n_2} - A_c| \tag{3.14}$$

and is simply the product of the characteristic polynomial of the plant and forward path controller.

Unity negative feedback is introduced as illustrated in Fig. 35 by the error equation (3.7). The state-variable model of the closed-loop system is obtained by substituting $e = r - y = r - Cx$ into (3.12), i.e.

$$\frac{dx(t)}{dt} = (A - BC)x(t) + Br(t)$$

$$y(t) = Cx(t) \tag{3.15}$$

The closed-loop characteristic polynomial hence takes the form

$$\rho_c(s) = |sI_n - A + BC| \tag{3.16}$$

3.2 Transfer Function Matrices

Although the state-variable models and characteristic polynomials derived in the previous section do, in principle, make possible the analysis of closed-loop system stability and dynamic responses, they are notationally clumsy and it is difficult to see the connection with single-input/single-output concepts. The picture does however become much clearer if we introduce the notion of *transfer function matrices.*

3.2.1 *Transfer function matrices and the impulse response matrix*

Consider the general l-input/m-output system

$$\frac{\mathrm{d}x(t)}{\mathrm{d}t} = Ax(t) + Bu(t), \qquad x(0) = x_0$$

$$y(t) = Cx(t) + Du(t) \tag{3.17}$$

and take Laplace transforms to yield the relations

$$sx(s) - x_0 = Ax(s) + Bu(s) \tag{3.18}$$

$$y(s) = Cx(s) + Du(s) \tag{3.19}$$

(The reader is reminded that, for notational simplicity, we are denoting the Laplace transform $\mathscr{L}f(t)$ of a function $f(t)$ by $f(s)$).

EXERCISE 3.2.1. Verify (3.18) by taking Laplace transforms of (3.17) element by element and reassembling the result into matrix form.

Rearranging (3.18) to solve for $x(s)$ yields

$$x(s) = (sI_n - A)^{-1}Bu(s) + (sI_n - A)^{-1}x_0 \tag{3.20}$$

and hence, substituting into (3.19),

$$y(s) = \{C(sI_n - A)^{-1}B + D\}u(s) + C(sI_n - A)^{-1}x_0 \tag{3.21}$$

These equations can be used to solve for $x(t)$ and $y(t)$ by taking inverse Laplace transforms. Our primary interest here, however, is the search for a generalization of the idea of transfer function.

The key to the generalization is seen by considering the case of zero initial conditions $x_0 = 0$, when (3.21) reduces to

$$y(s) = G(s)u(s) \tag{3.22}$$

Comparing this equation with (1.8) it is clear that the $m \times l$ matrix

$$G(s) = C(sI_n - A)^{-1}B + D \tag{3.23}$$

plays the role of the "transfer function" of the multivariable system. It is however a matrix if either or both of m and l are greater than one and is hence called the "system transfer function matrix". It is an $m \times l$ matrix whose elements are transfer functions (see later examples). The system is represented in the block form illustrated in Fig. 36 which has the same structure as the familiar single-input/single-output case illustrated in Fig. 2.

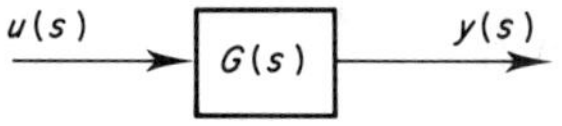

Fig. 36. Transfer function matrix representation of multivariable system dynamics.

Some insight into the meaning of (3.22) can be obtained by writing G in element form

$$G(s) = \begin{bmatrix} G_{11}(s) & G_{12}(s) \ldots & G_{1l}(s) \\ \vdots & & \vdots \\ G_{m1}(s) & \ldots\ldots\ldots & G_{ml}(s) \end{bmatrix} \tag{3.24}$$

and noting that (using the normal rules of matrix multiplication)

$$y_k(s) = G_{k1}(s)u_1(s) + \ldots + G_{kl}(s)u_l(s), \qquad 1 \leqslant k \leqslant m \tag{3.25}$$

i.e. the kth output transform is obtained as a linear combination of input transforms with transfer function coefficients. The transfer function $G_{ki}(s)$ in the kth row and the ith column of $G(s)$ hence describes the way that the input u_i affects the output y_k. In particular u_i has no effect on y_k if, and only if, $G_{ki}(s) \equiv 0$. In fact, the following very important result concerning system interaction follows immediately (see Section 1.1).

Theorem 3.2.1. *Taking the case of $m = l$, the system (3.17) is non-interacting if, and only if, its $m \times m$ transfer function matrix is diagonal for all s.*

EXERCISE 3.2.2. In the case of $m = l = 2$ verify that the system can be represented by the block diagram shown in Fig. 37.

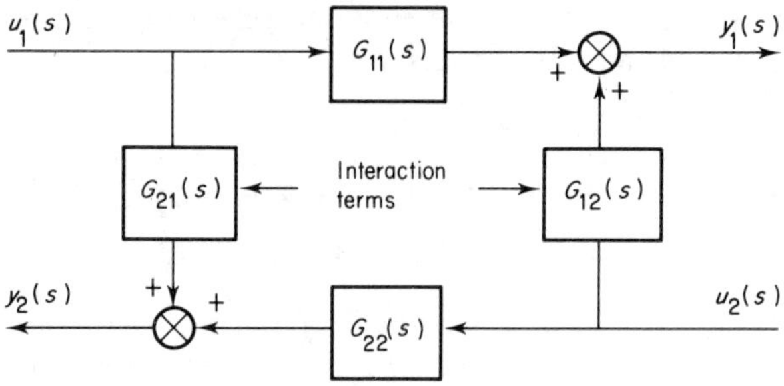

Fig. 37. Two-input/two-output system with interaction.

EXAMPLE 3.2.1. The transfer function matrix of the liquid level system (1.71) takes the form

$$G(s) = \begin{bmatrix} 1 & 0 \\ 0 & 1 \end{bmatrix} \begin{bmatrix} s + \dfrac{\beta}{a_1} & -\dfrac{\beta}{a_1} \\ -\dfrac{\beta}{a_2} & s + \dfrac{\beta}{a_2} \end{bmatrix}^{-1} \begin{bmatrix} a_1^{-1} & 0 \\ 0 & a_2^{-1} \end{bmatrix} + \begin{bmatrix} 0 & 0 \\ 0 & 0 \end{bmatrix}$$

$$= \frac{1}{s(a_1 a_2 s + \beta(a_1 + a_2))} \begin{bmatrix} a_2 s + \beta & \beta \\ \beta & a_1 s + \beta \end{bmatrix} \qquad (3.26)$$

indicating that the system is non-interacting only if $\beta = 0$.

EXERCISE 3.2.3. Verify that the mechanical system of Exercise 1.2.3 with $k_1 = k_2 = m_1 = m_2 = 1$ has the transfer function matrix

$$G(s) = \frac{1}{s^4 + 3s^2 + 1} \begin{bmatrix} s^2 + 1 \\ 1 \end{bmatrix} \qquad (3.27)$$

Note the dimensions of $G(s)$!

Finally, we note that the transfer function matrix $G(s)$ and the weighting matrix $H(t)$ defined by (1.165) and appearing in the convolution (1.167) satisfy the relation

$$G(s) = \mathscr{L}\{H(t)\} \qquad (3.28)$$

i.e. the system transfer function matrix is the Laplace transform of the system impulse response matrix. This is a familiar notion and is proved by considering the response of the system from zero initial conditions to an arbitrary impulsive input $u(t) = \alpha\delta(t)$ (see (1.168)). It follows directly that $u(s) = \mathscr{L}u(t) = \alpha$ and hence that $G(s)\alpha = y(s) = \mathscr{L}\{H(t)\}\alpha$ by (1.169). The relation (3.18) follows directly as α is arbitrary.

EXERCISE 3.2.4. Use (3.26) to show that the liquid level system with $a_1 = a_2 = \beta = 1$ has impulse response matrix

$$H(t) = \tfrac{1}{2}\begin{bmatrix} 1 + e^{-2t} & 1 - e^{-2t} \\ 1 - e^{-2t} & 1 + e^{-2t} \end{bmatrix} \tag{3.29}$$

3.2.2 Some block diagram algebra

As in the case of single-input/single-output systems the transfer function matrix representation of system dynamics leads to a simplicity of representation of the dynamics of *composite systems.* To demonstrate this, consider the configurations shown in Fig. 38 paralleling the equivalent classical configurations of Fig. 3.

In the *parallel* configuration of Fig. 38(a), the $l \times 1$ input vector $u(s)$ is fed simultaneously into q individual systems with transfer function matrices $G_i(s)$, $1 \leqslant i \leqslant q$, and $m \times 1$ output vectors $\hat{y}_i(s)$, $1 \leqslant i \leqslant q$, the output vector $y(s)$ from the system being the (matrix) sum of the subsystem outputs, i.e. using zero initial conditions $\hat{y}_i(s) = G_i(s)u(s)$, $1 \leqslant i \leqslant q$, and

$$\begin{aligned} y(s) &= \hat{y}_1(s) + \hat{y}_2(s) + \ldots + \hat{y}_q(s) \\ &= G_1(s)u(s) + G_2(s)u(s) + \ldots + G_q(s)u(s) \\ &= (G_1(s) + G_2(s) + \ldots + G_q(s))u(s) \end{aligned} \tag{3.30}$$

However, with zero initial conditions, we know that $y(s) = G(s)u(s)$ where $G(s)$ is the transfer function matrix of the composite system. It follows directly that we must have

$$G(s) = G_1(s) + G_2(s) + \ldots + G_q(s) \tag{3.31}$$

so that $G(s)$ is just the (matrix) sum of the individual subsystem transfer function matrices.

In the series configuration of Fig. 38(b), the output vector $\hat{y}_i(s)$ of a system with transfer function matrix $G_i(s)$ acts as input to a system with transfer function matrix $G_{i-1}(s)$. Assuming zero initial conditions,

$$y(s) = G_1(s)\hat{y}_2(s) = G_1(s)G_2(s)\hat{y}_3(s) = \ldots = G_1(s)G_2(s)\ldots G_q(s)u(s) \tag{3.32}$$

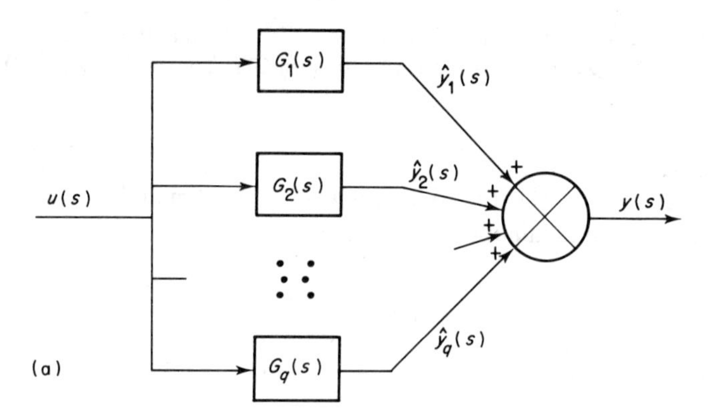

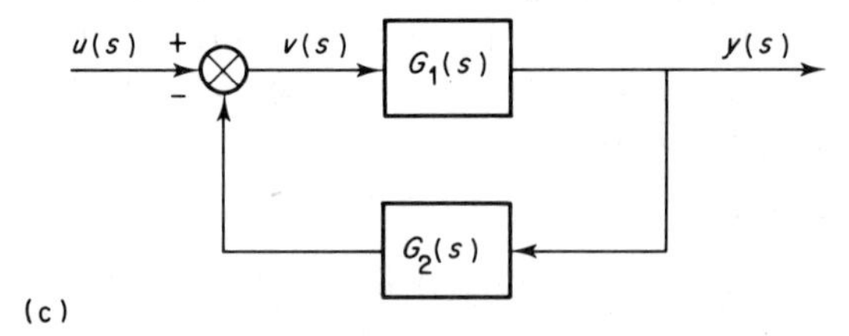

Fig. 38. Parallel, series and feedback multivariable systems.

so that the composite system transfer function matrix $G(s)$ has the form

$$G(s) = G_1(s)G_2(s) \ldots G_q(s) \tag{3.33}$$

i.e. $G(s)$ is just the (matrix) product of the individual subsystem transfer function matrices.

Finally, in the feedback system of Fig. 38(c), assuming zero initial conditions, it is clear that

$$y(s) = G_1(s)v(s), \qquad v(s) = u(s) - G_2(s)y(s) \tag{3.34}$$

and hence $y(s) = G_1(s)(u(s) - G_2(s)y(s)) = G_1 u - G_1 G_2 y$. Rearranging this equation in the form $y + G_1 G_2 y = (I + G_1 G_2)y = G_1 u$, it follows that

$$\begin{aligned} y(s) &= (I + G_1(s)G_2(s))^{-1}(I + G_1(s)G_2(s))y(s) \\ &= (I + G_1(s)G_2(s))^{-1}G_1(s)u(s) \end{aligned} \tag{3.35}$$

and hence that the composite system transfer function matrix

$$G(s) = (I + G_1(s)G_2(s))^{-1}G_1(s) \tag{3.36}$$

Comparing (3.31), (3.33) and (3.36) with (1.10), it is clear that all the well known relationships used in block diagram algebra generalize to the case of multivariable systems. All we have to do is to replace transfer functions by transfer function matrices, unity terms "1" by the unit matrix "I" of appropriate dimension and replace "division" by "matrix inversion". It is important to note, however, that in the multivariable case, the ordering of the transfer function matrices is vital in systems containing series or feedback connections. This is due to the fact that matrices do not (in general) commute. Taking, for example, (3.32) with $q = 2$, $G(s)$ cannot be written in the form $G_2(s)G_1(s)$.

EXERCISE 3.2.5. Use (3.34) to show that $v(s) = (I + G_2(s)G_1(s))^{-1}u(s)$ and hence that

$$G(s) = (I + G_1(s)G_2(s))^{-1}G_1(s) \equiv G_1(s)(I + G_2(s)G_1(s))^{-1} \tag{3.37}$$

Provide an alternative proof from the identity $(I + G_1G_2)G_1 = G_1(I + G_2G_1)$.

3.3 Poles and Zeros

The notion of transfer function matrix would not be complete without the notions of system poles and zeros. In the case of system poles, the generalization is quite straightforward but, in the case of system zeros, the generalization is technically rather tricky. For this reason the following development relies primarily on intuitive arguments to make the final definitions plausible.

Consider the l-input/m-output system (3.17) with transfer function matrix $G(s)$ given by (3.13) and write $(sI - A)^{-1} = \{\text{adj}\,(sI - A)\}/|sI - A|$ to yield

$$G(s) = \{C\,\text{adj}\,(sI - A)B + |sI - A|D\}/|sI - A| \tag{3.38}$$

Noting that $C\,\text{adj}\,(sI - A)B + |sI - A|D$ is simply an $m \times l$ matrix with elements that are *polynomials* in s and that $|sI - A|$ is a polynomial in s, it follows that every element of $G(s)$ is a transfer function expressed as the ratio of two polynomials. Neglecting the (technical) problem of cancellations it is clear that the denominator of each of these transfer functions is simply $|sI - A|$ and hence that the poles of each of the elements of $G(s)$ are the eigenvalues of the matrix A.
(Note: this connection between the poles of (elements of) G and the eigenvalues of A is not surprising bearing in mind known results for transfer functions and the stability results of Section 1.6. It perhaps also goes some way to explaining

the jargon of Section 2.3 where the problem of manipulation of eigenvalues is called the *pole*-allocation problem.)

In contrast to the above where the poles of the system are simply the poles of the transfer function elements of $G(s)$, the zeros of the system are not the zeros of the elements of G. The physical basis of their definition is suggested by considering the time-response of a single-input/single-output system with transfer function $g(s)$ to the input $u(t) = u_0 e^{s't}$,

$$y(t) = g(s')u(t) + y_0(t) \tag{3.39}$$

where $y_0(t)$ is a term describing the effect of initial conditions. By suitable choice of initial conditions it is possible to ensure that $y_0(t) \equiv 0$ and hence that $y(t) = g(s')u(t)$. In particular, neglecting the technical problem of cancellations, the response to a non-zero input is then identically zero if, and only if, s' is equal to a zero of $g(s)$.

Based on these observations, the following definition of zero is the one that best fits the spirit of this text: the complex number s' is a "zero" of the l-input/m-output ($l \leqslant m$) system (3.17) if, and only if, we can choose an initial condition x_0 and an $l \times 1$ input vector $u(t) = u_0 e^{s't}$ such that either $u_0 \neq 0$ and/or $x_0 \neq 0$ and the output response $y(t)$ is identically zero.

A general mathematical characterization of zeros is a sophisticated problem. Insight is obtained, however, by restricting attention to the case of $m = l$, writing the state vector $x(t) = x_0 e^{s't}$ and substituting into (3.17) to give

$$\frac{dx(t)}{dt} = s'x_0 e^{s't} = Ax_0 e^{s't} + Bu_0 e^{s't}$$

$$0 \equiv y(t) = Cx_0 e^{s't} + Du_0 e^{s't} \tag{3.40}$$

Cancelling the common terms $e^{s't}$ and rearranging yields

$$\begin{aligned} (s'I - A)x_0 - Bu_0 &= 0 \\ Cx_0 + Du_0 &= 0 \end{aligned} \tag{3.41}$$

or, in partitioned form

$$\begin{bmatrix} s'I - A & -B \\ C & D \end{bmatrix} \begin{bmatrix} x_0 \\ u_0 \end{bmatrix} = 0 \tag{3.42}$$

Noting that, by assumption, the column $(x_0^T, u_0^T)^T$ is non-zero it is clear that we require the determinant

$$\begin{vmatrix} s'I - A & -B \\ C & D \end{vmatrix} = 0 \tag{3.43}$$

suggesting (not proving!) that s' is a zero of the system (3.17) if, and only if, it is a zero of the (so-called) *zero polynomial*

$$z(s) = \begin{vmatrix} sI-A & -B \\ C & D \end{vmatrix} \tag{3.44}$$

This conclusion is, in fact, correct and the zeros of the system can be calculated by solving the equation $z(s') = 0$.

EXAMPLE 3.3.1. The liquid level system (1.71) has the zero polynomial

$$z(s) = \left|\begin{array}{cc|cc} s+a_1^{-1}\beta & -a_1^{-1}\beta & -a_1^{-1} & 0 \\ -a_2^{-1}\beta & s+a_2^{-1}\beta & 0 & -a_2^{-1} \\ \hline 1 & 0 & 0 & 0 \\ 0 & 1 & 0 & 0 \end{array}\right| = (a_1 a_2)^{-1} \tag{3.45}$$

and hence has no zeros. (Note, however, that some elements of $G(s)$ do have zeros!)

An important alternative formula to (3.44) is obtained by application of the identity described in the following exercise.

EXERCISE 3.3.1. Let M_{11}, M_{12}, M_{21} and M_{22} be $n \times n$, $n \times m$, $m \times n$ and $m \times m$ matrices respectively, and suppose that M_{11} is nonsingular. Verify that

$$\begin{bmatrix} I_n & 0 \\ -M_{21}M_{11}^{-1} & I_m \end{bmatrix} \begin{bmatrix} M_{11} & M_{12} \\ M_{21} & M_{22} \end{bmatrix} = \begin{bmatrix} M_{11} & M_{12} \\ 0 & M_{22} - M_{21}M_{11}^{-1}M_{12} \end{bmatrix} \tag{3.46}$$

Hence (by taking determinants!) deduce *Schur's formula*

$$\begin{vmatrix} M_{11} & M_{12} \\ M_{21} & M_{22} \end{vmatrix} = |M_{11}| \cdot |M_{22} - M_{21}M_{11}^{-1}M_{12}| \tag{3.47}$$

Applying Schur's formula to (3.44) yields

$$z(s) = |sI-A| \cdot |D + C(sI-A)^{-1}B| \equiv |sI-A| \cdot |G(s)| \tag{3.48}$$

indicating that the system zero polynomial can also be calculated as the product of the system characteristic polynomial and the *determinant* of the system transfer function matrix. It also indicates, in the case of $m = l = 1$, that $G(s) = z(s)/|sI - A|$ (i.e. $z(s)$ is a numerator polynomial of the system transfer function) and hence that the notions of zero in the single-input/single-output and multivariable cases do coincide.

EXERCISE 3.3.2. Verify that the single-input/single-output system with data

$$A = \begin{bmatrix} 0 & 1 & 0 \\ 0 & 0 & 1 \\ 0 & 1 & 1 \end{bmatrix}, \quad B = \begin{bmatrix} 0 \\ 0 \\ 1 \end{bmatrix} \quad \begin{matrix} C = (4 \quad 6 \quad 2) \\ \\ D = 0 \end{matrix} \tag{3.49}$$

has zero polynomial $z(s) = 2(s^2 + 3s + 2)$ and hence zeros at the points $s = -1$, $s = -2$. Check your result by calculation and inspection of the system transfer function.

Other important properties and equivalent definitions of zeros can be found in the reading list and in the problems at the end of the chapter.

3.4 Design Criteria and the Return-difference

The material described in Sections 3.2 and 3.3 make possible an alternative transfer function matrix approach to the analysis of the unity feedback system of Fig. 35 that complements the state variable methods of Section 3.1. The reader may be reassured to know that, apart from the necessity of "getting the matrices in the right order", the approach turns out to have exactly the same structure as familiar transfer function methods.

In the configuration of Fig. 35, suppose that the plant is described by the model (3.1) with $m \times l$ transfer function matrix

$$G(s) = C_p(sI_{n_1} - A_p)^{-1}B_p \tag{3.50}$$

and that the forward path controller either has the proportional form (3.2) with (taking Laplace transforms) constant $l \times m$ transfer function matrix

$$K(s) = K \tag{3.51}$$

or the dynamic form (3.10) with $l \times m$ transfer function matrix

$$K(s) = C_c(sI_{n_2} - A_c)^{-1}B_c + D_c \tag{3.52}$$

Taking the case of zero initial conditions it is clear that, in both cases,

$$y(s) = G(s)u(s), \qquad u(s) = K(s)e(s) \tag{3.53}$$

and hence that

$$y(s) = Q(s)e(s) \tag{3.54}$$

where the "forward path transfer function matrix" $Q(s) = G(s)K(s)$ (note the order!) must be the transfer function matrix of the forward path system (3.3) or (3.12), i.e. $Q(s) = C(sI - A)^{-1}B$.

Taking the Laplace transform of (3.7) yields

$$e(s) = r(s) - y(s) \tag{3.55}$$

and hence, using (3.54), $y = Q(r-y)$ or $y + Qy = (I_m + Q)y = Qr$ indicating that

$$y(s) = H_c(s)r(s) \tag{3.56}$$

where the "closed-loop transfer function matrix"

$$H_c(s) = \{I_m + Q(s)\}^{-1}Q(s) \tag{3.57}$$

must hence be the transfer function matrix of the closed-loop system (3.8) or (3.15).

EXERCISE 3.4.1. A more direct calculation of $H_c(s)$ should proceed directly from the closed-loop state-variable model (3.15), $H_c(s) = C(sI_n - A + BC)^{-1}B$. Use the matrix identity $(I_m + M_1M_2)M_1 = M_1(I_n + M_2M_1)$(valid for any $m \times n$ and $n \times m$ matrices M_1 and M_2 respectively) to verify that

$$\begin{aligned} H_c(s) &= C(sI - A + BC)^{-1}B = C(I_n + (sI_n - A)^{-1}BC)^{-1}(sI_n - A)^{-1}B \\ &\equiv (I_m + C(sI_n - A)^{-1}B)^{-1}C(sI_n - A)^{-1}B = (I_m + Q(s))^{-1}Q(s) \end{aligned} \tag{3.58}$$

hence proving (3.57) directly from the state-variable model. (Hint: take $M_1 = C$.)

The familiar structure of these results is revealed by using (3.53), (3.54) and (3.55) to show that the closed-loop system of Fig. 35 has the block representations shown in Fig. 39. Even (3.57) is the natural generalization of the single-input/single-output transfer function expression $Q/(1+Q)$, replacing the "1" by "I_m" and division by matrix inversion!

In general terms, the practical problem of unity negative feedback controller design for the plant $G(s)$ can be replaced by the theoretical problem of designing a forward path controller described by the transfer function matrix $K(s)$ such that the closed-loop system has "satisfactory" stability and transient performance

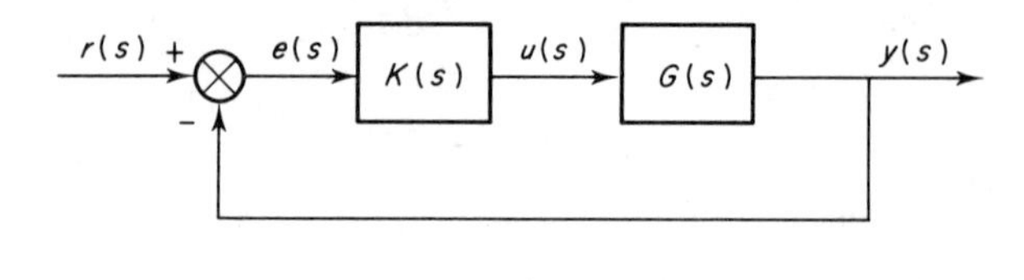

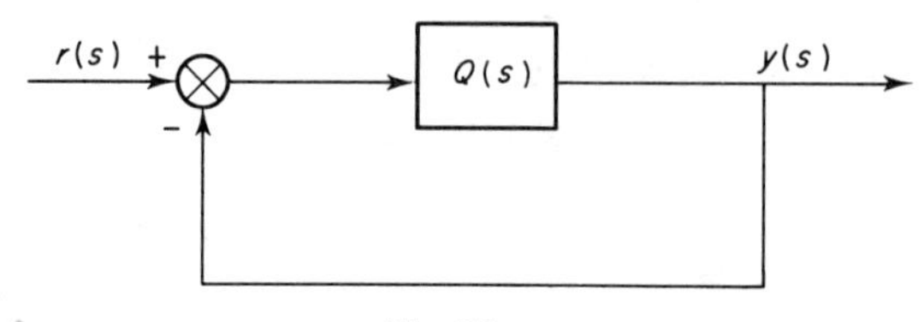

Fig. 39.

characteristics. The choice of $K(s)$ is equivalent to the choice of its ml transfer function elements describing the desired connections between the m error signals and the l plant inputs.

The definition of satisfactory stability and transient performance characteristics will vary from situation to situation. There are, however, a number of general criteria that commonly arise in practice, namely those of stability, steady-state and transient response and the essentially multivariable phenomenon of interaction. These are examined in more detail below.

Closed-loop stability. In general, the stability of the closed-loop system is absolutely essential. In principle, closed-loop stability can be investigated using the closed-loop characteristic polynomial $\rho_c(s)$ described in Section 3.1 but, for design purposes, it is useful to have a relationship between the forward path system transfer function matrix Q and closed-loop stability. More precisely, defining the *matrix return-difference* $T(s) = I_m + Q(s)$ it will be shown below that

$$\frac{\rho_c(s)}{\rho_0(s)} \equiv |T(s)| \tag{3.59}$$

i.e. the ratio of the closed-loop characteristic polynomial to the open-loop characteristic polynomial is identically equal to the *return-difference determinant* $|T(s)| = |I_m + Q(s)|$. This result is a natural generalization of the classical ($m = 1$) formula $\rho_c(s)/\rho_0(s) = 1 + Q(s)$ forming the basis of the well known Nyquist and root-locus analysis and design techniques. This is also true in the multivariable case, but the technical problems introduced in this case are beyond the scope of this text (see the reading list at the end of the chapter). In general terms, these difficulties are due to the complexity of $|T(s)|$ as a function of the elements of $K(s)$. These problems, together with some indications of solutions, are illustrated by example in the following sections.

The proof of (3.59) is based on the identity

$$|I_m + M_1 M_2| = |I_n + M_2 M_1| \tag{3.60}$$

valid for any $m \times n$ matrix M_1 and $n \times m$ matrix M_2 (see Exercise 3.4.2). In fact using, for example, (3.14) and (3.16), it is seen that

$$\begin{aligned}
\frac{\rho_c(s)}{\rho_o(s)} &\equiv \frac{|sI_n - A + BC|}{|sI_n - A|} \equiv |(sI_n - A)^{-1}| \cdot |sI_n - A + BC| \\
&\equiv |(sI_n - A)^{-1}\{sI_n - A + BC\}| \\
&\equiv |I_n + (sI_n - A)^{-1}BC| \\
&\equiv |I_m + C(sI_n - A)^{-1}B| \\
&\equiv |I_m + Q(s)| \equiv |T(s)|
\end{aligned} \tag{3.61}$$

as required.

EXERCISE 3.4.2. Prove (3.60). (Hint: if $m = n$ and M_1 is nonsingular then $|I_m + M_1 M_2| = |M_1(M_1^{-1} + M_2)| = |(M_1^{-1} + M_2)M_1| = |I_m + M_2 M_1|$. If $m = n$ and both M_1 and M_2 are singular, note that $M_1 + \epsilon I_m$ is nonsingular in some range $0 < \epsilon < \delta$ and apply the above procedure noting, by continuity, that your conclusions are valid for $\epsilon = 0$. If $m < n$ write $|I_m + M_1 M_2| = \left|I_n + \begin{bmatrix} M_1 \\ 0 \end{bmatrix} [M_2 \quad 0]\right|$.)

EXAMPLE 3.4.1. Consider the liquid level system (1.71) with data $a_1 = a_2 = \beta = 1$ and transfer function matrix of the form (see (3.26))

$$G(s) = \frac{1}{s(s+2)} \begin{bmatrix} s+1 & 1 \\ 1 & s+1 \end{bmatrix} \tag{3.62}$$

Suppose that the liquid level system is subject to unity proportional negative feedback with a diagonal controller

$$K(s) = K = \begin{bmatrix} k_1 & 0 \\ 0 & k_2 \end{bmatrix}$$

(k_1, k_2 real scalar gains) representing the situation where each vessel input is manipulated using measurements made on that vessel alone, i.e. $u_i(t)=k_i(r_i(t)-y_i(t))$, $i=1,2$. Verify that $|T(s)|=(s^2+(k_1+k_2+2)s+k_1k_2+(k_1+k_2))/s(s+2)$ and hence that the closed-loop system is stable, if and only if,

$$k_1+k_2+2>0, \qquad k_1k_2+k_1+k_2>0 \tag{3.63}$$

Check this result by direct calculation of the closed-loop characteristic polynomial from the matrices of the state-variable model.

Transient performance and interaction. These properties of the closed-loop system are described by the closed-loop transfer function matrix $H_c(s)$ given in (3.57). In general terms, transient performance is described by the gains, poles and zeros of all transfer function elements of H_c, whereas interaction is described by the *off-diagonal* terms (see Theorem 3.2.1) only.

As in a single-input/single-output systems design, closed-loop transient performance can be assessed by examination of the time response from zero initial conditions to specified demands such as steps, ramps or periodic block or sawtooth functions. For simplicity, we restrict our attention to the use of unit step functions (representing the demand for a change in set point) as this is probably the most important case in practice.

The multivariable equivalent of a general step demand is the $m \times 1$ demand vector $r(t)=(r_1(t), \ldots, r_m(t))^T$ where $r_i(t)=0$, $t \leqslant 0$, and $r_i(t)=\alpha_i$, $t>0$, $1 \leqslant i \leqslant m$. Taking Laplace transforms it is clear that $r_i(s)=s^{-1}\alpha_i$, $1 \leqslant i \leqslant m$, and hence that $r(s)=s^{-1}\alpha$. Now, although only one step response is required to assess the performance of a single-input/single-output system, an m-output system requires m independent step responses to assess all aspects of its performance. By far the most common choice of demand inputs are defined by the "unit vectors"

$$\tilde{e}_1=\begin{bmatrix}1\\0\\\cdot\\\cdot\\\cdot\\\cdot\\0\end{bmatrix}, \quad \tilde{e}_2=\begin{bmatrix}0\\1\\0\\\cdot\\\cdot\\\cdot\\0\end{bmatrix}, \ldots, \tilde{e}_m=\begin{bmatrix}0\\\cdot\\\cdot\\\cdot\\\cdot\\0\\1\end{bmatrix} \tag{3.64}$$

and the choice of $r(s)=s^{-1}\tilde{e}_i$, $1 \leqslant i \leqslant m$. It is worthwhile emphasizing the physical meaning of this definition by noting that the demand input $r(s)=s^{-1}\tilde{e}_i$ requires that

(a) a unit step change occurs in $y_i(t)$, and
(b) all outputs $y_k(t)$, $k \neq i$, are identically zero.

Such a demand input is termed a *unit step demand in* $y_i(t)$. In practice, of course, the output vector never follows the demand exactly and, as such, conventional performance criteria are stated as a relaxed form of the demand signal requirements. More precisely, the response $y(t)$ to a unit step demand $r(s) = s^{-1}\tilde{e}_i$ in output y_i should be such that

(i) the response $y_i(t)$ should possess "satisfactory" overshoot, risetime, settling time, steady state error etc.

(ii) The responses $y_k(t)$ $(k \neq i)$ describing the interaction effects should be "acceptably small".

The question "what constitutes satisfactory overshoot, risetime . . . ?" is a question that must be answered individually for each application considered, although, in general terms, it is anticipated that overshoot should not be too large, risetime not too long etc. Similar comments can be made about the question of what constitutes acceptably small interaction. In this case, acceptably small interaction does not necessarily imply small numerical magnitudes as a simple change of physical units can transform a numerically small measurement into a numerically large measurement, e.g. a (numerically) small displacement of 0.01 m is identical to the (numerically) large displacement of 10 mm. In the remainder of this text it is therefore assumed, for simplicity, that physical units are chosen such that interaction effects are acceptable if the responses $y_k(t)$ $(k \neq i)$ to a unit step demand in y_i satisfy $|y_k(t)| \ll 1$ for all $t > 0$ and $k \neq i$. These ideas are illustrated in Fig. 40. The responses of Fig. 40(a) and (b) are unacceptable due to excessive oscillation and interaction effects respectively. The responses of Fig. 40(c), however, satisfy the performance specifications, as outlined above.

Steady-state response and steady-state error. If the closed-loop system is stable, the steady state response to a unit step demand $r(s) = s^{-1}\tilde{e}_i$ in output y_i is calculated by applying the standard final-value theorem

$$\begin{aligned}\lim_{t \to \infty} y(t) &= \lim_{s \to 0} sy(s) = \lim_{s \to 0} sH_c(s)r(s) \\ &= \lim_{s \to \infty} H_c(s)\tilde{e}_i = H_c(0)\tilde{e}_i \end{aligned} \tag{3.65}$$

(Note: $\lim_{s \to 0} H_c(s) = H_c(0)$ as stability guarantees that H_c has no pole at $s = 0$.)

As $r(t) = \tilde{e}_i$, $t > 0$, it is also clear that the steady state error vector

$$\begin{aligned}\lim_{t \to +\infty} e(t) &= \lim_{t \to +\infty} (r(t) - y(t)) = \tilde{e}_i - H_c(0)\tilde{e}_i = \{I_m - H_c(0)\}\tilde{e}_i \\ &= i\text{th column of } I_m - H_c(0) \end{aligned} \tag{3.66}$$

and hence that steady state errors in response to a unit step demand in y_i are zero if, and only if, the ith column of $I_m - H_c(0)$ is zero. If we require that

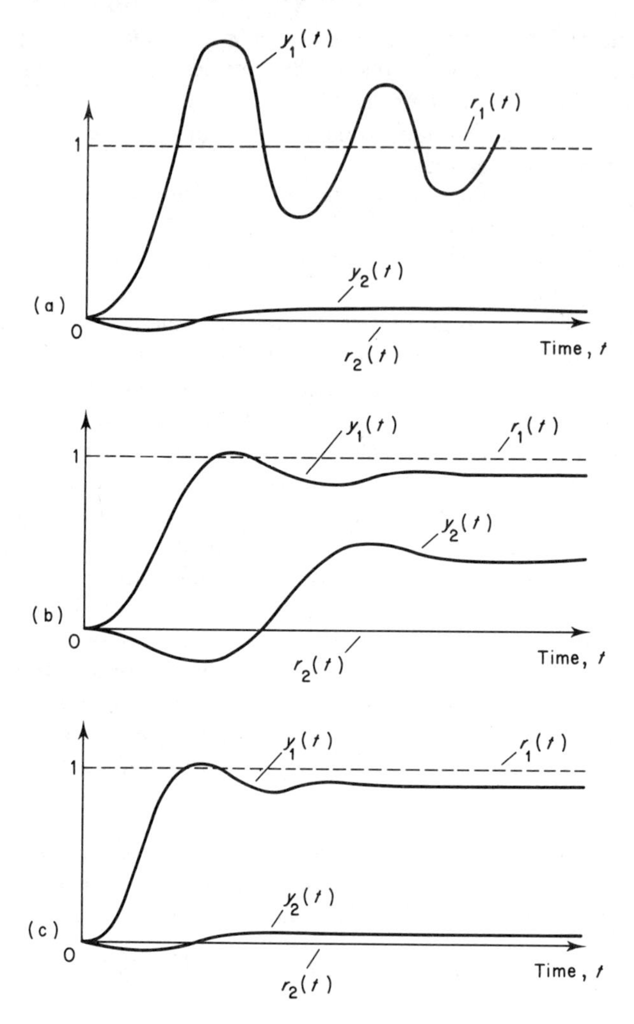

Fig. 40. Illustrative closed-loop responses of a two-output system to a unit step demand in y_1.

steady state errors in response to unit step demands in any output are zero it is clearly both necessary and sufficient that all columns of $I_m - H_c(0)$ are zero, i.e.

$$H_c(0) = I_m \tag{3.67}$$

(Compare this with the single-input/single-output requirement that the d.c. gain of the closed-loop transfer function is unity.)

EXAMPLE 3.4.2. Using the plant and proportional controller of Example 3.4.1 show that the closed-loop transfer function matrix is

$$H_c(s) = \frac{1}{(s^2 + (k_1 + k_2 + 2)s + k_1 + k_2 + k_1 k_2)} \times \begin{bmatrix} (s + 1 + k_2)k_1 & k_2 \\ k_1 & (s + 1 + k_1)k_2 \end{bmatrix} \tag{3.68}$$

It is then trivially verified that

$$H_c(0) = \frac{1}{(k_1 + k_2 + k_1 k_2)} \begin{bmatrix} (1 + k_2)k_1 & k_2 \\ k_1 & (1 + k_1)k_2 \end{bmatrix} \tag{3.69}$$

and hence that no finite choice of gains k_1, k_2 will produce a closed-loop system exhibiting zero steady state errors in response to unit step demand in y_1 or y_2. (Hint: evaluate the steady state error (3.66) for $i = 1, 2$, noting in each case that it is non-zero.) Note, however, that steady state errors can be made to be arbitrarily small by using high gains k_1, k_2 in both loops.

The analysis of closed-loop transient performance is a more complex matter. For simplicity, suppose that the loop gains k_1 and k_2 are identical, i.e. $k_1 = k_2 = k$ (where k is a convenient gain parameter) and consider the system response to a unit step demand in y_1,

$$y(s) = \begin{bmatrix} y_1(s) \\ y_2(s) \end{bmatrix} = H_c(s) \frac{1}{s} \begin{bmatrix} 1 \\ 0 \end{bmatrix} = \frac{k}{s(s + k)(s + k + 2)} \begin{bmatrix} (s + 1 + k) \\ 1 \end{bmatrix} \tag{3.70}$$

In particular,

$$\begin{aligned} y_1(t) &= \mathscr{L}^{-1} \frac{k(s + 1 + k)}{s(s + k)(s + k + 2)} \\ &= \frac{(1 + k)}{(2 + k)} - \tfrac{1}{2} e^{-kt} - \frac{k}{2(k + 2)} e^{-(k+2)t} \end{aligned} \tag{3.71}$$

showing that y_1 follows the demand rapidly with no overshoot and small steady state error if $k \gg 2$. The interaction term takes the form

$$\begin{aligned} y_2(t) &= \mathscr{L}^{-1} \frac{k}{s(s + k)(s + k + 2)} \\ &= \frac{1}{k + 2} - \tfrac{1}{2} e^{-kt} + \frac{k}{2(k + 2)} e^{-(k+2)t} \\ &= \frac{1}{k + 2} - \tfrac{1}{2} \{ e^{-kt} - e^{-(k+2)t} \} - \frac{e^{-(k+2)t}}{k + 2} \end{aligned} \tag{3.72}$$

A simple piece of calculus then indicates that

$$|y_2(t)| \leqslant \frac{1}{k+2} \tag{3.73}$$

and hence that the peak magnitude of the interaction term can be made to be arbitrarily small by the use of high gains $k \gg 2$. The resulting responses with the choice $k = 8$ are illustrated in Fig. 41 indicating interaction effects $|y_2(t)| < 0.1$, $t \geqslant 0$, no overshoot in y_1 and steady state errors of 0.1.

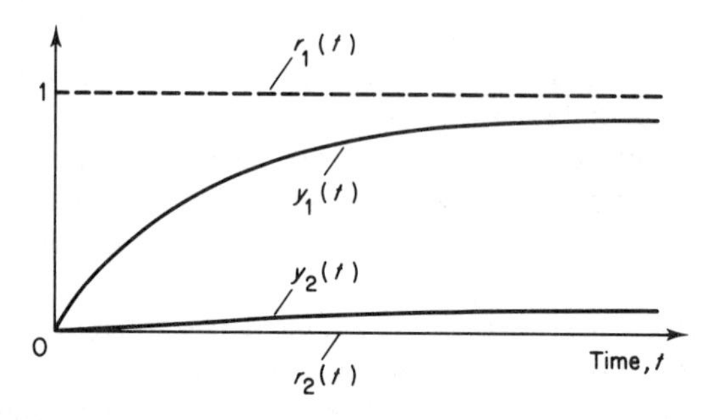

Fig. 41. Closed-loop response of liquid level system to a unit step demand in y_1.

Overall this analysis has shown that the proposed proportional controller is quite capable of producing a closed-loop system with excellent responses to unit step demands in y_1 if we use a high gain k. (Similar conclusions hold for the response to unit step demands in y_2 but this is left as an exercise for the reader.) The problem of why this controller rather than any other does this is not so easy to answer at this stage and is left for Section 3.6.

3.5 Closed-loop Interaction and Non-interacting Control

In this section (and Sections 3.6 and 3.7) we will restrict our attention to multivariable plant with the same number of inputs and outputs $m = l > 1$. It is important to recognize that the presence of interaction in the plant is a major source of difficulty in the design of the control system. In contrast, controller design for a non-interacting plant is a (relatively) straightforward task!

3.5.1 *Control of a non-interacting plant*

To illustrate these ideas, consider the non-interacting plant (see Theorem 3.2.1) described by the diagonal $m \times m$ transfer function matrix

$$G(s) = \begin{bmatrix} g_1(s) & 0 & \cdots & 0 \\ 0 & g_2(s) & & \vdots \\ \vdots & & & 0 \\ 0 & \cdots & 0 & g_m(s) \end{bmatrix} \quad (3.74)$$

By considering the relation $y(s) = G(s)u(s)$, it is clear that $y_i(s) = g_i(s)u_i(s)$, $1 \leqslant i \leqslant m$, and hence that our "multivariable" system consists of m independent single-input/single-output systems as illustrated in Fig. 42(a). It is also self-evident that each of these systems can be individually controlled by scalar unity feedback systems as shown in Fig. 42(b) with transfer functions

$$h_i(s) = \frac{g_i(s)k_i(s)}{1 + g_i(s)k_i(s)} \quad (3.75)$$

$u_1(s)$ $g_1(s)$ $y_1(s)$

$u_2(s)$ $g_2(s)$ $y_2(s)$

$u_m(s)$ $g_m(s)$ $y_m(s)$

(a)

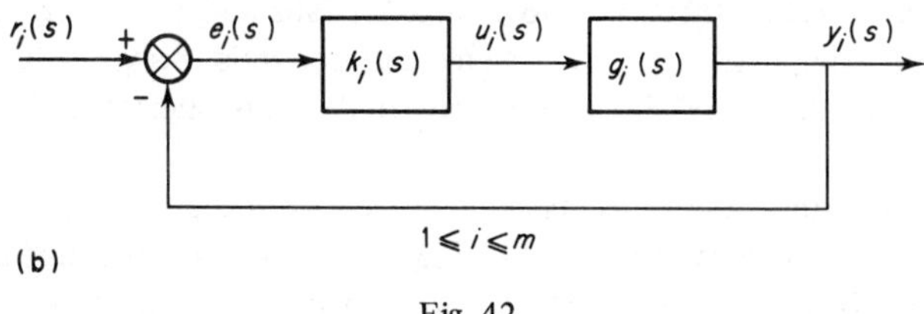

Fig. 42.

In particular the closed-loop system is non-interacting. It is also stable if, and only if, each of the closed-loop systems $h_i(s)$ is stable. We can state this in the multivariable systems language of previous sections by noting that $u_i(s) = k_i(s)e_i(s)$, $1 \leqslant i \leqslant m$, indicates that the equivalent forward path controller transfer function matrix is diagonal of the form

$$K(s) = \begin{bmatrix} k_1(s) & 0 & \cdots & 0 \\ 0 & k_2(s) & & \vdots \\ \vdots & & & 0 \\ 0 & \cdots & 0 & k_m(s) \end{bmatrix} \tag{3.76}$$

Also, it is easily verified that the closed-loop transfer function matrix is diagonal

$$H_c(s) = (I_m + G(s)K(s))^{-1}G(s)K(s) = \begin{bmatrix} h_1(s) & 0 & \cdots & 0 \\ 0 & & & \vdots \\ \vdots & & & 0 \\ 0 & \cdots & 0 & h_m(s) \end{bmatrix} \tag{3.77}$$

and the return-difference determinant

$$|T(s)| \equiv |I_m + G(s)K(s)| = \begin{vmatrix} 1+g_1k_1 & 0 & \cdots & 0 \\ 0 & & & \vdots \\ \vdots & & & 0 \\ 0 & \cdots & 0 & 1+g_mk_m \end{vmatrix}$$

$$= (1+g_1(s)k_1(s)) \ldots (1+g_m(s)k_m(s)) \tag{3.78}$$

3.5.2 *Non-interacting controllers*

If the plant possesses interaction effects (reflected by the presence of off-diagonal terms in $G(s)$) then the above analysis fails and there is a consequent increase in the complexity of the design exercise. It is, however, tempting to choose our controller $K(s)$ in such a way that similar design principles may apply.

The trick described below is initiated by constructing our controller $K(s)$ as two systems in series as illustrated in Fig. 43(a). The plant input $u(s)$ is the output from a *precompensator* with $m \times m$ transfer function matrix $K_p(s)$ and

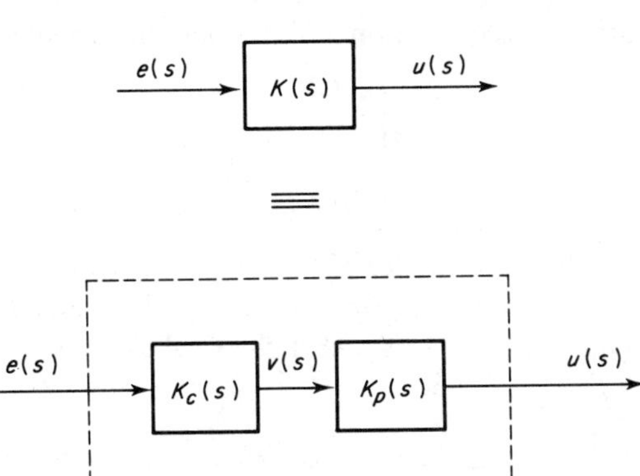

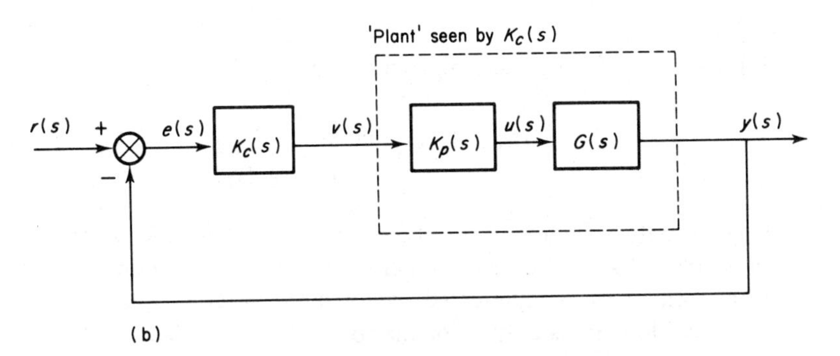

Fig. 43. Controller factorization and resulting feedback system.

$m \times 1$ input vector $v(s)$. The vector $v(s)$ is the output from a *compensator* with $m \times m$ transfer function matrix $K_c(s)$. From Section 3.2 it is clear that

$$K(s) = K_p(s)K_c(s) \tag{3.79}$$

The resulting closed-loop system is given in Fig. 43(b). Note that, from the point of view of the compensator K_c the "plant" takes the form $G(s)K_p(s)$ and that, if this plant were non-interacting, the design of $K_c(s)$ could proceed as in Section 3.5.1. Let us therefore choose the precompensator to make GK_p diagonal of the form

$$G(s)K_p(s) = \begin{bmatrix} g_1(s) & 0 \;.\;.\;. & 0 \\ 0 & g_2(s) & \vdots \\ 0. & \dots\dots & g_m(s) \end{bmatrix} \tag{3.80}$$

and, following the arguments of Section 3.5.1, choose the diagonal compensator

$$K(s) = \begin{bmatrix} k_1(s) & \dots & 0 \\ \vdots & & \vdots \\ 0 & \dots & k_m(s) \end{bmatrix} \tag{3.81}$$

as in (3.76). The resulting controller $K = K_p K_c$ is termed a "non-interacting controller". This is not due to its own structure (as K invariably has non-zero off-diagonal terms) but due to the fact that the final closed-loop system is non-interacting.

EXAMPLE 3.5.1. Consider the liquid level system (1.71) with transfer function matrix given by (3.26) and inverse transfer function matrix

$$G^{-1}(s) = \begin{bmatrix} sa_1 + \beta & -\beta \\ -\beta & sa_2 + \beta \end{bmatrix} \tag{3.82}$$

It is clear from (3.80) that the precompensator

$$K_p(s) = G^{-1}(s) \begin{bmatrix} g_1(s) & 0 \\ 0 & g_2(s) \end{bmatrix} \tag{3.83}$$

is completely specified by the choice of $g_1(s)$, $g_2(s)$. One approach to the choice of g_1 and g_2 is to examine the desired closed-loop responses from each vessel as represented by the diagonal terms $h_1(s)$, $h_2(s)$ of the closed-loop transfer function matrix

$$H_c(s) = \begin{bmatrix} h_1(s) & 0 \\ 0 & h_2(s) \end{bmatrix}, \qquad h_i(s) = \frac{g_i(s)k_i(s)}{1 + g_i(s)k_i(s)} \qquad (i = 1, 2) \tag{3.84}$$

For simplicity, suppose that we are looking for rapid first order responses to unit step demands with time constant $\tau = k^{-1}$ and zero steady state errors, i.e. we would like $h_i(s) = k/(s+k), i = 1, 2$. Solving these equations yields $g_i(s)k_i(s) = k/s$ $(i = 1, 2)$ suggesting the choice of $g_1(s) = g_2(s) = 1/s$ and $k_1 = k_2 = k$. The controller is now completely specified. The precompensator is obtained from (3.82) and (3.83) to be

$$K_p(s) = \begin{bmatrix} a_1 + \dfrac{\beta}{s} & -\dfrac{\beta}{s} \\ -\dfrac{\beta}{s} & a_2 + \dfrac{\beta}{s} \end{bmatrix} \tag{3.85}$$

and the compensator $K_c(s) = \begin{bmatrix} k & 0 \\ 0 & k \end{bmatrix}$. A block diagram of the resulting control system is illustrated in Fig. 44.

It is immediately apparent that, although the controller does achieve exact non-interaction and the desired closed-loop responses, it contains *dynamic* elements (in this case, integrators). The reader can easily verify that this will always be the case, independent of our choice of control parameters $g_i(s)$ and $k_i(s)$. This situation is a particularly simple example of the main problem with the use of non-interacting controllers, i.e. their dynamic complexity. In general terms, the more complex the plant, the more complicated is the resulting controller. If we note that, in general, controller complexity is proportional to economic cost it is easy to understand why there is a great incentive to achieve a satisfactory design using simple (preferably proportional) control elements. Note, however, if economic (or other) constraints insist that a proportional controller is used, then interaction will be present in the closed-loop system.

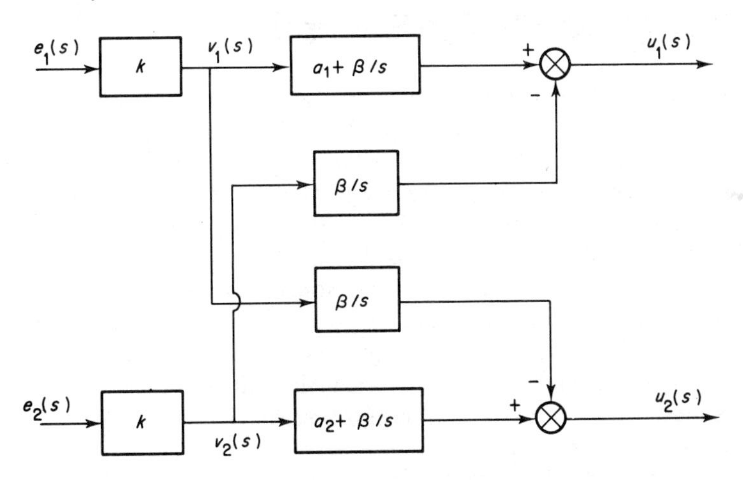

Fig. 44. Non-interacting controller for liquid level system.

EXERCISE 3.5.1. A system is described by the transfer function matrix

$$G(s) = \begin{bmatrix} \dfrac{2(s+1)}{s(s+2)} & \dfrac{s+1}{s(s+2)} \\ \dfrac{1}{s+1} & \dfrac{1}{s+1} \end{bmatrix} \quad (3.86)$$

EXERCISE 3.5.1. contd.

Verify that the choice of the constant precompensator

$$K_p(s) = \begin{bmatrix} 1 & -1 \\ -1 & 2 \end{bmatrix}$$

ensures that $G(s)K_p(s)$ is diagonal, with diagonal elements $g_1(s) = (s+1)/s(s+2)$ and $g_2(s) = 1/(s+1)$. Choosing the constant compensator

$$K_c(s) = \begin{bmatrix} k_1 & 0 \\ 0 & k_2 \end{bmatrix}$$

verify that the closed-loop system is stable for all choice of gains $k_1 > 0$, $k_2 > 0$.

3.6 Illustrative Design Example

If it were not for the undesirably high complexity of non-interacting controllers the analysis of the preceding section would be an almost complete solution to the design problem for multivariable feedback systems. It has, however, been necessary to undertake a major research effort to provide design techniques that generate dynamically simple controllers satisfying the (relaxed) performance specifications described in Section 3.4. The need to design dynamically simple controllers does, in general, mean that the closed-loop system will possess some interaction properties but, with careful design, these can, in many cases, be reduced to acceptable levels. It is the purpose of this section to illustrate this possibility by consideration of a simple example. Although simple, the example contains most of the fundamental concepts (in a simple form, of course) used in more advanced study (see reading list).

Consider the liquid level system (1.71) with data $a_1 = a_2 = \beta = 1$ and the problem of designing a unity feedback control system that regulates the liquid levels (in the absence of demands) to the specified equilibrium points and has the property that a demanded increase in level in one vessel is achieved rapidly, with little overshoot and steady state error and with little dynamic effect on the level in the other vessel. In other words, we want a stable closed-loop system exhibiting fast responses with small interaction effects and steady state errors in response to unit step demands in either of the system outputs. This has been achieved using a dynamic non-interacting controller in Example 3.5.1. In this section, we attempt to achieve these objectives using a *proportional* control system.

The basic data on which we must base the design is the plant transfer function matrix (obtained from (3.26))

$$G(s) = \frac{1}{s(s+2)}\begin{bmatrix} s+1 & 1 \\ 1 & s+1 \end{bmatrix} \tag{3.87}$$

and the open-loop characteristic polynomial (see Section 3.1.1)

$$\rho_0(s) = \begin{vmatrix} s+1 & -1 \\ -1 & s+1 \end{vmatrix} = s(s+2) \tag{3.88}$$

The question is, of course, where to start? It has been noted in Section 3.5.1 that, if the interaction terms were not present, the design would be straightforward. It has also been seen in Example 3.5.1 that interaction can be removed by a dynamic (non-interacting) controller but we cannot accept this solution if we insist on a proportional control element. The key to the way out of these difficulties can be motivated by analogy with the techniques of Section 1.5.2 where problems due to the off-diagonal terms of the "A matrix" were eliminated by *transforming* it to diagonal form. Let us therefore examine the possibility of diagonalizing $G(s)$ using similarity transformations.

EXERCISE 3.6.1. Verify that the matrix

$$M = \begin{bmatrix} \alpha & \beta \\ \beta & \alpha \end{bmatrix}$$

has the characteristic polynomial $s^2 - 2\alpha s + \alpha^2 - \beta^2 = (s-(\alpha+\beta)) \times (s-(\alpha-\beta))$, eigenvalues $\lambda_1 = \alpha+\beta$, $\lambda_2 = \alpha-\beta$ and eigenvector matrix $T = \begin{bmatrix} 1 & -1 \\ 1 & 1 \end{bmatrix}$. Deduce that

$$M = T\begin{bmatrix} \alpha+\beta & 0 \\ 0 & \alpha-\beta \end{bmatrix}T^{-1} \tag{3.89}$$

Noting that $G(s)$ has the structure defined in the above exercise with $\alpha = (s+1)/s(s+2)$ and $\beta = 1/s(s+2)$, it is clear that it has eigenvalues

$$g_1(s) = \frac{(s+1)}{s(s+2)} + \frac{1}{s(s+2)} = \frac{1}{s}$$

$$g_2(s) = \frac{(s+1)}{s(s+2)} - \frac{1}{s(s+2)} = \frac{1}{s+2} \tag{3.90}$$

and that

$$G(s) = T\begin{bmatrix} g_1(s) & 0 \\ 0 & g_2(s) \end{bmatrix} T^{-1} \tag{3.91}$$

i.e. the liquid level system can be regarded as a non-interacting system "sandwiched" between two constant (i.e. no dynamics) systems as illustrated in Fig. 45. In particular, the transfer functions $g_1(s)$ and $g_2(s)$ are simply first order systems (in contrast to the elements of $G(s)$ which are second order systems).

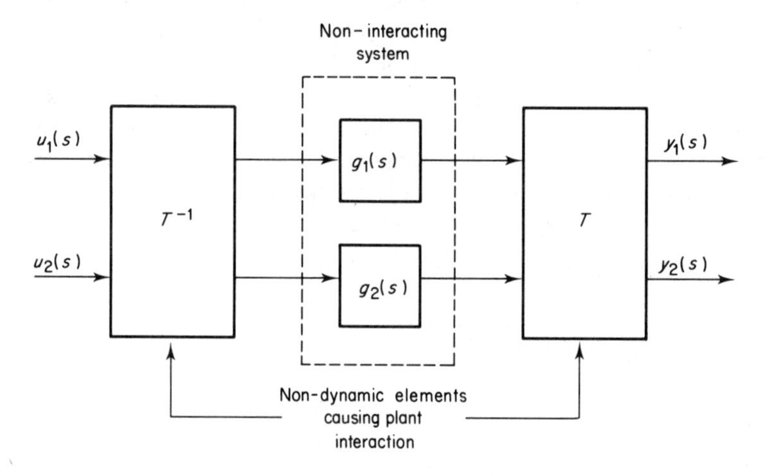

Fig. 45. Structural decomposition of liquid level system.

It is clear that design would be a simple matter in the absence of the terms T and T^{-1} in (3.91). Let us ignore them for the moment and design proportional feedback controllers for the scalar transfer functions $g_1(s)$ and $g_2(s)$ of the form shown in Fig. 46. The closed-loop transfer functions are

$$h_i(s) = \frac{g_i(s)k_i}{1 + g_i(s)k_i}, \qquad i = 1,2 \tag{3.92}$$

or, more precisely,

$$h_1(s) = \frac{k_1}{s + k_1}, \qquad h_2(s) = \frac{k_2}{s + 2 + k_2} \tag{3.93}$$

from which we deduce that the scalar closed-loop systems are stable if, and only if,

$$k_1 > 0, \qquad k_2 + 2 > 0 \tag{3.94}$$

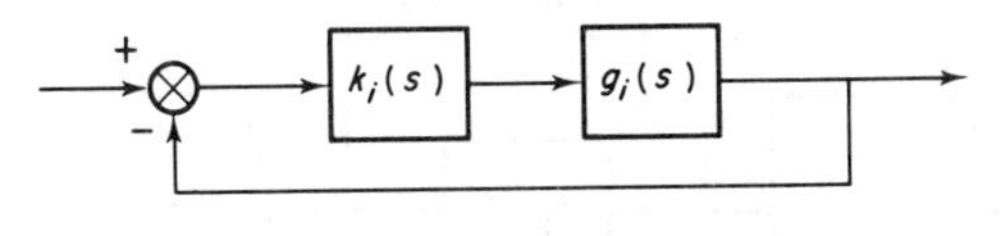

Fig. 46

Fast response speeds and small steady state errors are also guaranteed if the gains k_1, k_2 are both large.

The pressing question is now, of course, what is the relevance of the above analysis to the real, interacting multivariable system? More precisely, can we convert the designs k_1 and k_2 for g_1 and g_2 into a controller design for the real plant? The trick is to use the controller transfer function matrix (c.f. (3.91)) with the same eigenvector matrix as the plant and eigenvalues k_1 and k_2, i.e.

$$K(s) = T\begin{bmatrix} k_1 & 0 \\ 0 & k_2 \end{bmatrix}T^{-1} \tag{3.95}$$

Note that $K(s)$ is independent of s and hence a *proportional* controller as required. The forward path transfer function matrix

$$Q(s) = G(s)K(s) = T\begin{bmatrix} g_1(s)k_1 & 0 \\ 0 & g_2(s)k_2 \end{bmatrix}T^{-1} \tag{3.96}$$

indicates that the closed-loop transfer function matrix is

$$\begin{aligned} H_c(s) &= (I+Q)^{-1}Q = TT^{-1}(I+Q)^{-1}TT^{-1}QTT^{-1} \\ &= T(T^{-1}(I+Q)T)^{-1}T^{-1}QTT^{-1} \\ &= T(I+T^{-1}QT)^{-1}T^{-1}QTT^{-1} \\ &= T\begin{bmatrix} h_1(s) & 0 \\ 0 & h_2(s) \end{bmatrix}T^{-1} \end{aligned} \tag{3.97}$$

where we have used (3.96) to substitute for $T^{-1}QT$. It follows from (3.97) that the closed-loop system has the structure shown in Fig. 47 and is identical to the plant with h_i replacing g_i, $i = 1, 2$. Figure 47 also suggests the following intuitive points

(a) The closed-loop system possesses interaction effects and these interaction effects are due totally to the nondynamic terms T and T^{-1}.

(b) The stability and dynamic performance of the closed-loop system is dominated by the chosen dynamics for the scalar feedback system $h_1(s)$ and $h_2(s)$.

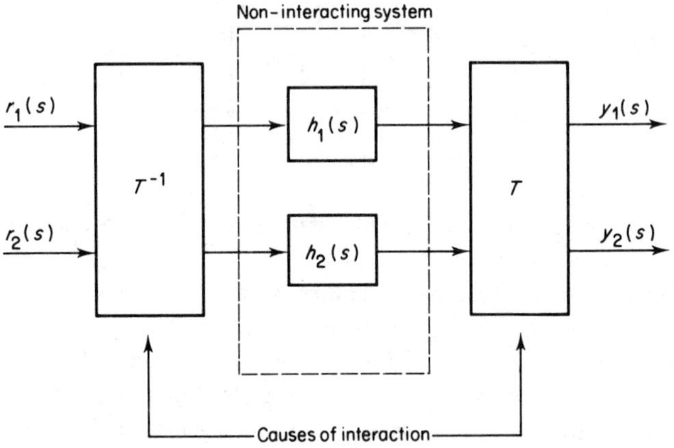

Fig. 47. Closed-loop liquid level system.

These observations are valid and we can strengthen (b) by noting the system return-difference determinant

$$
\begin{aligned}
|I_2 + Q(s)| &= \left| I + T \begin{bmatrix} g_1 k_1 & 0 \\ 0 & g_2 k_2 \end{bmatrix} T^{-1} \right| \\
&= \left| T \left(I + \begin{bmatrix} g_1 k_1 & 0 \\ 0 & g_2 k_2 \end{bmatrix} \right) T^{-1} \right| \\
&= \begin{vmatrix} 1 + g_1(s) k_1 & 0 \\ 0 & 1 + g_2(s) k_2 \end{vmatrix} \\
&= \frac{(s + k_1)(s + k_2 + 2)}{s(s + 2)}
\end{aligned} \tag{3.98}
$$

Using (3.88) it follows (Section 3.4) that the closed-loop characteristic polynomial takes the form $\rho_c(s) = (s + k_1)(s + k_2 + 2)$ and hence that the closed-loop (multivariable) system is stable if, and only if, the gains k_1 and k_2 satisfy (3.94), i.e. the closed-loop (multivariable) system is stable if, and only if, the scalar systems $h_1(s)$ and $h_2(s)$ are stable. We could not hope for a better correspondence that that!

The problem of stabilization is now completely solved by choosing gains k_1 and k_2 to satisfy (3.94). The problem of suppressing steady state errors is also easily solved by observing from (3.97) that

$$H_c(0) = \begin{bmatrix} \frac{k_2+1}{k_2+2} & \frac{1}{k_2+2} \\ \frac{1}{k_2+2} & \frac{k_2+1}{k_2+2} \end{bmatrix} \tag{3.99}$$

If $k_2 \gg 2$, it is clear that $H_c(0) \simeq I_2$ and steady state errors in response to unit step demands will be small. The problem of suppressing interaction effects to a satisfactory level is not so easily solved, however, and requires the introduction of new, and essentially multivariable, design rules. The problem is probably best stated graphically by regarding the gains k_1 and k_2 as coordinates in a two-dimensional "gain-space", when the "point" (k_1, k_2) satisfying the stability conditions (3.94) must lie in the "stable region" in Fig. 48. There are hence an infinite number of choices of (k_1, k_2) some aggravating and others easing the interaction problem in the closed-loop system. In the following paragraphs, we will consider three choices represented by the points $a(k_1 \gg k_2)$, $b(k_2 \gg k_1)$ and $c(k_1 \simeq k_2)$ illustrated in Fig. 48.

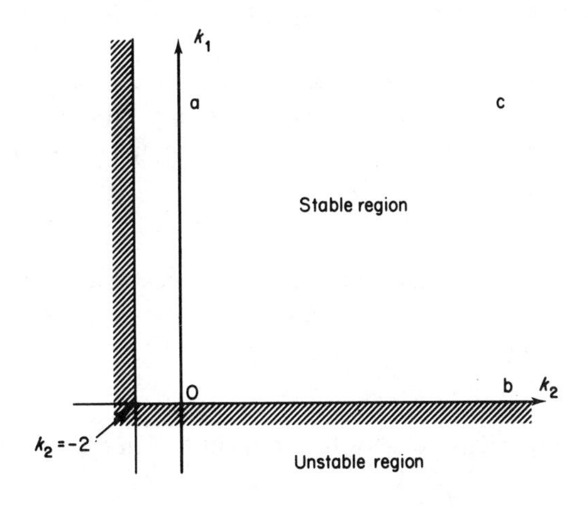

Fig. 48. Stable region in gain space.

Consider the response of the closed-loop system from zero initial conditions to a unit step demand in the level $y_2(t)$ in vessel two. Using (3.97) it follows that

$$y(s) = H_c(s)\begin{bmatrix} 0 \\ \frac{1}{s} \end{bmatrix} = \frac{1}{s(s+k_1)(s+k_2+2)}\begin{bmatrix} \frac{1}{2}(k_1-k_2)s+k_1 \\ \frac{1}{2}(k_1+k_2)s+k_1(1+k_2) \end{bmatrix} \tag{3.100}$$

and hence that

$$y_1(t) = \tfrac{1}{2}\left\{\frac{2}{2+k_2} - e^{-k_1 t} + \frac{k_2}{2+k_2} e^{-(k_2+2)t}\right\} \tag{3.101}$$

$$y_2(t) = \tfrac{1}{2}\left\{\frac{2(1+k_2)}{(2+k_2)} - e^{-k_1 t} - \frac{k_2}{2+k_2} e^{-(k_2+2)t}\right\} \tag{3.102}$$

Taking the case of point "a" represented by the condition $k_1 \gg k_2$, the magnitude of interaction effects can be assessed by considering small values of time t in the range $0 < t \ll 1/(k_2+2)$. In this interval the first and third terms in both (3.101) and (3.102) remain essentially constant whereas the term $-\frac{1}{2}e^{-k_1 t}$ rapidly decays to zero. This has the effect of causing both outputs to rise to values of approximately 0.5. Clearly transient interaction effects (i.e. $y_1(t)$) have peak magnitude of order 0.5 or greater.

Taking now the case of point "b" represented by the condition $k_2 \gg k_1$ and considering time $0 < t \ll 1/k_1$, the first and second terms in (3.101) and (3.102) remain essentially constant whilst the third terms decay rapidly to zero. This has the effect of causing $y_2(t)(y_1(t))$ to rise (drop) to a value of approximately 0.5 (-0.5). Clearly transient interaction effects again have peak magnitude of order 0.5 or greater.

Interaction effects of the magnitude described would have to be tolerated in practice if it were not possible to do better. Insight into the "best" choice of k_1 and k_2 is obtained by writing the interaction term $y_1(t)$ in the form

$$y_1(t) = \frac{1}{2+k_2}(1 - e^{-(k_2+2)t}) + \tfrac{1}{2}(e^{-(k_2+2)t} - e^{-k_1 t}) \tag{3.103}$$

The first term has a peak magnitude $1/(2+k_2)$ which tends to zero if the gain k_2 is large whereas the second term has no such property. We can, however, eliminate this second term by choosing

$$k_2 + 2 = k_1 \tag{3.104}$$

(a condition corresponding roughly to the point "c" in Fig. 47) when the responses are

$$y_1(t) = \frac{1}{k_1}(1 - e^{-k_1 t})$$

$$y_2(t) = \frac{k_1 - 1}{k_1}(1 - e^{-k_1 t}) \tag{3.105}$$

It is immediately apparent that closed-loop response speeds increase as k_1 increases, that the steady state errors also decrease and that the peak numerical magnitude of the interaction effect $\max_{t \geq 0} |y_1(t)| = 1/k_1$ becomes arbitrarily small. The responses for the choice $k_1 = 10$ (and hence $k_2 = 8$) are illustrated in Fig. 49 and suggest that the design is highly satisfactory.

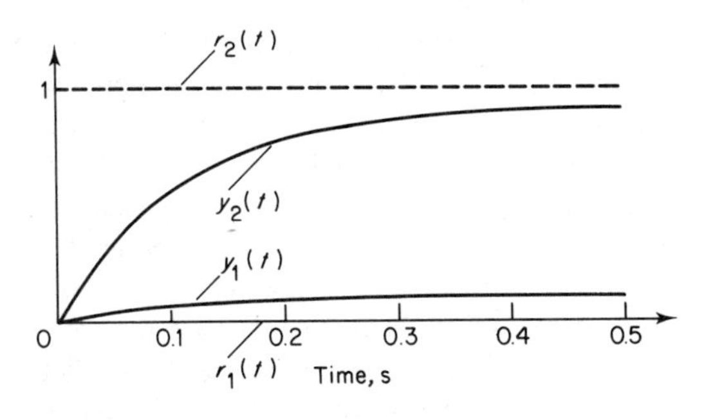

Fig. 49.

The controller (transfer function) matrix is obtained by substituting (3.104) into (3.96)

$$K = T\begin{bmatrix} k_1 & 0 \\ 0 & k_2 \end{bmatrix} T^{-1} = \begin{bmatrix} 1 & -1 \\ 1 & 1 \end{bmatrix} \begin{bmatrix} k_1 & 0 \\ 0 & k_1 - 2 \end{bmatrix} \tfrac{1}{2} \begin{bmatrix} 1 & 1 \\ -1 & 1 \end{bmatrix}$$

$$= \begin{bmatrix} k_1 - 1 & 1 \\ 1 & k_1 - 1 \end{bmatrix} \tag{3.106}$$

and the forward path controller is hence interacting (note the off-diagonal terms!) and has the block structure indicated in Fig. 50.

In summary the above example indicates that it is possible to achieve satisfactory designs for multivariable feedback systems without the need to use dynamically complex non-interacting controllers. The approach used above was to use eigenvalue analysis of the plant transfer function matrix $G(s)$ to decompose the system into three components, one of which is non-interacting. Controllers were then designed for the non-interacting element and were subsequently transformed into a desirable controller for the whole system. Although this procedure was highly systematic, there are two problems that are of an essentially multivariable nature.

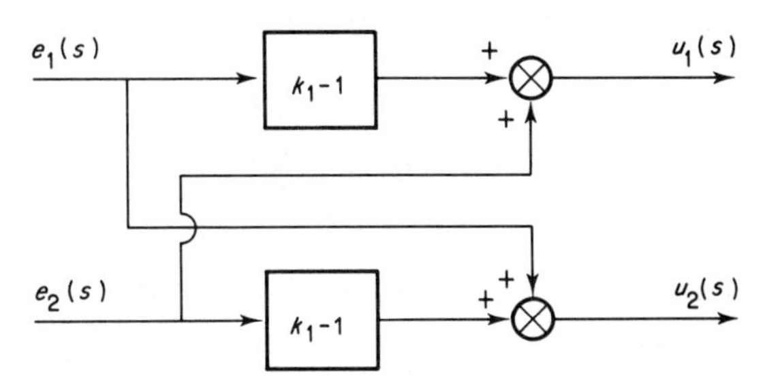

Fig. 50. Proportional controller for liquid level system.

(a) The reduction in interaction requires the use of high controller gains. This is a general principle but as high gains can, in more complex systems, introduce instability, it cannot always be realized in practice.

(b) The reduction in interaction also needed careful consideration to be given to the relative magnitude of gains k_1 and k_2. Our analysis lead us to the choice of k_1 and k_2 related by (3.104). The principle underlying this choice can be exposed by noting that (3.104) is equivalent to the requirement that $h_1(s)$ and $h_2(s)$ have identical time-constants or, more generally, the requirement is that $h_1(s)$ and $h_2(s)$ have highly similar step responses. In fact, in such a situation, we can write $h_1(s) \simeq h_2(s)$ and note from (3.97) that

$$H_c(s) \simeq Th_1(s)I_2T^{-1} = h_1(s)\begin{bmatrix} 1 & 0 \\ 0 & 1 \end{bmatrix} \tag{3.107}$$

indicating that the closed-loop transfer function matrix is approximately diagonal and hence the closed-loop system approximately non-interacting!

3.7 Multivariable First Order Lags

Despite its simplicity, many of the techniques and observations generated in the control analysis of the liquid level system in the last section do, in fact, carry over to more general complex cases. It is the purpose of this section to underline this statement by consideration of *unity, negative proportional feedback control* of the linear, time-invariant plant (notation as in Section 3.1.1).

$$\frac{dx_p(t)}{dt} = A_p x_p(t) + B_p u(t)$$

$$y(t) = C_p x_p(t) \tag{3.108}$$

in the special case when $m = l = n_1$ (i.e. the system has the same number of states, inputs and outputs) and both the $m \times m$ matrices B_p and C_p are non-singular. The plant transfer function matrix $G(s) = C_p(sI_m - A_p)^{-1}B_p$ then has an inverse of the form

$$\begin{aligned} G^{-1}(s) &= (C_p(sI_m - A_p)^{-1}B_p)^{-1} \\ &= B_p^{-1}(sI_m - A_p)C_p^{-1} \\ &= sA_0 + A_1 \end{aligned} \tag{3.109}$$

where the $m \times m$ matrices A_0, A_1 are given by the formulae

$$A_0 = B_p^{-1}C_p^{-1} = (C_pB_p)^{-1}, \qquad A_1 = -B_p^{-1}A_pC_p^{-1} \tag{3.110}$$

A scalar first order lag has transfer function $g(s) = 1/(a_0s + a_1)$ with inverse

$g^{-1}(s) = a_0 s + a_1$. The observation that (3.109) reduces to this form in the case of $m = 1$ (when A_0 and A_1 are scalars) suggests that it represents a generalization of the notion of a first order lag. This is, in fact, the case and such a system is termed an $m \times m$ *multivariable first order lag.*

EXERCISE 3.7.1. Verify that the liquid level system (1.71) is a first order lag (independent of plant data) with inverse transfer function matrix

$$G^{-1}(s) = \begin{bmatrix} sa_1 + \beta & -\beta \\ -\beta & sa_2 + \beta \end{bmatrix}$$

$$\equiv s\underbrace{\begin{bmatrix} a_1 & 0 \\ 0 & a_2 \end{bmatrix}}_{A_0} + \underbrace{\begin{bmatrix} \beta & -\beta \\ -\beta & \beta \end{bmatrix}}_{A_1} \tag{3.111}$$

EXERCISE 3.7.2. Verify that the first order lag (3.108) can be written in the form

$$\frac{dy(t)}{dt} = -A_0^{-1}A_1 y(t) + A_0^{-1}u(t) \tag{3.112}$$

Hence deduce that the plant characteristic polynomial is

$$\rho_0(s) = |sI_m + A_0^{-1}A_1| \tag{3.113}$$

Check this result by noting that $A_p = -C_p^{-1}A_0^{-1}A_1C_p$ and hence that A_p and $-A_0^{-1}A_1$ have the same eigenvalues.

As in Section 3.6 the problem in controller design is where to start. In fact the key to the solution of the problem is to use eigenvalue and eigenvector methods (as in Section 3.6) together with the use of *constant precompensation.* More precisely, we will borrow some ideas from non-interacting control (Section 3.5.2) and regard the forward path proportional controller $K(s)$ as a cascaded precompensator $K_p(s)$ and compensator $K_c(s)$ as illustrated in Fig. 43(a). For reasons that may become apparent later we will make the (non-unique) choice

$$K_p(s) = A_0 \tag{3.114}$$

when the "plant seen by the compensator" (see Fig. 43(b)) has transfer function matrix

$$G(s)K_p(s) = (sA_0 + A_1)^{-1}A_0 = (sI_m + A_0^{-1}A_1)^{-1} \tag{3.115}$$

The trick now is to examine the eigenstructure of $A_0^{-1}A_1$. More precisely, suppose that $A_0^{-1}A_1$ has eigenvector matrix T and eigenvalues $\lambda_1, \lambda_2, \ldots, \lambda_m$ satisfying

$$T^{-1}A_0^{-1}A_1T = \begin{bmatrix} \lambda_1 & 0 & \cdots & 0 \\ 0 & \lambda_2 & & \vdots \\ \vdots & & & 0 \\ 0 & \cdots & 0 & \lambda_m \end{bmatrix} \tag{3.116}$$

It follows that

$$\begin{aligned} G(s)K_p(s) &= (sI_m + TT^{-1}A_0^{-1}A_1TT^{-1})^{-1} \\ &= (T(sI_m + T^{-1}A_0^{-1}A_1T)T^{-1})^{-1} \\ &= T\begin{bmatrix} g_1(s) & 0 & \cdots & 0 \\ 0 & g_2(s) & & \vdots \\ \vdots & & & 0 \\ 0 & \cdots & 0 & g_m(s) \end{bmatrix}T^{-1} \end{aligned} \tag{3.117}$$

where the scalar transfer functions $g_j(s)$ take the *first order form*

$$g_j(s) = \frac{1}{s+\lambda_j}, \qquad 1 \leqslant j \leqslant m \tag{3.118}$$

i.e. the first order lag can be regarded as a non-interacting system "sandwiched" between two constant systems (as illustrated in Fig. 45 for the case of $m = 2$), the constant elements T and T^{-1} being the sole cause of interaction in the system.

Following the arguments of Section 3.6, let us now design proportional feedback systems for $g_1(s), \ldots, g_m(s)$ as shown in Fig. 46 with transfer functions

$$h_i(s) = \frac{g_i(s)k_i}{1+g_i(s)k_i} = \frac{k_i}{s+\lambda_i+k_i}, \qquad 1 \leqslant i \leqslant m \tag{3.119}$$

and choose the compensator (c.f. (3.95))

$$K_c(s) = T\begin{bmatrix} k_1 & 0 & \cdots & 0 \\ 0 & k_2 & & \vdots \\ \vdots & & & 0 \\ 0 & \cdots & 0 & k_m \end{bmatrix}T^{-1} \tag{3.120}$$

The resulting controller

$$K(s) = K_p(s)K_c(s) = A_0T \begin{bmatrix} k_1 & & \\ & k_2 & \\ & & \ddots & \\ & & & k_m \end{bmatrix} T^{-1} \tag{3.121}$$

is a proportional controller as required yielding the forward path transfer function matrix

$$Q(s) = G(s)K_p(s)K_c(s) = T \begin{bmatrix} g_1(s)k_1 & 0 & \cdots & 0 \\ 0 & g_2(s)k_2 & & \vdots \\ & & & 0 \\ 0 & \cdots & 0 & g_m(s)k_m \end{bmatrix} T^{-1} \tag{3.122}$$

and the closed loop transfer function matrix (see (3.97))

$$H_c(s) = T \begin{bmatrix} h_1(s) & 0 & \cdots & 0 \\ 0 & h_2(s) & & \vdots \\ \vdots & & & 0 \\ 0 & \cdots & & h_m(s) \end{bmatrix} T^{-1} \tag{3.123}$$

In general, (3.123) indicates that the closed-loop system possesses interaction effects due to the constant elements T and T^{-1}. At this point we will learn from the experience of Section 3.6 and attempt to reduce closed-loop interaction by choosing the gains $k_1, k_2, \ldots, k_m$ to ensure that the scalar feedback systems $h_1(s), \ldots, h_m(s)$ have similar dynamic characteristics. More precisely we will choose k_i such that $h_i(s)$ has a time constant k^{-1} (independent of i), i.e.

$$k = \lambda_i + k_i, \qquad 1 \leqslant i \leqslant m \tag{3.124}$$

The resulting controller is obtained by substituting into (3.121) and using (3.116)

$$K = A_0T \begin{bmatrix} k-\lambda_1 & \cdots & 0 \\ \vdots & & \vdots \\ 0 & \cdots & k-\lambda_m \end{bmatrix} T^{-1} = A_0(kI_m - A_0^{-1}A_1)$$

$$= kA_0 - A_1 \tag{3.125}$$

(which is a nice, easy formula to remember!) and the closed-loop transfer function matrix takes the form

$$H_c(s) = T\begin{bmatrix} \frac{k-\lambda_1}{s+k} & \cdots & 0 \\ \vdots & & \vdots \\ 0 & \cdots & \frac{k-\lambda_m}{s+k} \end{bmatrix} T^{-1}$$

$$= \frac{k}{s+k}(I_m - k^{-1}A_0^{-1}A_1) \tag{3.126}$$

The stability, steady state accuracy and transient performance of the closed loop system are analysed below.

(a) The stability of the closed-loop system is assessed by calculating the return-difference determinant using (3.113), (3.114), (3.115) and (3.125)

$$\begin{aligned} |I_m + G(s)K(s)| &\equiv |G(s)K_p(s)| \cdot |(G(s)K_p(s))^{-1} + K_p^{-1}(s)K(s)| \\ &\equiv |(GK_p)^{-1} + K_p^{-1}K|/|(GK_p)^{-1}| \\ &\equiv \frac{|sI_m + A_0^{-1}A_1 + A_0^{-1}(kA_0 - A_1)|}{|sI_m + A_0^{-1}A_1|} \\ &\equiv (s+k)^m/\rho_0(s) \end{aligned} \tag{3.127}$$

i.e. the closed-loop characteristic polynomial $\rho_c(s) = (s+k)^m$ and hence the closed-loop system is stable if, and only if,

$$k > 0 \tag{3.128}$$

(b) The steady-state accuracy of the closed-loop system in response to unit step demands is assessed by noting that

$$H_c(0) = I_m - k^{-1}A_0^{-1}A_1 \rightarrow I_m \qquad (\text{as } k \rightarrow +\infty) \tag{3.129}$$

i.e. (Section 3.4) steady state errors can be made to be arbitrarily small by the use of suitably large "gain" k.

(c) The transient interaction effects of the closed-loop system in response to unit step demands can be assessed by noting that

$$H_c(s) = \frac{k}{s+k}(I_m - k^{-1}A_0^{-1}A_1) \rightarrow \frac{k}{s+k}I_m \qquad (\text{as } k \rightarrow +\infty) \tag{3.130}$$

i.e. the closed-loop transfer function matrix is approximately diagonal at high gains k indicating that closed-loop interaction effects can be made arbitrarily small.

All of these points are illustrated in the following example. Before continuing, however, it is worthwhile pointing out that the eigenvector techniques introduced in Section 3.6 combined with the use of precompensation do make possible the design of simple (proportional) controllers for multivariable first order lags. Again the approach was based on the search for a non-interacting element within the system and the reduction of closed-loop interaction required (a) the use of high gains and (b) a certain dynamic "compatibility" between the scalar feedback systems $h_1(s), \ldots, h_m(s)$ (expressed by equation (3.124)). These observations are simply special cases of more general principles (see reading list).

EXAMPLE 3.7.1. Despite its apparent complexity, the application of the above theory requires a knowledge of the controller (3.125) and the closed-loop transfer function matrix (3.126) only.

Consider the design of a unity negative proportional feedback system for the control of the first order lag

$$\frac{dx_p(t)}{dt} = \begin{bmatrix} 3 & 2 \\ -11 & -8 \end{bmatrix} x_p(t) + \begin{bmatrix} 1 & -2 \\ -1 & 3 \end{bmatrix} u(t)$$

$$y(t) = \begin{bmatrix} 2 & 1 \\ 1 & 1 \end{bmatrix} x_p(t) \tag{3.131}$$

and suppose that it is required that all steady state errors and peak interaction effects in response to unit step demands in any output are $\leqslant 0.1$.

The first step is to calculate A_0 and A_1 from (3.110)

$$A_0 = \begin{bmatrix} 1 & 1 \\ 0 & 1 \end{bmatrix}, \qquad A_1 = \begin{bmatrix} 3 & 7 \\ 2 & 4 \end{bmatrix} \tag{3.132}$$

from which

$$A_0^{-1} A_1 = \begin{bmatrix} 1 & 3 \\ 2 & 4 \end{bmatrix} \tag{3.133}$$

The use of the controller (3.125) yields the closed-loop transfer function matrix of (3.126), i.e.

$$H_c(s) = \frac{k}{s+k} \begin{bmatrix} 1-k^{-1} & -3k^{-1} \\ -2k^{-1} & 1-4k^{-1} \end{bmatrix} \tag{3.134}$$

The response to a unit step demand in $y_1(t)$ is simply

$$y(t) = \mathscr{L}^{-1} H_c(s) \frac{1}{s} \begin{bmatrix} 1 \\ 0 \end{bmatrix} = (1 - e^{-kt}) \begin{bmatrix} 1-k^{-1} \\ -2k^{-1} \end{bmatrix} \tag{3.135}$$

indicating that the interaction effects and steady state errors are $\leqslant 0.1$ if, and only if, $2k^{-1} \leqslant 0.1$ and $k^{-1} \leqslant 0.1$. A similar analysis of the response to a unit step demand in $y_2(t)$ indicates that we need $3k^{-1} \leqslant 0.1$ and $4k^{-1} \leqslant 0.1$. All of these inequalities must be satisfied, i.e. we need $k > 40$. The resulting controller is simply

$$K = kA_0 - A_1 = \begin{bmatrix} k-3 & k-7 \\ -2 & k-4 \end{bmatrix} \tag{3.136}$$

EXERCISE 3.7.3. If, in Example 3.7.1, we replace the performance specifications by the requirement that steady state errors in response to unit step demands in y_1 (y_2) are less than 0.2 (0.15) and that transient interaction effects are less than 0.02 (0.1) verify that we need $k \geqslant 100$.

3.8 Stability and the Forward Path System Frequency Response Matrix

In the classical situation of a single-input/single-output system or plant with transfer function $g(s)$ subjected to unity negative feedback with forward path controller transfer function $k(s)$, it is well known that the stability of the closed-loop system can be assessed by examining the frequency response loci $q(j\omega)$, $\omega \geqslant 0$, of the forward path transfer function $q(s) = g(s)k(s)$. The technique used is, of course, the application of the *Nyquist stability criterion.* It is the purpose of this section to outline a generalization of this well known result to the situation of an l-input/m-output plant $G(s)$ subjected to unity negative feedback with $l \times m$ forward path controller $K(s)$. Our attention is limited to the derivation and interpretation of one particular stability theorem. More generalizations and their use in design can be found in the reading list.

The results of Section 3.4 indicate that the stability of the closed-loop system of Fig. 39 is described by the relation

$$\frac{\rho_c(s)}{\rho_0(s)} \equiv |I_m + Q(s)| \tag{3.137}$$

where $\rho_c(s)$ ($\rho_0(s)$) is the closed (open)-loop characteristic polynomial and $Q(s) = G(s)K(s)$ is the forward path transfer function matrix. The forward path frequency response matrix at a real frequency ω is simply $Q(j\omega)$ and can be interpreted (Section 1.7.3) as a representation of the steady state response of the forward path system to sinusoidal inputs of frequency ω. The natural notion of the frequency response *plot* of the forward path system is a graphical representation of the "value" of $Q(j\omega)$ as the frequency ω increases from 0 to $+\infty$. It is important to note, however, that in the multivariable case $(m > 1) Q(j\omega)$ has m^2 elements and hence the frequency response plot of Q will consist of the m^2 plots of the frequency responses of its elements $Q_{11}, Q_{12}, \ldots, Q_{1m}, Q_{21}, \ldots, Q_{mm}$. This collection of loci in the complex plane is commonly termed the forward path system *Nyquist Array* (a generalization of the well known Nyquist diagram for a single-input/single-output system).

The use of the Nyquist array of Q in the assessment of the stability of the closed-loop system is based on the same piece of complex-variable theory that yields the classical Nyquist stability criterion and requires a consideration of the gain and place characteristics of the elements of $Q(s)$ on a *closed contour* in the complex plane. More precisely, let D be the familiar Nyquist contour in the complex plane consisting of the imaginary axis $s = j\omega$, $-R \leqslant \omega \leqslant R$ and the semi-circle in the right half complex plane of radius $|s| = R$. The radius R is supposed to be large enough to ensure that all zeros of the open-loop and closed-loop characteristic polynomials that lie in the right-half complex plane also lie in the interior of D. If, as sometimes happens, either ρ_c or ρ_0 has a zero at some point $s_0 = j\omega_0$ on the imaginary axis, the D contour is "indented" (using an infinitessimally small semi-circle) into the left-half complex-plane. These ideas are illustrated in Fig. 51.

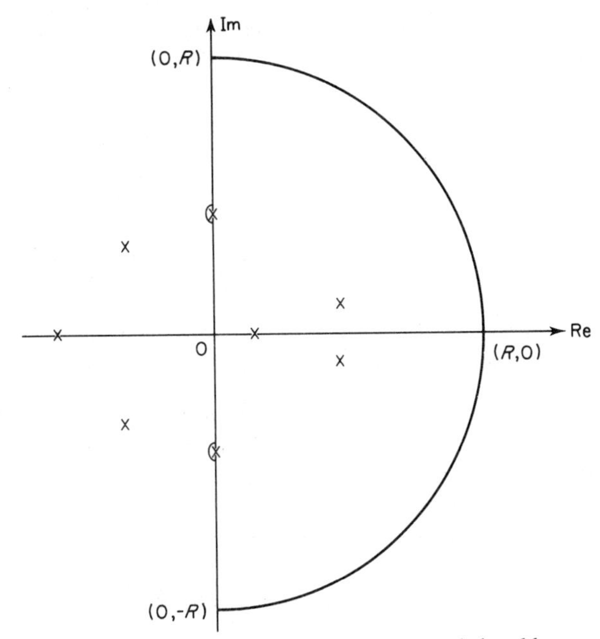

Fig. 51. The Nyquist "D-contour". x = open- and closed-loop poles.

Consider now the situation when we move from an arbitrary point s_0 on the D contour around the contour once only in a clockwise manner. If at each point we evaluate and plot in the complex plane the numerical value of the return-difference determinant $|I + Q(s)|$, we will generate a closed contour Γ in the complex plane as illustrated in Fig. 52. (Note that, as D avoids all zeros of ρ_c and ρ_0, Γ is finite and does not pass through the origin of the complex plane.)

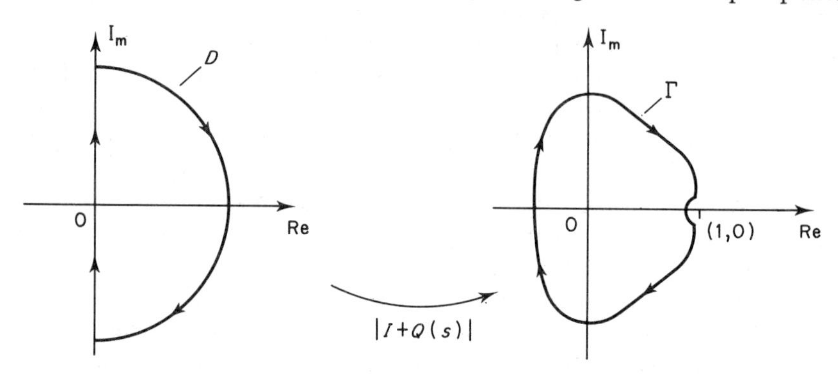

Fig. 52. The return-difference and the contour Γ.

Suppose now that, as we move round D once in a clockwise manner, the equivalent point on Γ encircles the *origin* of the complex plane n_T times in a clockwise manner (anti-clockwise encirclements being regarded as negative clockwise encirclements). It follows directly from (3.137) that n_T is equal to the number of clockwise encirclements of the origin generated by $\rho_c(s)/\rho_0(s)$ as s moves round D in a clockwise manner. Applying standard arguments (see, for example, Raven, 1978, p. 440) it follows immediately that

$$n_c - n_0 = n_T \tag{3.138}$$

where

n_0 = number of zeros of $\rho_0(s)$ inside the D-contour (i.e. the number of unstable poles in the forward path system)

n_c = number of zeros of $\rho_c(s)$ inside the D-contour (i.e. the number of unstable poles in the closed-loop system).

Normally, n_0 is known from physical arguments or direct evaluation of the roots of the open-loop characteristic polynomial and n_T can be obtained by inspection of Γ. Equation (3.138) then allows the calculation of n_c. Noting that the closed-loop system is asymptotically stable if, and only if, $n_c = 0$, we obtain the following generalization of the Nyquist stability criterion.

Theorem 3.8.1. *The closed-loop system is asymptotically stable if, and only if,*

$$n_0 + n_T = 0 \tag{3.139}$$

EXERCISE 3.8.1. Suppose that the forward path system generating the contour Γ in Fig. 52 is asymptotically stable (i.e. $n_0 = 0$). Deduce that $n_T = 1$ and hence that the closed-loop system is unstable and possesses one pole in the right-half complex plane.

In principle the evaluation of n_T requires a plot of Γ and, hence, a knowledge of the behaviour of $|I + Q(s)|$ (and, therefore, $Q(s)$) at all points of the D contour. It is important to note, however, that it is only necessary, in practice, to know the behaviour of $|I + Q(s)|$ (and, therefore, $Q(s)$) with s on the positive imaginary axis $\{s = j\omega, 0 \leqslant \omega \leqslant R\}$, i.e. it is only necessary to know the Nyquist array of Q. This follows from the observations

(a) The transfer function matrix $Q(s)$ of the forward path system is obtained from (3.3) or (3.12) to be $Q(s) = C(sI - A)^{-1}B$ and, hence, because of the inverse,

$$\lim_{|s| \to +\infty} Q(s) = 0 \tag{3.140}$$

This leads naturally to the observation that, on the semicircular part of D, $|I + Q(s)| \simeq |I| = 1$ if R is very large. If we formally consider R to be infinitely large, the contribution to Γ from the semicircular part of D reduces to the single point $(1, 0)$.

(b) As the data in the matrices A, B, C are real, it is easily verified that

$$Q(\bar{s}) \equiv \overline{Q(s)} \tag{3.141}$$

and hence that $|I + Q(-j\omega)| = \overline{|I + Q(j\omega)|}$. That is, the contribution to Γ from the negative imaginary axis can be obtained by complex conjugation of the contribution from the positive imaginary axis.

EXAMPLE 3.8.1. Consider the problem of assessing the stability of a unity negative feedback system with asymptotically stable forward path system with transfer function matrix

$$Q(s) = \frac{1}{(s+1)(s+3)} \begin{bmatrix} 7s + 11, & -12s - 16 \\ 6s + 8, & -11s - 13 \end{bmatrix} \tag{3.142}$$

Examination of the poles of $Q(s)$ indicates that the fastest open-loop time-constant is 1/3 s. This suggests that a plot of $|I + Q(j\omega)|$ with ω in the range $0 < \omega < 12$ will provide a reasonable representation of its behaviour on the whole of the imaginary axis. This is shown in Fig. 53 where the rest of the contour Γ has been inserted in dotted lines by applying the observations (a) and (b) above. It is immediately apparent that there is one complete clockwise encirclement $n_T = 1$. Noting

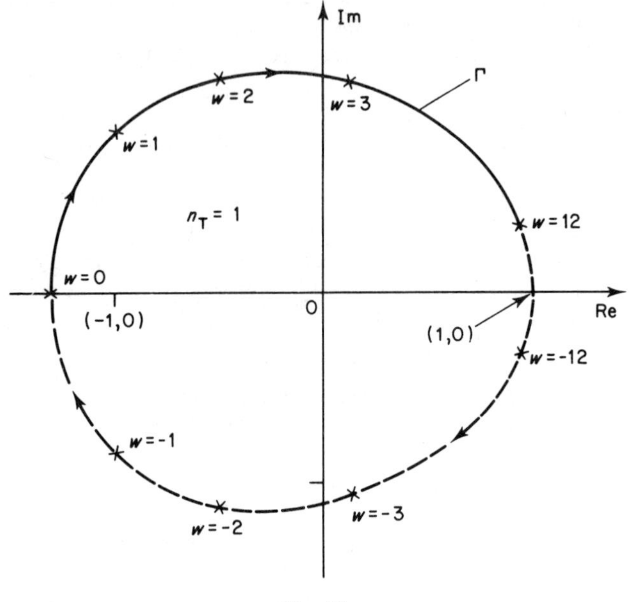

Fig. 53.

that $n_0 = 0$, it follows directly from (3.138) that $n_c = n_0 + n_T = 1$ indicating one closed-loop pole in the right-half plane. This is easily verified by noting that

$$|I + Q(s)| = \frac{(s+2)(s-2)}{(s+1)(s+3)} \tag{3.143}$$

Problems

(1) A single-input/two-output plant is described by the model

$$\frac{dx_p(t)}{dt} = \begin{bmatrix} -1 & 1 \\ 1 & 0 \end{bmatrix} x_p(t) + \begin{bmatrix} 0 \\ 1 \end{bmatrix} u(t)$$

$$y(t) = \begin{bmatrix} 1 & 2 \\ 1 & 1 \end{bmatrix} x_p(t)$$

If the plant is to be controlled by unity negative feedback with control input $u(t) = k_1 e_1(t) + k_2 e_2(t)$, show that the state variable model of the closed-loop system takes the form

$$\frac{dx_p(t)}{dt} = \begin{bmatrix} -1 & 1 \\ 1-k_1-k_2 & -2k_1-k_2 \end{bmatrix} x_p(t) + \begin{bmatrix} 0 & 0 \\ k_1 & k_2 \end{bmatrix} r(t)$$

together with the output equation. Deduce that the closed-loop system is stable if, and only if,

$$1+2k_1+k_2 > 0, \qquad 3k_1+2k_2-1 > 0$$

(Hint: apply Routh's test to the closed-loop characteristic polynomial.)

(2) Using the plant of example one above, suppose that the forward path controller is extended to include integral control action $u(s) = (1 + (1/Ts))(k_1 e_1(s) + k_2 e_2(s))$. Show that the controller has a state variable model

$$\frac{dx_c(t)}{dt} = \frac{1}{T}(k_1 e_1(t) + k_2 e_2(t)), \qquad u(t) = x_c(t) + k_1 e_1(t) + k_2 e_2(t)$$

and hence that the forward path system has the state-variable model

$$\frac{dx(t)}{dt} = \begin{bmatrix} -1 & 1 & 0 \\ 1 & 0 & 1 \\ 0 & 0 & 0 \end{bmatrix} x(t) + \begin{bmatrix} 0 & 0 \\ k_1 & k_2 \\ \frac{k_1}{T} & \frac{k_2}{T} \end{bmatrix} e(t)$$

$$y(t) = \begin{bmatrix} 1 & 2 & 0 \\ 1 & 1 & 0 \end{bmatrix} x(t)$$

Finally, show that the closed-loop system has the model

$$\frac{dx(t)}{dt} = \begin{bmatrix} -1 & 1 & 0 \\ 1-k_1-k_2 & -2k_1-k_2 & 1 \\ -\frac{k_1}{T}-\frac{k_2}{T} & -\frac{2k_1}{T}-\frac{k_2}{T} & 0 \end{bmatrix} x(t) + \begin{bmatrix} 0 & 0 \\ k_1 & k_2 \\ \frac{k_1}{T} & \frac{k_2}{T} \end{bmatrix} r(t)$$

together with the output equation above. Investigate the stability of this system.

(3) Verify that the forward path controller $K(s)$ where elements are each proportional plus integral transfer functions $K_{ik}(s) = \alpha_{ik} + \beta_{ik}/s$ can be written in the matrix form

$$K(s) = K_1 + \frac{1}{s} K_2$$

where K_1 and K_2 are $l \times m$ constant matrices with elements of the form $(K_1)_{ik} =$

α_{ik}, $(K_2)_{ik} = \beta_{ik}$, $1 \leqslant i \leqslant l$, $1 \leqslant k \leqslant m$. Show that this "multivariable proportional plus integral controller" can be represented by the model

$$\frac{\mathrm{d}x_c(t)}{\mathrm{d}t} = e(t), \qquad u(t) = K_2 x_c(t) + K_1 e(t)$$

which has the characteristic polynomial $|sI_m| = s^m$. Finally confirm that the system is always controllable and is also observable if $m \leqslant l$ and rank $K_2 = m$.

(4) Verify directly from (3.23) that the system

$$\frac{\mathrm{d}x(t)}{\mathrm{d}t} = \begin{bmatrix} 0 & 1 \\ -2 & -3 \end{bmatrix} x(t) + \begin{bmatrix} 0 \\ 1 \end{bmatrix} u(t), \qquad y(t) = [1 \quad 0] x(t)$$

has the transfer function $g(s) = 1/(s^2 + 3s + 2)$.

(5) Verify that a system with A, B, C, D matrices

$$A = \begin{bmatrix} -2 & 3 \\ 1 & -4 \end{bmatrix}, \qquad B = \begin{bmatrix} 1 \\ 0 \end{bmatrix}, \qquad C = \begin{bmatrix} 1 & 1 \\ 0 & 1 \end{bmatrix}, \qquad D = \begin{bmatrix} 0 \\ 0 \end{bmatrix}$$

has the transfer function matrix

$$G(s) = \frac{1}{s^2 + 6s + 5} \begin{bmatrix} s + 5 \\ 1 \end{bmatrix}$$

Hence show that the output responses to a unit step input are simply $y_1(t) = 1 - \mathrm{e}^{-t}$, $y_2(t) = 0.2 - 0.25\,\mathrm{e}^{-t} + 0.05\,\mathrm{e}^{-5t}$.

(6) Prove that the characteristic polynomial of the series configuration of Fig. 38(b) is simply the product of the characteristic polynomials of the individual subsystems $G_j(s)$, $1 \leqslant j \leqslant q$. Hence show that its zero polynomial is simply the product of the zero polynomials of the individual subsystems. (Hint: use (3.32) and (3.48).)

(7) In Fig. 38(c), prove that the zero polynomial of the feedback system is simply the product of the zero polynomial of the system G_1 and the characteristic polynomial of the system G_2. (Hint: write down the zero polynomial in terms of the matrices in the state-variable models of G_1 and G_2.)

(8) Using the plant and controller of problem (1) verify that the forward path transfer function matrix is

$$Q(s) = \frac{1}{s^2 + s - 1} \begin{bmatrix} 2s + 3 \\ s + 2 \end{bmatrix} [k_1 \quad k_2]$$

and hence that the return-difference determinant is $|T(s)| = (s^2 + (1 + 2k_1 + k_2)s + 3k_1 + 2k_2 - 1)/(s^2 + s - 1)$. Hence verify the stability predictions found in problem (1).

(9) Verify that the non-interacting controller for the system

$$G(s) = \frac{1}{s^2 + s + 1} \begin{bmatrix} s + 1 & -s \\ -1 & s + 1 \end{bmatrix}$$

that generates the closed-loop system with "nice" transfer function matrix

$$H_c(s) = \frac{k}{s + k} \begin{bmatrix} 1 & 0 \\ 0 & 1 \end{bmatrix}$$

is the proportional plus integral controller

$$K(s) = k \begin{bmatrix} 1 + \frac{1}{s} & 1 \\ \frac{1}{s} & 1 + \frac{1}{s} \end{bmatrix}$$

(10) Given the system

$$G(s) = \frac{1}{(s - 2)(s + 4)} \begin{bmatrix} s + 1 & 3 \\ 3 & s + 1 \end{bmatrix}$$

investigate the response of a unity negative feedback system for the control of this plant using the proportional controller

$$K(s) = \frac{1}{2} \begin{bmatrix} k_1 + k_2 & k_1 - k_2 \\ k_1 - k_2 & k_1 + k_2 \end{bmatrix}$$

where k_1, k_2 are real scalars. In particular investigate the closed-loop response to a unit step demand in y_1 in the cases (a) $k_1 \gg k_2$ and (b) $k_1 \ll k_2$. Use your results to motivate the choice of $k_1 - 2 = k_2 + 4 = k$. In this case verify that the closed-loop transfer function matrix

$$H_c(s) = \frac{k}{s+k}\begin{bmatrix} 1 - \frac{1}{k} & \frac{3}{k} \\ \frac{3}{k} & 1 - \frac{1}{k} \end{bmatrix}$$

and deduce that the peak interaction effects in response to unit step demands are less than 0.1 if, and only if, $k > 30$. (Hint: follow the procedure of Section 3.6 noting that the results of Exercise 3.6.1 apply to this plant.)

(11) Design a proportional output feedback system for the plant

$$G(s) = \frac{1}{(s+2)(s+4)}\begin{bmatrix} s+3 & 1 \\ 1 & s+3 \end{bmatrix}$$

to ensure stable responses to unit step demands and peak interaction effects less than 0.05. (Hint: follow the procedure of Section 3.6.)

(12) Verify that the system

$$G(s) = \frac{1}{s^2+3s+3}\begin{bmatrix} (s+2) & -(s+1) \\ -1 & (s+2) \end{bmatrix}$$

is a first order lag with the matrices

$$A_0 = \begin{bmatrix} 1 & 1 \\ 0 & 1 \end{bmatrix}, \qquad A_1 = \begin{bmatrix} 2 & 1 \\ 1 & 2 \end{bmatrix}$$

and characteristic polynomial $s^2 + 3s + 3$. Show that a suitable proportional controller for G takes the form

$$K = \begin{bmatrix} k-2 & k-1 \\ -1 & k-2 \end{bmatrix}$$

and hence that the closed-loop transfer function matrix is

$$H_c(s) = \frac{k}{s+k}\begin{bmatrix} 1 - \frac{1}{k} & \frac{1}{k} \\ -\frac{1}{k} & 1 - \frac{2}{k} \end{bmatrix}$$

Hence verify that interaction effects in response to unit step demands are less than 0.14 for all $t > 0$ if, and only if, $k > 7.143$ but that the steady state error in the second output in response to a unit step demand in the second output is less than 0.1 if, and only if, $k > 20$.

(13) Suppose that the l-input linear, time-invariant system $\mathrm{d}x(t)/\mathrm{d}t = Ax(t) + Bu(t)$ is subjected to constant state feedback $u(t) = r(t) - Fx(t)$. Derive the stability relation

$$\frac{\rho_c(s)}{\rho_0(s)} \equiv |I_l + F(sI_n - A)^{-1}B|$$

(Hint: we use a development similar to equation (3.61).) Hence, derive a state feedback law for the system of equation (2.39) to allocate the closed-loop poles to the positions $\mu_1 = \mu_2 = -1 + j$, $\mu_3 = \mu_4 = -1 - j$ and compare your answer with the results of Exercise 2.3.2. (Hint: derive the closed-loop characteristic polynomial from the above relation and solve the equations $\rho_c(\mu_k) = 0$, $1 \leqslant k \leqslant 4$ for the elements of F).

Remarks and Further Reading

This chapter has not claimed to provide a complete treatment of the analysis and design of multivariable feedback control systems using output feedback. It is hoped, however, that the reader will find that he has seen the basic groundwork required for more advanced study (Rosenbrock, 1974; Owens, 1978; Munro, 1979a; Postlethwaite and MacFarlane, 1979; MacFarlane, 1980) and he has also seen some of the techniques that can be used applied to a real (albeit simple) physical example. More precisely, the chapter describes the natural generalization of transfer function, poles and zeros, block diagram algebra and unity negative feedback systems to the multivariable case. The generalizations take their most natural and simplest form when expressed in matrix notation and have an uncanny similarity to their scalar counterparts. The differences between scalar and multivariable systems only really crop up in the discussion of performance specifications where the essentially multivariable phenomenon of interaction appears. It cannot be emphasized too much that interaction can be a major source of design problems. In fact, almost all of the design techniques discussed in the references are constructed with the aim of reducing interaction in the closed-loop system. More detailed comments on individual topics follow.

(1) The unity negative feedback systems introduced in this chapter can be generalized to the case of non-unity feedback (a common situation when measurement dynamics must be taken into account in the design exercise). Many of the fundamental ideas of the characteristic polynomial and the

relationship between the closed-loop characteristic polynomial and a return-difference determinant carry over to this case (see Rosenbrock, 1974; Owens, 1978). More generally, the ideas of state and output feedback control can often be combined to advantage yielding closed-loop systems possessing two feedback loops – an "inner" state feedback loop and an "outer" unity output feedback loop. (See, for example, Owens, 1978; MacFarlane *et al.*, 1978.)

(2) The material in Sections 3.1 and 3.2 are to be regarded as complementary. The material in Section 3.1 lays the foundations but its primary importance in practice lies in the use of the derived state variable models of the closed-loop system for *simulation* purposes. In contrast the material of Section 3.2 lays the foundations required for the *design* of the forward path control system.

(3) The calculation of the plant transfer function matrix can be a difficult task. An introduction to simple numerical methods can be found in Owens (1978).

(4) Although the intuitive foundation of the ideas of the zero of a linear multivariable system is fairly straightforward, the rigorous theoretical foundations are not so simple. In fact, it is necessary to introduce a host of different types of zero. A review of some aspects of the theory of zeros is given by MacFarlane and Karcanius (1976) but the reader can also obtain information from Owens (1978), Rosenbrock (1970) and Postlethwaite and MacFarlane (1979). Some elementary numerical methods can also be found in Kouvaritakis and MacFarlane (1976) and Owens (1979c) and the connections between the systems zeros and its root-locus plot are outlined in Owens (1978).

(5) The list of design criteria described in Section 3.4 is hardly exhaustive. This is partly due to the fact that each individual design exercise tends to throw up specific performance requirements that cannot be easily fitted into a general treatment. In contrast, in applications where the primary purpose is the stabilization of an unstable plant, the requirements of high steady state accuracy and low interaction effects in response to a unit step demand are irrelevant. Finally, in many industries where safety is of prime importance, it is necessary to design the feedback system to be reliable even in the presence of component failures. This property of *integrity* is amenable to mathematical analysis in certain well defined situations (see Owens, 1978).

(6) The non-interacting controller technique outlined in Section 3.5.2 was

introduced by Boksenbrom and Hood (1949). It is largely neglected now but is a useful conceptual beginning for multivariable studies.

(7) The example described in Section 3.6 contains many of the fundamental ingredients of more advanced techniques. It illustrates the physical fact that the suppression of closed-loop interaction requires the use of high controller gains. From a theoretical point of view, however, the two most important generalizations are as follows.

(a) Most of the available design techniques attempt to reduce the need to consider interaction terms in the design exercise or to transform the problem into an equivalent problem where interaction effects are zero or small enough to be neglected. In the example of Section 3.6, the liquid level system was converted into a non-interacting system by the use of *similarity* transformations. This is not the only technique (see Owens, 1978; MacFarlane, 1980) but the idea of attempting to eliminate the presence of interaction in the design process is now so widely accepted it can be taken as a principle of multivariable design.

(b) Having reduced the design exercise to a sequence of classical single-input/single-output designs, there is normally some compatibility relationship between these designs required to ensure that closed-loop interaction effects are small.

Perhaps the most natural generalization of Section 3.6 is the notion of the commutative controller (MacFarlane, 1970a). This technique suffers from problems, however, and the more successful techniques are based on the Nyquist-type stability criterion outlined in Section 3.8. An account of the fundamental importance of the return-difference can be found in MacFarlane (1970b), Owens (1978) and Postlethwaite and MacFarlane (1979) and, more recently, MacFarlane (1979, 1980) and the special issue of the Institution of Electrical Engineers Control and Science Record, edited by Harris and Owens (1979). This last reference describes most of the available techniques (with the exception of Rosenbrock (1974)) including the characteristic locus design method (Kouvaritakis), the method of dyadic expansion (Owens), multivariable root-locus theory (Postlethwaite, Owens), sequential design techniques (Mayne), optimization methods (Heunis *et al.*) and other numerical approaches (Zakian). Some unifying threads can be found in Owens (1979e).

(8) The idea of a multivariable first order lag can be extended to the second and higher order cases (Owens, 1978). Their importance lies in their use as approximate plant models for controller design (Edwards and Owens, 1977; Owens, 1978; Owens and Chotai, 1980, 1981).

4. Discrete Output Feedback

It is clear from Chapter three that many of the familiar and highly successful transfer function methods for the analysis and design of scalar feedback systems can be generalized to cope with multivariable systems. It is the purpose of this chapter to demonstrate that these general observations also hold for sampled-data systems described by linear, time-invariant discrete state-variable models. The methods used are those of the z-transform (of course!) and the use of matrix notation. The development will be a close parallel of that of Chapter three.

4.1 Discrete Feedback Systems

The purpose of this section is to repeat the analysis of Sections 3.1–3.4 for the case of discrete systems control. In many cases the treatment is so close to that of the continuous case that results will be stated with only an outline proof. It is expected that the interested reader will be able to fill in the detail using the techniques of the previous chapter.

4.1.1 State-variable models of unity feedback systems

Consider the l-input/m-output plant described by the linear, time-invariant discrete model

$$\begin{aligned} x_{k+1}^p &= \Phi_p x_k^p + \Delta_p u_k \\ y_k &= C_p x_k^p \qquad k \geqslant 0 \end{aligned} \tag{4.1}$$

where $\{u_0, u_1, u_2, \ldots\}$ is a sequence of $l \times 1$ input vectors, $\{x_0^p, x_1^p, \ldots\}$ is the sequence of $n_1 \times 1$ state vectors evolving from the initial condition x_0^p and $\{y_0, y_1, y_2, \ldots\}$ is the corresponding sequence of output vectors. As in Section 3.1 we will suppose that there is no "D-term" in the output equation.

Suppose now that the input sequence is generated as the output of a *forward path controller* with an input sequence $\{e_0, e_1, e_2, \ldots\}$ of $m \times 1$ error vectors. There are two distinct cases:

Proportional control. The case of proportional control is represented by the matrix relation

$$u_k = Ke_k, \qquad k \geq 0 \tag{4.2}$$

(c.f. equation (3.2)) when the input vector u_k at the kth sample instant is derived from the error vector e_k at the kth sample instant by multiplication by the $l \times m$ matrix K of constant gains. The forward path system describes the relation between $\{e_0, e_1, e_2, \ldots\}$ and $\{y_0, y_1, y_2, \ldots\}$ and is obtained from (4.1) and (4.2) by eliminating u_k

$$x_{k+1} = \Phi x_k + \Delta e_k, \qquad y_k = Cx_k, \qquad k \geq 0 \tag{4.3}$$

where $x_k = x_k^p$ $(k \geq 0)$, $\Phi = \Phi_p$, $\Delta = \Delta_p K$ and $C = C_p$. The corresponding open-loop characteristic polynomial is

$$\rho_0(z) = |zI - \Phi| \tag{4.4}$$

and is obviously equal to the characteristic polynomial of the plant.

Unity negative feedback is introduced by the matrix equation

$$e_k = r_k - y_k, \qquad k \geq 0 \tag{4.5}$$

where $\{r_0, r_1, r_2, \ldots\}$ is the sequence of $m \times 1$ demand vectors, i.e. r_k is the ideal or demanded output vector at the kth sample instant. The resulting feedback system has the structure shown in Fig. 35 and has a state-variable model obtained by substituting (4.5) into (4.3), i.e.

$$x_{k+1} = (\Phi - \Delta C)x_k + \Delta r_k, \qquad y_k = Cx_k, \qquad k \geq 0 \tag{4.6}$$

This closed-loop system has the closed-loop characteristic polynomial

$$\rho_c(z) = |zI - \Phi + \Delta C| \tag{4.7}$$

describing its stability.

Dynamic control elements. In a similar manner to Section 3.1.2, the introduction of dynamic effects into the forward path controller can be represented by using a forward path controller described by the linear model

$$x_{k+1}^c = \Phi_c x_k^c + \Delta_c e_k, \qquad u_k = C_c x_k^c + D_c e_k, \qquad k \geq 0 \tag{4.8}$$

where $\{x_0^c, x_1^c, x_2^c, \ldots\}$ is a sequence of $n_2 \times 1$ control state vectors describing the dynamic aspects of the control element. The forward path system is obtained by introducing the $n \times 1$ state vector $(n = n_1 + n_2)$

$$x_k = \begin{bmatrix} x_k^p \\ x_k^c \end{bmatrix}, \qquad k \geqslant 0 \tag{4.9}$$

and combining equations (4.1) and (4.8). The resulting system has the form of (4.3) with

$$\Phi = \begin{bmatrix} \Phi_p & \Delta_p C_c \\ 0 & \Phi_c \end{bmatrix}, \qquad \Delta = \begin{bmatrix} \Delta_p D_c \\ \Delta_c \end{bmatrix}, \qquad C = (C_p, 0) \tag{4.10}$$

and has the open-loop characteristic polynomial $\rho_0(z) = |zI - \Phi|$ as in (4.4). Unity negative feedback is introduced as described by (4.5) and leads to the closed-loop system of the form of (4.6) with closed-loop characteristic polynomial as in (4.7).

4.1.2 z-transforms and z-transfer function matrices

The state-variable models described above are a necessary and very important part of systems analysis and design. In particular they are a necessary basis for simulation studies (and hence performance assessment) of the closed-loop system. It is, however, difficult to see how the state-variable models can be used in design and, in particular, how the formulation links up with the single-input/single-output concepts. The picture does, however, become clearer if we introduce the notion of *z-transfer function matrix.*

The first step is the extension of the notion of the z-transform of a sequence of numbers to that of the *z-transform* of a sequence of matrices: the z-transform of a sequence $\{f_0, f_1, f_2, \ldots\}$ of $r_1 \times r_2$ constant matrices is the $r_1 \times r_2$ matrix function of the complex variable z defined by

$$\begin{aligned} f(z) &= f_0 + \frac{1}{z} f_1 + \frac{1}{z^2} f_2 + \ldots \\ &= \sum_{k=0}^{\infty} z^{-k} f_k \end{aligned} \tag{4.11}$$

and is defined for all z such that the infinite series converges absolutely.

EXAMPLE 4.1.1. The z-transform of the sequence $\{\alpha, \alpha, \alpha, \alpha, \alpha, \ldots\}$ of identical $n \times 1$ vectors is the $n \times 1$ vector function of z

$$\alpha(z) = \alpha + z^{-1}\alpha + z^{-2}\alpha + \ldots = \frac{1}{1 - z^{-1}}\alpha = \frac{z}{z-1}\alpha \tag{4.12}$$

defined whenever $|z| > 1$.

EXAMPLE 4.1.2. The z-transform of the sequence $\{x_0, Ax_0, A^2x_0, \ldots\}$ is the function

$$\begin{aligned} x_0(z) &= x_0 + z^{-1}Ax_0 + z^{-2}A^2x_0 + \ldots \\ &= (I + z^{-1}A + z^{-2}A^2 + \ldots)x_0 = (zI - A)^{-1}zx_0 \end{aligned} \tag{4.13}$$

defined for $|z| >$ maximum of the moduli of the eigenvalues of A (see problem (1) at the end of the chapter).

EXERCISE 4.1.1 (linearity of z-transformation). If $f(z)$ is the z-transform of the matrix sequence $\{f_0, f_1, f_2, \ldots\}$ and λ is any scalar, verify that $\lambda f(z)$ is the z-transform of the sequence $\{\lambda f_0, \lambda f_1, \lambda f_2, \ldots\}$. Let $f(z)$, $g(z)$ and $h(z)$ be the z-transforms of the matrix sequences $\{f_0, f_1, f_2, \ldots, \}$, $\{g_0, g_1, g_2, \ldots\}$ and $\{f_0 + g_0, f_1 + g_1, f_2 + g_2, \ldots\}$ respectively. Prove that $h(z) = f(z) + g(z)$.

EXERCISE 4.1.2. Verify that the z-transform of the sequence $\{x_0, \lambda x_0, \lambda^2 x_0, \ldots\}$ with λ a non-zero complex scalar is simply $(z/(z - \lambda))x_0$ and is defined for $|z| > |\lambda|$.

EXERCISE 4.1.3. If $f(z)$ is the z-transform of the sequence $\{f_0, f_1, f_2, \ldots\}$ prove that the z-transform of the "forward-shifted" sequence $\{0, f_0, f_1, f_2, \ldots\}$ is simply $z^{-1}f(z)$.

The second step in the analysis is the representation of systems dynamics in transform terms. For this purpose, consider the general l-input/m-output system of (1.188), i.e.

$$\begin{aligned} x_{k+1} &= \Phi x_k + \Delta u_k \\ y_k &= Cx_k + Du_k, \qquad k \geqslant 0 \end{aligned} \tag{4.14}$$

Defining

$$\begin{aligned} x(z) &= x_0 + z^{-1}x_1 + z^{-2}x_2 + \ldots \\ u(z) &= u_0 + z^{-1}u_1 + z^{-2}u_2 + \ldots \end{aligned} \tag{4.15}$$

to be the z-transforms of the state and input sequences respectively, it follows from (1.190) that

$$\begin{aligned} x(z) &= x_0 + z^{-1}(\Phi x_0 + \Delta u_0) + z^{-2}(\Phi^2 x_0 + \Phi\Delta u_0 + \Delta u_1) \\ &\quad + z^{-3}(\Phi^3 x_0 + \Phi^2 \Delta u_0 + \Phi\Delta u_1 + \Delta u_2) + \ldots \\ &= (I_n + z^{-1}\Phi + z^{-2}\Phi^2 + \ldots)x_0 + z^{-1}\Delta u_0 + z^{-2}(\Phi\Delta u_0 + \Delta u_1) \\ &\quad + z^{-3}(\Phi^2\Delta u_0 + \Phi\Delta u_1 + \Delta u_3) + \ldots \end{aligned}$$

$$= (I_n + z^{-1}\Phi + z^{-2}\Phi^2 + \ldots)x_0$$

$$+ \frac{1}{z}(I_n + z^{-1}\Phi + z^{-2}\Phi^2 + \ldots)\Delta(u_0 + z^{-1}u_1 + z^{-2}u_2 + \ldots) \quad (4.16)$$

or, using the results of problem (1) at the end of the chapter

$$x(z) = (zI_n - \Phi)^{-1}zx_0 + (zI - \Phi)^{-1}\Delta u(z) \quad (4.17)$$

Letting

$$y(z) = y_0 + z^{-1}y_1 + z^{-2}y_2 + \ldots \quad (4.18)$$

denote the z-transform of the output sequence, it is easily seen that $y(z) = Cx_0 + Du_0 + z^{-1}(Cx_1 + Du_1) + \ldots = C(x_0 + z^{-1}x_1 + z^{-2}x_2 + \ldots) + D(u_0 + z^{-1}u_1 + z^{-2}u_2 + \ldots) = Cx(z) + Du(z)$ and, hence, substituting into (4.17),

$$y(z) = \{C(zI_n - \Phi)^{-1}\Delta + D\}u(z) + C(zI_n - \Phi)^{-1}zx_0 \quad (4.19)$$

This equation is the z-transform equivalent of the state equations (4.14). It is also the sampled-data system equivalent of the Laplace transform relation (3.21) for continuous systems.

Considering the case of zero initial conditions $x_0 = 0$, equation (4.19) reduces to

$$y(z) = G(z)u(z) \quad (4.20)$$

where the $m \times l$ matrix function of the complex variable z defined by

$$G(z) = C(zI_n - \Phi)^{-1}\Delta + D \quad (4.21)$$

is the $m \times l$ z-transfer function matrix of the system. It is also the multivariable generalization of the classical notion of z-transfer function and suggests the block system representation illustrated in Fig. 54 (compare with Fig. 36 for the continuous case).

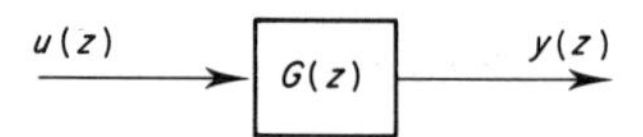

Fig. 54. Transform representation of discrete system input-output dynamics.

EXAMPLE 4.1.3. The discrete single-input/two-output system

$$x_{k+1} = \begin{bmatrix} -\frac{1}{2} & \frac{1}{2} \\ \frac{1}{2} & -\frac{1}{2} \end{bmatrix} x_k + \begin{bmatrix} 1 \\ 0 \end{bmatrix} u_k$$

$$y_k = \begin{bmatrix} 1 & 0 \\ 1 & 1 \end{bmatrix}, \qquad k \geqslant 0 \quad (4.22)$$

has z-transfer function matrix

$$G(z) = \frac{1}{z(z+1)}\begin{bmatrix} z+\frac{1}{2} \\ z+1 \end{bmatrix} \tag{4.23}$$

The z-transform of the output response from zero-initial conditions to the unit step input $u_k = 1, k \geqslant 0$, is simply $y(z) = G(z)u(z)$ and $u(z) = z/(z-1)$. Substituting into (4.23) yields the relations

$$y_1(z) = \frac{(z+\frac{1}{2})}{(z+1)(z-1)}, \qquad y_2(z) = \frac{1}{z-1} \tag{4.24}$$

It is clear from this and the results of Example 4.1.1 and Exercise 4.1.3 that the response of the second output is just the sequence $\{0, 1, 1, 1, 1, \ldots\}$. The response of the first output can be obtained by (a) expanding $y_1(z)$ as a power series in z^{-1} and examining the coefficients or (b) expanding $y_1(z)$ in the partial fraction form

$$y_1(z) = \frac{(\frac{1}{4})}{z+1} + \frac{(\frac{3}{4})}{z-1} \tag{4.25}$$

and, using the result of Exercises 4.1.1, 4.1.2 and 4.1.3, it follows that the response of the first output is just the sequence $\{0, \frac{1}{4}+\frac{3}{4}, -\frac{1}{4}+\frac{3}{4}, \frac{1}{4}+\frac{3}{4}, \ldots\} = \{0, 1, \frac{1}{2}, 1, \frac{1}{2}, 1, \ldots\}$.

Finally, before we move on to the business of applying the above concepts to the representation of feedback systems in transform notation, there are a number of important concepts to be established.

Interaction. Writing $G(z)$ in element form

$$G(z) = \begin{bmatrix} G_{11}(z) & \ldots & G_{1l}(z) \\ \vdots & & \vdots \\ G_{m1}(z) & \ldots & G_{ml}(z) \end{bmatrix} \tag{4.26}$$

then relation (4.20) takes the form of (3.25),

$$y_k(z) = G_{k1}(z)u_1(z) + \ldots + G_{kl}(z)u_l(z), \qquad 1 \leqslant k \leqslant m \tag{4.27}$$

indicating that the z-transfer function $G_{ki}(z)$ describes the way that the ith input affects the kth output. In particular, in the case of $m = l$, we obtain the following parallel to Theorem 3.2.1.

Theorem 4.1.1. *Taking $m = l$, the system (4.14) is non-interacting if, and only if, its $m \times m$ z-transfer function matrix is diagonal for all z.*

Block diagram algebra. Comparing (4.20) with its equivalent (equation (3.22)) for the continuous system case it is clear that all the results of Section 3.2.2 carry over directly to the case of discrete systems. In other words, block diagram algebra can be used in the analysis of discrete system dynamics using z-transfer function matrices.

Poles. In a similar manner to (3.38) it is easily verified that

$$G(z) = \{C \operatorname{adj}(zI_n - \Phi)\Delta + |zI_n - \Phi| D\}/|zI_n - \Phi| \tag{4.28}$$

and hence that the poles of $G(z)$ are simply the eigenvalues of Φ. The anticipated relationship between the poles of G and system stability is hence established (Section 1.8.2).

Zeros. The concept of zero in the discrete case is, in many ways, identical to that in the continuous case (Section 3.3) with a small change due to discrete system structure. More precisely, the complex number λ is a zero of the discrete system (4.14) in the case of $l \leqslant m$ if, and only if, we can choose an initial condition x_0 and an $l \times 1$ input sequence $u_k = u_0\lambda^k$, $k \geqslant 0$, such that either $u_0 \neq 0$ and/or $x_0 \neq 0$ and the output sequence $y_k = 0, k \geqslant 0$.

A simple characterization of discrete systems zeros can be obtained in the case of $m = l$ by writing $x_k = \lambda^k x_0$, $k \geqslant 0$. Substituting into (4.14) yields the equations

$$\begin{aligned}(\lambda I_n - \Phi)x_0 - Bu_0 &= 0 \\ Cx_0 + Du_0 &= 0\end{aligned} \tag{4.29}$$

Noting that this equation has identical structure to (3.41), it follows, using a similar argument to that following that equation, that λ is a zero of the discrete system (4.14) if, and only if, it is a zero of the so-called "zero-polynomial"

$$z(\lambda) = \begin{vmatrix} \lambda I_n - \Phi & -\Delta \\ C & D \end{vmatrix} \tag{4.30}$$

An alternative form of $z(\lambda)$ is obtained from Schur's formula (Exercise 3.3.1), namely

$$z(\lambda) = |\lambda I - \Phi|\,|D + C(\lambda I - \Phi)^{-1}\Delta| = |\lambda I - \Phi| \cdot |G(\lambda)| \tag{4.31}$$

Finally as the polynomials (4.30) and (3.44) have identical structure it follows that all of the properties of continuous system zeros described in the problems and reading list of Chapter three will also apply to discrete systems zero.

4.1.3 *Feedback systems, performance specifications and the return-difference*

In this section, we construct the z-transfer function matrix version of the state-variable analysis of Section 4.1.1. More precisely, in the configuration of

Fig. 35, suppose that the plant (4.1) is described by the z-transfer function matrix

$$G(z) = C_p(zI_{n_1} - \Phi_p)^{-1}\Delta_p \tag{4.32}$$

and that the forward path controller has z-transfer function matrix $K(z)$. There are two cases:

(a) *Proportional control.* Equation (4.2) can be written in the form $u(z) = Ke(z)$ and hence

$$K(z) = K \tag{4.33}$$

(b) *Dynamic control element.* The forward path controller (4.8) has the z-transfer function matrix

$$K(z) = C_c(zI_{n_2} - \Phi_c)^{-1}\Delta_c + D_c \tag{4.34}$$

Taking the case of zero initial conditions, it is clear that, in both cases,

$$y(z) = G(z)u(z), \qquad u(z) = K(z)e(z) \tag{4.35}$$

and hence that

$$y(z) = Q(z)e(z) \tag{4.36}$$

where the forward path z-transfer function matrix $Q(z) = G(z)K(z)$ must be the transfer function matrix of the forward path system (4.3), i.e.

$$Q(z) = G(z)K(z) = C(zI - \Phi)^{-1}\Delta \tag{4.37}$$

Taking z-transforms of (4.5) yields

$$e(z) = r(z) - y(z) \tag{4.38}$$

and, hence, using (4.36), $y = Q(r - y)$ or $y + Qy = (I_m + Q)y = Qr$ yielding

$$y(z) = H_c(z)r(z) \tag{4.39}$$

where the "closed-loop z-transfer function matrix"

$$H_c(z) = (I_m + Q(z))^{-1}Q(z) \tag{4.40}$$

is, hence, the transfer function matrix of the closed-loop system (4.6). Note the formal similarity to (3.57).

EXERCISE 4.1.4. Using the technique of Exercise 3.4.1, verify that $H_c(z) = C(zI - \Phi + \Delta C)^{-1}\Delta$.

EXERCISE 4.1.5. Use equations (4.35), (4.36) and (4.38) to prove that the closed-loop system can be represented in block form as illustrated in Fig. 55(a) or (b).

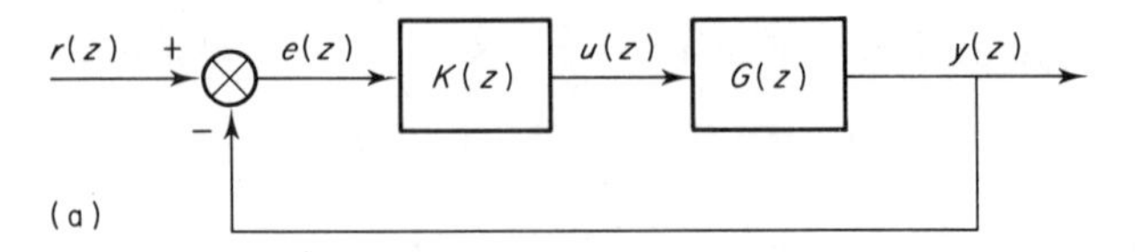

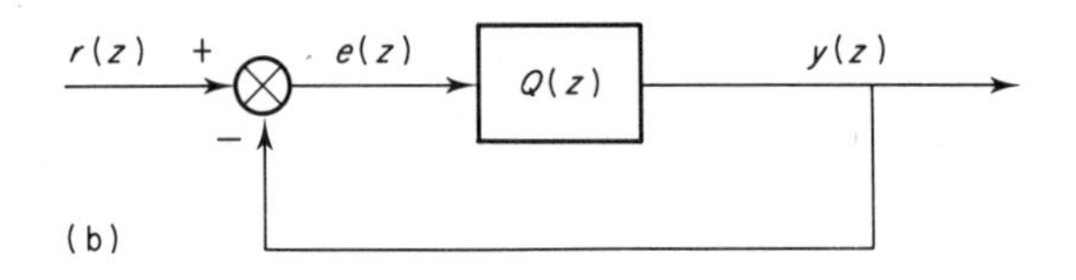

Fig. 55.

Equations (4.39) and (4.40) are the foundations required for the study of the behaviour of the closed-loop system using z-transform methods, the problem of the design of the forward path controller to ensure "satisfactory" stability and performance of the closed-loop system being regarded as the choice of the $l \times m$ transfer function matrix $K(z)$.

The formal definition of performance specifications follows the same lines as that of Section 3.4 and is, of course, subject to the same limitations. The details are outlined below.

Closed-loop stability. As in the continuous case (equations (3.59)), closed-loop stability can be described in terms of the forward path transfer function matrix. More precisely, the closed-loop characteristic polynomial $\rho_c(z)$ (equation (4.7)) and open-loop characteristic polynomial $\rho_0(z)$ (equation 4.4)) are related by

$$\frac{\rho_c(z)}{\rho_0(z)} \equiv |T(z)| \tag{4.41}$$

where $T(z) = I_m + Q(z)$ is the *matrix return-difference.* The proof of this relation follows the development of equation (3.61) and is left as an exercise for the reader.

Transient performance and interaction. As in the continuous case, closed-loop system transient performance is assumed to be assessed by examination of the time response from zero initial conditions to specified step demands. A general step demand is represented by the sequence $\{\alpha, \alpha, \alpha, \alpha, \ldots\}$ of identical $m \times 1$ vectors $r_k = \alpha, k \geqslant 0$. The corresponding demand transform is $r(z) = (z/(z-1)\alpha)$ and the m independent step responses required to assess all aspects of closed-loop system performance are obtained by choosing $\alpha = \tilde{e}_1, \tilde{e}_2, \ldots, \tilde{e}_m$ (equation

(3.64)) sequentially. The demand $r(z) = (z/(z-1))\tilde{e}_i$ is termed a "unit step demand in the ith output".

Steady state response and steady state error. If the closed-loop system is stable, the steady state response to the unit step demand $r(z) = (z/(z-1))\tilde{e}_i$ is calculated by applying the standard final-value theorem

$$\lim_{k \to +\infty} y_k = \lim_{z \to 1} \frac{(z-1)}{z} y(z) = \lim_{z \to 1} \frac{(z-1)}{z} H_c(z) r(z)$$

$$= \lim_{z \to 1} H_c(z)\tilde{e}_i = H_c(1)\tilde{e}_i \tag{4.42}$$

(Note: $\lim_{z \to 1} H_c(z) = H_c(1)$ as stability guarantees that H_c has no pole at $z = 1$.) It is left as an exercise for the reader to show that the steady state error vector

$$\lim_{k \to +\infty} e_k = (I_m - H_c(1))\tilde{e}_i \tag{4.43}$$

and hence that steady state errors in response to unit step demands in any output are zero if, and only if,

$$H_c(1) = I_m \tag{4.44}$$

(Note: compare this relation with (3.67).)

EXERCISE 4.1.6. Verify (4.42) by writing $x_\infty = \lim_{k \to +\infty} x_k$, deducing from (4.6) that $\lim_{k \to +\infty} y_k = C(I_n - \Phi + \Delta C)^{-1} \Delta \tilde{e}_i$ and using the results of Exercise 4.1.4.

4.2 Discrete Multivariable First Order Lags

On paper there is an obvious formal similarity in the open- and closed-loop relationship defining the stability and performance properties of both continuous and discrete-time systems. This is verified by a comparison of equations (3.22) and (4.20), (3.54) and (4.36), (3.56) and (4.39), (3.57) and (4.40) and, finally, (3.59) and (4.41). It follows from this observation that many of the techniques and principles used in the design of continuous output feedback systems are suitable, with little modification, for application to the design of discrete output feedback systems. One major difference, however, is that whereas in the continuous case, controller gain is a limiting factor in attainable closed-loop performance, the equivalent parameter in the discrete-time case is sampling rate. It is not possible to present a general analysis of this problem but it is possible to illustrate the principles and possibilities by the detailed analysis of discrete first order lags described below. The reader may find it useful to compare the material with that of Section 3.7.

An m-input/m-output discrete time plant is said to be a "discrete multivariable first order lag" if, and only if, it can be described by a linear time-invariant state-variable model of the form

$$\begin{aligned} x_{k+1}^p &= \Phi_p x_k^p + \Delta_p u_k \\ y_k &= C_p x_k^p, \qquad k \geqslant 0 \end{aligned} \tag{4.45}$$

where the system dimension $n = m$ and the (square!) $m \times m$ matrices C_p and Δ_p are both nonsingular. The plant z-transfer function matrix $G(z) = C_p (zI_m - \Phi_p)^{-1} \Delta_p$ then has inverse of the form

$$G^{-1}(z) = (C_p(zI_m - \Phi_p)^{-1}\Delta_p)^{-1} = (z-1)A_0 + A_1 \tag{4.46}$$

where the $m \times m$ matrices A_0 and A_1 are given by the formulae

$$A_0 = \Delta_p^{-1} C_p^{-1} = (C_p \Delta_p)^{-1}, \qquad A_1 = \Delta_p^{-1}(I_m - \Phi_p)C_p^{-1} \tag{4.47}$$

and hence can be computed directly from the matrices occuring in the state-variable model. A scalar discrete first order lag has transfer function $g(z) = 1/(a_0(z-1) + a_1)$ with inverse $g^{-1}(z) = a_0(z-1) + a_1$. Hence the motivation for the terminology of "first-order lag" in the (more general) multivariable case (4.46).

EXAMPLE 4.2.1. There is a strong connection between the notions of continuous and discrete multivariable first order lags. More precisely, if a *continuous first order lag* (Section 3.7)

$$\frac{dx_p(t)}{dt} = A_p x_p(t) + B_p u(t), \qquad y(t) = C_p x_p(t) \tag{4.48}$$

is to be controlled by piecewise constant inputs using sampled output data (all signals being synchronized with sampling interval $h > 0$), then the design could proceed by computing a discrete state-variable model of the form of (4.45) with (equation (1.204))

$$\Phi_p = e^{A_p h}, \qquad \Delta_p = \int_0^h e^{A_p(h-t)} B \, dt \tag{4.49}$$

It is clear that $n = m$ and hence, provided Δ_p is nonsingular, the discrete model derived from a continuous multivariable first-order lag is a discrete multivariable first-order lag!

Taking, for example, the liquid level system (1.71) with data $a_1 = a_2 = \beta = 1$, then

$$A_p = \begin{bmatrix} -1 & 1 \\ 1 & -1 \end{bmatrix}, \qquad B_p = C_p = \begin{bmatrix} 1 & 0 \\ 0 & 1 \end{bmatrix} \tag{4.50}$$

and, using the results of Example 1.7.2,

$$e^{At} = \tfrac{1}{2}\begin{bmatrix} 1+e^{-2t} & 1-e^{-2t} \\ 1-e^{-2t} & 1+e^{-2t} \end{bmatrix} \tag{4.51}$$

Substitution into (4.49) indicates that the matrices defining a discrete model of the liquid level system take the form

$$\Phi_p = \tfrac{1}{2}\begin{bmatrix} 1+e^{-2h} & 1-e^{-2h} \\ 1-e^{-2h} & 1+e^{-2h} \end{bmatrix} \tag{4.52}$$

$$\Delta_p = \tfrac{1}{4}\begin{bmatrix} 1-e^{-2h}+2h & e^{-2h}-1+2h \\ e^{-2h}-1+2h & 1-e^{-2h}+2h \end{bmatrix} \tag{4.53}$$

Taking, for illustrative purposes, a sampling interval of $h = \frac{1}{20}$, the discrete state-variable model of liquid level system dynamics takes the form

$$x_{k+1}^p = \begin{bmatrix} 0.952 & 0.048 \\ 0.048 & 0.952 \end{bmatrix} x_k^p + \begin{bmatrix} 0.049 & 0.001 \\ 0.001 & 0.049 \end{bmatrix} u_k$$

$$y_k = \begin{bmatrix} 1 & 0 \\ 0 & 1 \end{bmatrix} x_k^p, \qquad k \geqslant 0 \tag{4.54}$$

and, using (4.47), the system has an inverse z-transfer function matrix of the form of (4.46) with

$$A_0 = \begin{bmatrix} 20.5 & -0.4 \\ -0.4 & 20.5 \end{bmatrix}, \qquad A_1 = \begin{bmatrix} 1 & -1 \\ -1 & 1 \end{bmatrix} \tag{4.55}$$

EXERCISE 4.2.1. Using the construction and notation of Example 3.7.1 verify that

$$|\Delta_p| = \left(\prod_{k=1}^{m} \left(\int_0^h e^{\lambda_k(h-t)}dt\right)\right)|B| \tag{4.56}$$

where $\lambda_1, \lambda_2, \ldots, \lambda_m$ are the eigenvalues of A_p. (Hint: express A_p in its diagonal or Jordan form.) Hence, deduce that Δ_p is singular if, and only if, no

EXERCISE 4.2.1. contd.

non-zero imaginary eigenvalue of A_p is an integer multiple of $2\pi j/h$. Provide a physical interpretation of this result.

EXERCISE 4.2.2. Verify that the first order lag (4.45) can be written in the form

$$y_{k+1} = (I - A_0^{-1}A_1)y_k + A_0^{-1}u_k, \qquad k \geqslant 0 \tag{4.57}$$

Hence deduce that the plant characteristic polynomial is

$$\rho_0(s) = |(z-1)I_m + A_0^{-1}A_1| \tag{4.58}$$

and check this result directly from (4.47). Deduce that the discrete first order lag has no zeros. (Hint: use (4.31).)

Noting the formal similarity between (4.46) and (3.109) it is clear that the principle of control analysis outlined in Section 3.7. for continuous first order lags will carry through to the discrete case. The theoretical development is outlined below, the details being left to the conscientious reader.

Suppose that $A_0^{-1}A_1$ has eigenvector matrix T and eigenvalues $1-\eta_1$, $1-\eta_2, \ldots, 1-\eta_m$ satisfying

$$T^{-1}A_0^{-1}A_1T = \begin{bmatrix} 1-\eta_1 & \cdots & & 0 \\ \vdots & 1-\eta_2 & & \vdots \\ & & \ddots & \\ 0 & \cdots & & 1-\eta_m \end{bmatrix} \tag{4.59}$$

and choose the forward path proportional controller with z-transfer function matrix

$$K(z) = A_0T\begin{bmatrix} k_1 & \cdots & 0 \\ \vdots & k_2 & \vdots \\ 0 & \cdots & k_m \end{bmatrix}T^{-1} \tag{4.60}$$

It follows that

$$Q(z) = G(z)K(z) = T\begin{bmatrix} q_1(z) & 0 & \cdots & 0 \\ 0 & q_2(z) & & \vdots \\ 0 & \cdots & & q_m(z) \end{bmatrix}T^{-1} \tag{4.61}$$

where the scalar z-transfer functions are

$$q_i(z) = \frac{k_i}{z-\eta_i}, \qquad 1 \leqslant i \leqslant m \tag{4.62}$$

Finally, the closed-loop z-transfer function matrix is

$$H_c(z) = T\begin{bmatrix} h_1(z) & 0 \ldots & 0 \\ 0 & h_2(z) & \vdots \\ \vdots & & \vdots \\ 0 & \ldots\ldots & h_m(z) \end{bmatrix}T^{-1} \tag{4.63}$$

where

$$h_i(z) = \frac{q_i(z)}{1+q_i(z)} = \frac{k_i}{z-\eta_i+k_i}, \qquad 1 \leqslant i \leqslant m \tag{4.64}$$

Using the experience of Section 3.6 and the success of the principle in Section 3.7 we will now attempt to reduce interaction effects in the closed-loop system by choosing the gains $k_1, k_2, \ldots, k_m$ to ensure that the scalar feedback systems $h_i(z)$, $1 \leqslant i \leqslant m$, have similar dynamic characteristics. More precisely, we will choose the k_i such that each h_i has identical poles, i.e.

$$k = \eta_i - k_i, \qquad 1 \leqslant i \leqslant m \tag{4.65}$$

resulting in the controller (c.f. (3.125))

$$K(z) = A_0(1-k) - A_1 \tag{4.66}$$

The corresponding closed-loop transfer function matrix can be obtained by substituting (4.65) into (4.63) or, more directly, noting that $H_c = (I+GK)^{-1}GK = (G^{-1}+K)^{-1}K$ and substituting from (4.46) and (4.66)

$$H_c(z) = \frac{(1-k)}{(z-k)}\left\{I_m - \frac{1}{(1-k)}A_0^{-1}A_1\right\} \tag{4.67}$$

The stability, steady state accuracy and transient performance of the closed-loop system are analysed below.

(a) The *stability* of the system is described by the return-difference determinant

$$\begin{aligned} |I_m + G(z)K(z)| &\equiv |G(z)| \cdot |G^{-1}(z) + K(z)| \\ &\equiv \frac{|G^{-1}(z)+K(z)|}{|G^{-1}(z)|} \\ &\equiv (s-k)^m/\rho_0(z) \end{aligned} \tag{4.68}$$

where $\rho_0(z)$ takes the form described in Exercise 3.7.2, i.e. the closed-loop characteristic polynomial is simply $\rho_c(z) = (z-k)^m$ and hence the closed-loop system is stable if, and only if,

$$|k| < 1 \tag{4.69}$$

(b) The *steady state accuracy* of the closed-loop system can be assessed by noting that

$$H_c(1) = I_m - \frac{1}{(1-k)} A_0^{-1} A_1 \tag{4.70}$$

In particular it is clear that steady state errors in response to unit step demands are small if, and only if, the elements of $A_0^{-1}A_1$ are small when compared with $1-k$.

(c) The *transient interaction effects* of the closed-loop system in response to unit step demands can be assessed in general terms by noting that $H_c(z) \simeq (1-k)/(z-k)I_m$ is approximately diagonal if the elements of $A_0^{-1}A_1$ are small compared to $1-k$ and, hence (theorem 4.1.1), closed-loop interaction effects are small.

It is clear from the above that the quality of the closed-loop transient and steady state response depends critically upon the magnitude of the elements of the matrix $(1-k)^{-1}A_0^{-1}A_1$, the quality of the performance improving as the maximum of the moduli of the elements decreases. Unfortunately, we cannot make this maximum arbitrarily small by increasing the magnitude of k due to the stability constraint (4.69). We must therefore conclude that the attainable performance is limited by the structure of the matrix $A_0^{-1}A_1$. Fortunately, we do have some control here as $A_0^{-1}A_1$ depends explicitly upon our choice of sample interval h. More precisely, if we assume that the discrete first-order plant is derived from a continuous first-order plant of the form of (4.48), then equations (4.47) and (4.49) indicate that

$$A_0^{-1}A_1 = C_p(I_m - e^{A_p h})C_p^{-1} \tag{4.71}$$

and hence that

$$\lim_{h \to 0} A_0^{-1}A_1 = 0 \tag{4.72}$$

as $e^{A_p h} = I_m$ at $h = 0$. That is, for a given choice of k satisfying (4.69) a reduction in sample interval h (or, equivalently, an increase in sampling rate h^{-1}) will reduce both steady state errors and interaction effects in the closed-loop system response to unit step demands.

EXERCISE 4.2.3. By expressing $e^{A_p h}$ in terms of the eigenvalues $\lambda_1, \lambda_2, \ldots, \lambda_m$ and eigenvector matrix T of A_p, verify that $A_0^{-1}A_1$ is small if

$$\lambda_i h \ll 1, \qquad 1 \leqslant i \leqslant m \tag{4.73}$$

What does this mean?

EXERCISE 4.2.4. Verify that an increase in sampling rate with k held fixed does, in fact, correspond to an increase in the "real-time" response speed of the closed-loop system.

EXAMPLE 4.2.2. To illustrate the application of the above theory, consider the design of a unity negative proportional feedback system for the control of the liquid level system (1.71) with data $a_1 = a_2 = \beta = 1$ using sampled output data with sampling interval $h = \frac{1}{20}$. The discrete plant model has been derived in Example 3.7.1 and, using (4.55), it is clear that

$$A_0^{-1}A_1 = \begin{bmatrix} 0.048 & -0.048 \\ -0.048 & 0.048 \end{bmatrix} \tag{4.74}$$

Suppose that we wish the closed-loop system to respond rapidly to unit step demands and with small steady state errors and interaction effects. With this in mind, choose a controller of the form of (4.66).

$$K(z) = \begin{bmatrix} 19.5 - 20.5k & 0.6 + 0.4k \\ 0.6 + 0.4k & 19.5 - 20.5k \end{bmatrix}, \qquad |k| < 1 \tag{4.75}$$

when, it is known (equation (3.67)) that the closed-loop z-transfer function matrix

$$H_c(z) = \frac{(1-k)}{(z-k)} \begin{bmatrix} 1 - \dfrac{0.048}{(1-k)} & \dfrac{0.048}{(1-k)} \\ \dfrac{0.048}{(1-k)} & 1 - \dfrac{0.048}{(1-k)} \end{bmatrix} \tag{4.76}$$

Note the presence of off-diagonal interaction effects in the closed-loop system.

The transient performance of the closed-loop system is assessed by examination of the response from zero initial conditions to unit step demands. Taking, for example, the demand signal $r(z) = \dfrac{z}{(z-1)}\begin{bmatrix} 1 \\ 0 \end{bmatrix}$, the system response is

$$y(z) = H_c(z)r(z) = \frac{z(1-k)}{(z-1)(z-k)} \begin{bmatrix} 1 - \dfrac{0.048}{(1-k)} \\ \dfrac{0.048}{(1-k)} \end{bmatrix} \tag{4.77}$$

Taking inverse z-transforms by expanding $z(1-k)/(z-k)(z-1) = 1/(z-1) - k/(z-k)$ and using Exercises 4.1.2 and 4.1.3 yields the response

$$y_i = (1-k^i)\begin{bmatrix} 1 - \dfrac{0.048}{(1-k)} \\ \dfrac{0.048}{(1-k)} \end{bmatrix}, \qquad i \geqslant 0 \tag{4.78}$$

It is now clear that our choice of k is critical to closed-loop performance. Steady state errors, for example, in both outputs are seen to have modulus $0.048/(1-k)$ and, hence, if steady state errors are to have modulus less than a, k must lie in the range defined by

$$0.048 \leqslant a(1-k) \tag{4.79}$$

In a similar manner, the peak transient interaction effect has modulus $0.048/(1-k)$ if $k \geqslant 0$ and 0.048 if $k < 0$. Hence, if peak transient interaction effects are to be less than a specified value b, k must satisfy the inequality

$$b \geqslant \begin{cases} 0.048/(1-k) & \text{if} \quad k \geqslant 0 \\ 0.048 & \text{if} \quad k < 0 \end{cases} \tag{4.80}$$

The limitations in performance are now apparent as, from (4.79), there is no solution for k in the range $-1 < k < 1$ if $a < 0.024$ and, from (4.80), there is no solution for k in the required range if $b < 0.048$. That is, steady state errors and peak interaction effects cannot be reduced in peak magnitude below the values of 0.024 and 0.048 respectively, unless faster sampling rates are introduced.

To illustrate the possibilities, suppose that $a = b = 0.1$. It is clear that both (4.79) and (4.80) reduce to $k < 0.52$ which, when combined with the stability condition $|k| < 1$, reduces to $-1 < k < 0.52$. Choosing, for simplicity, $k = -0.15$, the resulting closed-loop responses of the liquid levels are illustrated in Fig. 56. The reader should repeat the analysis for the case of a unit step demand in level in the second vessel.

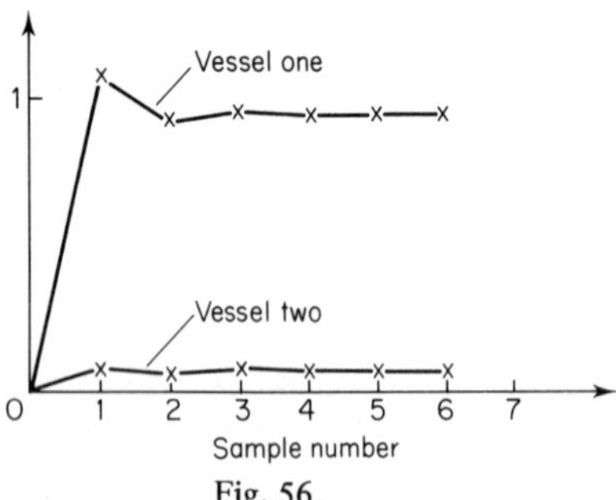

Fig. 56.

EXERCISE 4.2.5. Using the development of Example 3.7.1, verify that the liquid level system with $a_1 = a_2 = \beta = 1$ generates the matrix

$$A_0^{-1}A_1 = \frac{(1-e^{-2h})}{2}\begin{bmatrix} 1 & -1 \\ -1 & 1 \end{bmatrix} \tag{4.81}$$

Repeat the analysis of Example 3.7.2 and verify that steady state errors and interaction effects in response to unit step demands cannot be reduced below the values of $(1-e^{-2h})/4$ and $(1-e^{-2h})/2$ respectively. Show that these bounds tend to zero as h becomes infinitessimally small.

4.3 Frequency Domain Criterion for Stability

As in Section 3.8, the use of frequency domain concepts can be used as the basis of simple graphical methods of assessing the stability of the discrete closed-loop system illustrated in Fig. 55. The basis of the stability analysis is again the return-difference relationship (4.41) expressed in the form

$$\frac{\rho_c(z)}{\rho_0(z)} \equiv |I_m + Q(z)| \tag{4.82}$$

where $\rho_c(z)$ ($\rho_0(z)$) is the closed (open)-loop characteristic polynomial and $Q(z) = G(z)K(z)$ is the forward path z-transfer function matrix. However, the Nyquist D-contour (Fig. 51) used in the analysis of continuous systems, is replaced by the familiar "unit circle" in the complex plane. If, as sometimes happens, one or both of the characteristic polynomials has a zero at some point on the unit circle, then the contour is "indented" (using an infinitessimally small semi-circle). These ideas are illustrated in Fig. 57.

If we now move from an arbitrary point s_0 on the contour around the contour once only in a clockwise manner and, at each point, we evaluate the numerical value of $|I_m + Q(z)|$, then we will generate a closed contour Γ in the complex plane as illustrated in Fig. 58. Suppose that this contour encircles the origin of the complex plane n_T times in a clockwise manner (anti-clockwise encirclements being counted negative). Applying standard encirclement arguments to (4.82) yields the relation

$$\begin{aligned} n_T = {} & \text{(number of zeros of } \rho_c \text{ inside the unit circle)} \\ & - \text{(number of zeros } \rho_0 \text{ inside the unit circle)} \end{aligned} \tag{4.83}$$

or, if

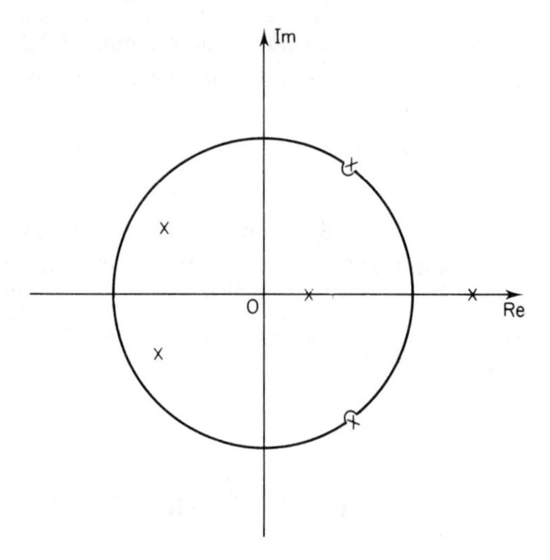

Fig. 57. Unit circle contour for discrete system stability studies. x = open- and closed-loop poles.

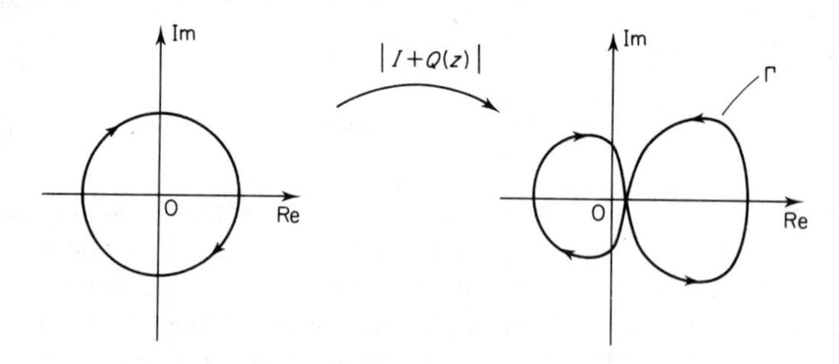

Fig. 58. The return-difference and the contour Γ.

n_c = number of zeros of ρ_c outside the unit circle

n_0 = number of zeros of ρ_0 outside the unit circle, and

n = dimension of the state-variable model of Q,

then it is clear that $n_T = (n - n_c) - (n - n_0)$ or

$$n_T = n_0 - n_c \tag{4.84}$$

and we hence obtain the following generalization of the Nyquist stability criterion for discrete/sampled-data systems.

Theorem 4.3.1. *The closed-loop system is asymptotically stable if, and only if,* $n_T = n_0$.

The proof of this result follows directly from the observation (Section 1.8.2) that stability is equivalent to the requirement that $n_c = 0$, i.e. ρ_c has no zero outside the unit circle.

Problems

(1) Let

$$\Phi_d = \begin{bmatrix} \eta_1 & 0 & \dots & 0 \\ 0 & \eta_2 & & \vdots \\ \vdots & & & 0 \\ 0 & \dots & 0 & \eta_n \end{bmatrix}$$

be a diagonal $n \times n$ matrix. Verify, by expanding its elements as a power series that

$$(zI_n - \Phi_d)^{-1} = z^{-1} I_n + z^{-2} \Phi_d + z^{-3} \Phi_d^2 + \dots$$

the series converging absolutely for $|z| > \max \{|\eta_1|, |\eta_2|, \dots, |\eta_n|\}$. More generally, let Φ be an $n \times n$ matrix possessing eigenvalues $\eta_1, \eta_2, \dots, \eta_n$ and a nonsingular eigenvector matrix T. It follows that we can write $\Phi = T^{-1} \Phi_d T$. Prove that

$$(zI_n - \Phi)^{-1} = z^{-1} I_n + z^{-2} \Phi + z^{-3} \Phi^2 + \dots$$

by verifying the identities $T\Phi_d^k T^{-1} = \Phi^k$ and $(zI_n - \Phi)^{-1} = T(zI_n - \Phi_d)^{-1} T^{-1}$. (Note: these results are also true if Φ possesses only a Jordan canonical form.)

(2) A discrete system described by the model

$$x_{k+1} = \begin{bmatrix} 0 & 1 & 0 \\ 0 & 0 & 1 \\ 0 & 0 & 0 \end{bmatrix} x_k + \begin{bmatrix} 0 \\ 0 \\ 1 \end{bmatrix} u_k$$

$$y_k = \begin{bmatrix} 1 & 0 & 0 \\ 0 & 1 & 0 \end{bmatrix} x_k, \qquad k \geqslant 0$$

is to be subjected to the feedback law

$$u_k = [k_1 \quad k_2](r_k - y_k)$$

Verify that the closed-loop system is described by the model

$$x_{k+1} = \begin{bmatrix} 0 & 1 & 0 \\ 0 & 0 & 1 \\ -k_1 & -k_2 & 0 \end{bmatrix} x_k + \begin{bmatrix} 0 & 0 \\ 0 & 0 \\ k_1 & k_2 \end{bmatrix} r_k, \qquad k \geq 0$$

and that the open (resp. closed)-loop characteristic polynomials take the form z^3 (resp. $z^3 + k_2 z + k_1$). Check these results by computing the z-transfer function matrices

$$G(z) = \begin{bmatrix} \frac{1}{z^3} \\ \frac{1}{z^2} \end{bmatrix}, \qquad K(z) = [k_1 \quad k_2]$$

and verifying that $|I_2 + Q(z)| = (z^3 + k_2 z + k_1)/z^3$.

(3) Verify that the forward path controller $K(z)$ whose elements are proportional plus summation systems of the form

$$K_{ik}(z) = \alpha_{ik} + \beta_{ik} \frac{z}{z-1}$$

can be written in the form $K(z) = K_1 + z/(z-1)K_2$ where K_1 and K_2 are constant matrices with elements $(K_1)_{ik} = \alpha_{ik}$, $(K_2)_{ik} = \beta_{ik}$. Show that this "proportional plus summation controller" can be represented by the model

$$x_{k+1}^c = x_k^c + e_k, \qquad u_k = (K_1 + K_2)e_k + K_2 x_k^c, \qquad k \geq 0$$

which has the characteristic polynomial $(z-1)^m$. Finally, confirm that the system is always controllable and is also observable if $m \leq l$ and rank $K_2 = m$.

(4) Show that the z-transfer function of the discrete system

$$x_{k+1} = \begin{bmatrix} 0.25 & 0.25 \\ 0.25 & 0.25 \end{bmatrix} x_k + \begin{bmatrix} 0 \\ 1 \end{bmatrix} u_k, \qquad y_k = (1 \quad 0) x_k$$

takes the form $G(z) = 0.25/z(z-0.5)$. Hence, show that the output response from zero initial conditions to a unit step input $u_k = 1$ $(k \geq 0)$ is

$$y_k = \tfrac{1}{2}(1 - (0.5)^{k-1}), \; k \geq 1$$

(5) Formulate and prove the equivalent results of problems (6) and (7) in Chapter three in the case of discrete systems.

(6) In the light of Theorem 4.1.1 and the analysis of Section 3.5, show that a "non-interacting" controller for a m-input/m-output discrete plant with $m \times m$

z-transfer function matrix $G(z)$ takes the form

$$K(z) = G^{-1}(z)\begin{bmatrix} q_1(z) & 0 & \dots & 0 \\ 0 & & & \vdots \\ \vdots & & & 0 \\ 0 & \dots & 0 & q_m(z) \end{bmatrix}$$

where $q_1(z), q_2(z), \dots, q_m(z)$ are scalar z-transfer functions. Show also that the closed-loop system z-transfer function matrix is diagonal with diagonal elements $h_i(z) = q_i(z)/(1 + q_i(z))$, $1 \leqslant i \leqslant m$.

(7) Using the results of problem (6), choose a (non-interacting) controller for the system

$$x_{k+1} = \begin{bmatrix} 0.25 & 0.25 \\ 0.25 & 0.25 \end{bmatrix} x_k + \begin{bmatrix} 1 & 1 \\ 0 & 1 \end{bmatrix} u_k$$

$$y_k = \begin{bmatrix} 1 & 0 \\ 1 & 1 \end{bmatrix} x_k$$

such that the closed-loop system has z-transfer function matrix

$$H_c(z) = \begin{bmatrix} \dfrac{1-k_1}{z-k_1} & 0 \\ 0 & \dfrac{1-k_2}{z-k_2} \end{bmatrix}$$

Verify that this controller is a proportional plus summation controller in the sense of problem (3).

(8) Verify that the system

$$G(z) = \frac{1}{(z-0.7)(z-0.9)}\begin{bmatrix} z-0.8 & z-0.7 \\ 0.1 & z-0.7 \end{bmatrix}$$

is a first-order lag with matrices

$$A_0 = \begin{bmatrix} 1 & -1 \\ 0 & 1 \end{bmatrix}, \qquad A_1 = \begin{bmatrix} 0.3 & -0.3 \\ -0.1 & 0.2 \end{bmatrix}$$

and characteristic polynomial $(z-0.7)(z-0.9)$. Using the proportional controller

$$K(z) = A_0(1-k) - A_1 = \begin{bmatrix} 0.7-k & k-0.7 \\ 0.1 & 0.8-k \end{bmatrix}$$

show that the closed-loop z-transfer function matrix takes the form

$$H_c(z) = \frac{(1-k)}{(z-k)} \begin{bmatrix} 1-\dfrac{0.2}{1-k} & \dfrac{0.1}{1-k} \\ \dfrac{0.1}{1-k} & 1-\dfrac{0.2}{1-k} \end{bmatrix}$$

Investigate the steady state and interaction effects in the system as k varies in the interval $-1 < k < 1$.

(9) Suppose that the continuous $m \times m$ first order lag (4.48) subjected to synchronized piecewise constant inputs and sampled outputs gives rise to the discrete $m \times m$ first order lag (4.45) with inverse z-transfer function matrix of the form of (4.46). Verify that

$$\begin{aligned} A_0^{-1}A_1 &= C_p(I_m - e^{A_p h})C_p^{-1} \\ &= I_m - e^{C_p A_p C_p^{-1} h} \end{aligned}$$

If the system is to be controlled by the proportional controller $K(z) = A_0(1-k) - A_1$, consider how this knowledge of the dependence of $A_0^{-1}A_1$ on h could be used to choose the sample interval h. Apply your results to the continuous system

$$\frac{dx(t)}{dt} = \begin{bmatrix} 0 & 1 \\ 0 & 1 \end{bmatrix} x(t) + \begin{bmatrix} 1 & 1 \\ 1 & 0 \end{bmatrix} u(t)$$

$$y(t) = \begin{bmatrix} 0 & 1 \\ 1 & 1 \end{bmatrix} x(t)$$

to choose the sample interval h such that the resulting closed-loop system (assumed to be dead-beat, i.e. $k = 0$) has peak interaction effects in the response from zero initial conditions to unit step demands of less than 0.1.

(10) Formulate and prove the discrete system equivalent of the results of problem (13) in Chapter three.

Remarks and Further Reading

This chapter has been a complete parallel of Chapter three and the reader is encouraged to compare the two chapters in some detail. Overall he will find that the material (in formal mathematical terms) is virtually identical and from this he must (correctly) conclude that many of the design techniques outlined in the reading list at the end of that chapter will carry over, with little change, to the discrete case. There are two important differences that are worthy of mention

(i) the Nyquist D-contour is replaced by the unit circle in frequency response stability studies;

(ii) from Chapter three we know that for continuous systems, the loop gains are parameters that have an important effect on system performance. In contrast, in the case of discrete systems, the most important parameter is probably sample rate, as illustrated by the analysis of first order lags in Section 4.2.

Finally, the reader can brush up his knowledge of z-transforms by reference to Cappelini *et al.* (1978), Cadzow (1973), Lindorff (1965), Bishop (1975) or Jury (1964). More information on discrete first order lags and their use as approximate plant models for design purposes can be found in Owens (1979) and Owens and Chotai (1980, 1981).

5. Unconstrained Optimal Control: an Introduction

It is an important and valuable characteristic of *frequency domain* methods of controller design that, given time and experience, the designer can very often use pole, zero and frequency response information to design systematically a feedback system with satisfactory stability and transient performance properties *in the time-domain*. It is clear, however, that these techniques cannot deal exactly with control problems involving one or more of the following features:

(i) "hard" control constraints on the magnitude of the elements of the plant input vector, e.g. saturation constraints of the form $|u_k(t)| \leqslant M_k$, $1 \leqslant k \leqslant l$, that typically occur in valve actuators;

(ii) limits on the time available to complete the control task. For example, in military pursuit systems, a missile must *rendezvous* with its quarry before the quarry reaches its target!

(iii) tight accuracy requirements;

(iv) the desire to improve aspects of performance that are difficult to analyse in the frequency domain. For example, how would one improve fuel consumption for a vehicle on a specified journey?

In fact, these more precise constraints and criteria of system performance require that the design process be undertaken in the time domain!

This chapter introduces the mathematical machinery of OPTIMAL CONTROL as a means of approaching the solution of the above problems. The philosophy of the approach is that the design problem is formulated mathematically in such a manner that it is directly solvable by a digital computer. This has obvious advantages at the design stage but it does mean that the engineer, for the time being at least, forfeits all decisions on the structure and complexity of the resulting control scheme. In fact, in general, he cannot even guarantee that the controller takes a feedback form! Leaving these difficulties to one side, however, the information generated by optimal control analysis can yield, at minimum, valuable insight into control potential.

Although the treatment described below and in following chapters is not comprehensive, the standard of mathematical rigour is, on the whole, high and most of the essential results forming the cornerstones of applications studies are included. The reader is referred to the reading list for applications and further developments.

5.1 Formulation of the Control Problem

We begin our discussion of optimal control at the foundation of all control theory: the formulation of the control problem. The essential elements of the control problem are:

(i) a *dynamical system* to be controlled;
(ii) a *specified objective* for the system;
(iii) a set of *admissible controllers*;
(iv) a *means of measuring performance* to test the effectiveness of any given control strategy.

The formulation of the control problem is the translation of these elements into a mathematical framework. We will examine them, one by one.

The dynamical system. This is assumed, in general, to be represented by a state variable model of the general form (1.26)–(1.28), although our attention will be restricted to the case of the linear time-invariant system

$$\frac{\mathrm{d}x(t)}{\mathrm{d}t} = Ax(t) + Bu(t), \qquad x(0) = x_0$$

$$y(t) = Cx(t) + Du(t) \tag{5.1}$$

The system objective. The objective of the control system is to complete some specified task. We assume that this is expressed as a combination of

(a) constraints on the output or state variables at specified times;
(b) limits to the time available to complete the control task.

For example, the objective of an aircraft is, say, to fly from London to Paris with a journey time of between 25 and 30 minutes. In this case the system objective is to transfer the aircraft from a given initial condition to a given final condition in the specified available time.

Admissible controllers. Control input signals in practice are obtained from devices capable of providing only a limited amount of energy. The possible inputs to the system are hence limited! The class of controls that can be considered for the given problem is termed the set of admissible controls. We will restrict attention in the remainder of this text to admissible controls defined by constraints on the relationship between, and the magnitude of, the elements of the input vector $u(t)$ at each point in time. For example, if each input $u_k(t)$ represents the opening of a control valve, and the valve strokes are limited, it is clear that all admissible inputs must satisfy constraints of the form

$$-M_k^{(1)} \leqslant u_k(t) \leqslant M_k^{(2)}, \qquad 1 \leqslant k \leqslant l \tag{5.2}$$

when $M_k^{(i)}$ are scalars, $1 \leqslant k \leqslant l, i = 1, 2$, or, in the case of $M_k^{(1)} = M_k^{(2)} (= M_k)$,

$$|u_k(t)| \leqslant M_k, \qquad 1 \leqslant k \leqslant l \tag{5.3}$$

For notational purposes in our general discussion (where the form of the constraints need not be specified) the presence of constraints will be indicated by the symbolism

$$u(t) \in \Omega(t) \tag{5.4}$$

which is to be interpreted as meaning that the elements of the input vector at time t are subject to constraints represented by the symbol $\Omega(t)$.

Performance assessment. In contrast to frequency domain design methods where performance is assessed by visual inspection of the transient characteristics of the closed-loop system on completion of the design, the optimal control methodology demands that the means of assessing performance is quantified in the form of a *performance index* or *cost functional* at the beginning of the design exercise. More precisely, the design engineer is asked to construct a numerical index of performance $J(u)$ whose value reflects the quality of any admissible control function $u(t)$ in accomplishing the system objective. This performance index is assumed to have the property that its numerical value decreases as the quality of control action increases, i.e. if $u^{(1)}(t)$ and $u^{(2)}(t)$ are admissible controls accomplishing the system objective, then the relation

$$J(u^{(1)}) > J(u^{(2)}) \tag{5.5}$$

implies that the control input $u^{(2)}$ is "better" than the input $u^{(1)}$.

Given the above the search for the "best" controller or, equivalently, the controller producing the highest performance is a well defined mathematical problem, namely, that of calculating the controller that produces the smallest value of J. More precisely, we have the following *optimal control problem*:

> Find an admissible controller $u^*(t)$ that simultaneously ensures the completion of the system objective and the minimization of the performance index $J(u)$, i.e. symbolically,
>
> $$J(u^*) = \min J(u) \tag{5.6}$$
>
> where the minimization is performed over all admissible inputs $u(t)$ that ensure the completion of the system objective.

The admissible controller u^* is called an "optimal controller" for the system.

The remainder of this section is devoted to the construction of illustrative physical examples of the abstract concepts outlined above.

EXAMPLE 5.1.1 (minimum fuel control of a land vehicle). Consider the land vehicle illustrated in Fig. 59 moving on a horizontal straight road from left to right from a point a at rest at the time $t = 0$ and arriving at the point b at rest at a specified time $t = T$. The control problem is simply to drive from a to b in the specified time and, in such a manner that the total fuel consumed in the journey is minimized.

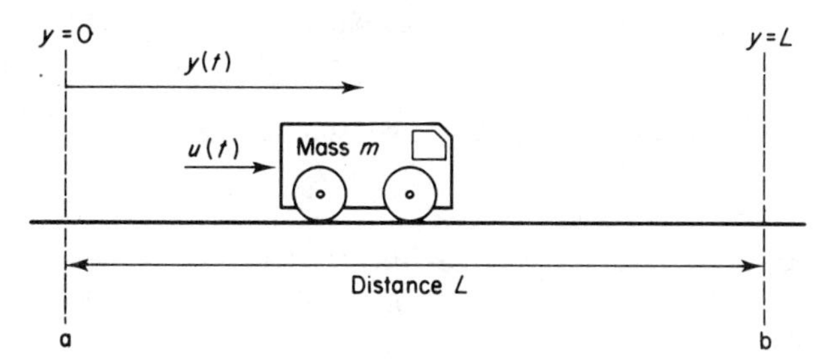

Fig. 59. A simple land vehicle system.

The *system model* has the simple form

$$m\frac{d^2y(t)}{dt} = u(t) \tag{5.7}$$

where $u(t)$ is the net force acting on the vehicle (regarded as an input). Using the state variables $x_1(t) = y(t)$ and $x_2(t) = dy(t)/dt$, this model has the state-variable form

$$\frac{dx(t)}{dt} = \begin{bmatrix} 0 & 1 \\ 0 & 0 \end{bmatrix} x(t) + \begin{bmatrix} 0 \\ \frac{1}{m} \end{bmatrix} u(t) \tag{5.8}$$

The system initial condition is

$$x(0) = \begin{bmatrix} 0 \\ 0 \end{bmatrix} \tag{5.9}$$

and the *control objective* is to drive the system to the condition

$$x(T) = \begin{bmatrix} L \\ 0 \end{bmatrix} \tag{5.10}$$

at the fixed time $t = T$. It is assumed that the available acceleration is limited by *constraints* on the input of the form

$$-M_1 \leqslant u(t) \leqslant M_2, \qquad 0 \leqslant t \leqslant T \tag{5.11}$$

where M_1 and M_2 are constants. Equation (5.11) defines the class of admissible controllers.

In general, there are an infinite number of ways to drive from point a to point b in the specified time using an admissible control strategy. In mathematical terms there are an infinite number of control functions $u(t)$ satisfying (5.11) that generate solutions of (5.8) satisfying the boundary conditions (5.9) and (5.10). The optimal control problem is to choose, from this infinity of candidates, a controller that minimizes the fuel consumption. To do this, suppose that the scalar function

$$L_f(x(t), u(t), t) = \text{rate of fuel consumption at time } t \text{ with input } u(t) \text{ and states } x(t) \tag{5.12}$$

is known. It is then clear that the total fuel consumed in the period $0 \leqslant t \leqslant T$ is simply the integral of L. More precisely, the performance index required is simply

$$J(u) = \text{total fuel consumed} = \int_0^T L_f(x(t), u(t), t)\mathrm{d}t \tag{5.13}$$

It is now self-evident that the problem of choosing a driving strategy to minimize fuel consumption on the specified journey is equivalent to the optimal control problem of finding a controller $u^*(t)$ satisfying (5.8)–(5.11) and minimizing the performance index (5.13). A solution to this particular problem is deferred to the next chapter in Section 6.2.3.

EXAMPLE 5.1.2 (minimum energy control of a mine-winder). Consider the mine-winder illustrated in Fig. 60. The objective of the control system is to lower the cage in a controlled way from rest at the top of the shaft at time $t = 0$ down to a rest position at the bottom of the shaft at a specified time $t = T$. It is required that the controller operates in such a manner as to minimize total energy losses in the winding engine in this period.

A simplified system model has the form

$$m\frac{\mathrm{d}^2 y(t)}{\mathrm{d}t^2} = u_0(t) - mg \tag{5.14}$$

where g is the acceleration due to gravity and $u_0(t)$ is the cage force

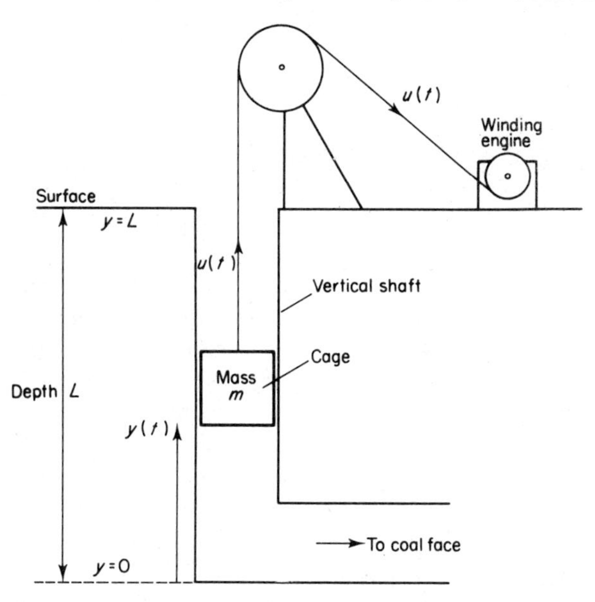

Fig. 60. Schematic of a simple mine-winder.

developed by the winding engine. Choosing state variables as in Example 5.1.1 yields the state variable model

$$\frac{\mathrm{d}x(t)}{\mathrm{d}t} = \begin{bmatrix} 0 & 1 \\ 0 & 0 \end{bmatrix} x(t) + \begin{bmatrix} 0 \\ \frac{1}{m} \end{bmatrix} (u_0(t) - mg) \tag{5.15}$$

which is identical to (5.8) with $u(t)$ replaced by $u_0(t) - mg$. The system initial and final conditions take the form

$$x(0) = \begin{bmatrix} L \\ 0 \end{bmatrix}, \qquad x(T) = \begin{bmatrix} 0 \\ 0 \end{bmatrix} \tag{5.16}$$

and it is assumed that the admissible controllers are limited by boundedness constraints of the form given in (5.11).

The problem of minimum energy losses in the winding engine is now formulated as an optimal control problem by defining the (assumed known) scalar function.

$$L_e(x(t), u_0(t), t) = \text{rate of energy dissipation in the winding gear at time } t, \text{ with input } u_0(t) \text{ and the system in the state } x(t). \tag{5.17}$$

and the performance index

$$J(u_0) = \text{total energy dissipated} = \int_0^T L_e(x(t), u_0(t), t)\mathrm{d}t \qquad (5.18)$$

The optimal controller $u_0^*(t)$ satisfies (5.11), (5.15) and (5.16) and simultaneously minimizes the performance index (5.18). This problem will be examined later in this chapter and in the next chapter.

EXAMPLE 5.1.3 (a problem involving compromise). The reader should not find it difficult to accept that most design situations contain the need to reach a compromise between two conflicting design objectives. The problem of formulation of an optimal problem is no exception to this rule. We will illustrate this situation by consideration of the first order system

$$\frac{\mathrm{d}x(t)}{\mathrm{d}t} = -x(t) + u(t) \qquad (5.19)$$

with the initial condition $x(0) = 1$. There are no other constraints on the system state. In contrast to the previous two examples, the system objective is rather ill defined: the objective is to control the system on a fixed time-interval $0 \leqslant t \leqslant T$ in such a manner that the state $x(t)$ remains "small" and yet the control input required is "not too large".

The problem of making $x(t)$ "small" could, we suppose, be represented by the introduction of the performance index

$$J_1(u) = \frac{1}{2}\int_0^T (x(t))^2\,\mathrm{d}t \qquad (5.20)$$

Certainly, if $J_1(u)$ is small, then $x(t)$ is small (in a least-squares sense) in the interval of interest. In contrast, the need to use control inputs that are "not too large" could be represented by the introduction of a performance index

$$J_2(u) = \frac{1}{2}\int_0^T (u(t))^2\,\mathrm{d}t \qquad (5.21)$$

It is important to recognize the fact that we cannot simultaneously minimize $J_1(u)$ and $J_2(u)$ as the minimization of $J_1(u)$ requires large control signals, whereas the minimization of $J_2(u)$ requires small control signals.

The way out of this dilemma is to attempt the construction of a performance index whose minimization will produce a compromise between the two conflicting objectives. A standard approach is to construct the compromise performance index $J(u)$ as a *convex combination* of $J_1(u)$ and $J_2(u)$, namely

$$J(u) = \lambda J_1(u) + (1-\lambda)J_2(u) = \frac{1}{2}\int_0^T (\lambda x^2 + (1-\lambda)u^2)\mathrm{d}t \quad (5.22)$$

where λ is some scalar parameter in the range $0 \leqslant \lambda \leqslant 1$. If $\lambda = 1$, then $J(u) = J_1(u)$ and, if $\lambda = 0$, we have $J(u) = J_2(u)$, i.e. by suitable choice of λ we can minimize either $J_1(u)$ or $J_2(u)$. A choice of λ between the values of 0 and 1 will generate a compromise between these two extremes. An actual choice of λ must be made on a trial and error basis.

The above examples are designed only to give a flavour of the source and structure of optimal control problems. In the first two cases, the formulation is relatively straightforward but, in the case of Example 5.1.3, the reader will note that the formulation of the optimal control problem can be a difficult matter involving an element of uncertainty that can only be resolved by "old-fashioned" trial-and-error methods.

The remainder of this chapter is devoted primarily to the detailed analysis of two general optimal control problems of particular interest in applications. The analysis is important in this sense but the development is also designed to generate many of the fundamental mathematical concepts of optimal control in a simple context. An important restriction, however, is that the control input vector is assumed not to be subject to any constraints. The inclusion of constraints is left to the next chapter.

5.2 The Linear Quadratic Control Problem

The linear quadratic problem discussed in this section is probably the most important general problem for the purposes of applications studies as it is the only optimal control problem that gives rise to an optimal control law of linear state feedback form. The general problem can be defined as follows: given a linear, time-invariant system

$$\frac{\mathrm{d}x(t)}{\mathrm{d}t} = Ax(t) + Bu(t) \quad (5.23)$$

with a specified initial condition $x(0) = x_0$, calculate the control input vector $u(t)$ on the time interval $0 \leqslant t \leqslant t_f$ (t_f being a fixed terminal time) that minimizes the quadratic performance index

$$J(u) = \frac{1}{2}x^T(t_f)Fx(t_f) + \frac{1}{2}\int_0^{t_f} \{x^T(t)Qx(t) + u^T(t)Ru(t)\}\mathrm{d}t \quad (5.24)$$

It is assumed that the control input vector is unconstrained, i.e. any control input is an admissible controller.

The above problem is, in essence, a generalization of the problem of Example 5.1.3, the aim of the optimal controller being to reach a compromise between the separate objectives:

(i) ensuring that the elements of the state vector at the terminal time $t = t_f$ are "small";

(ii) ensuring that the elements of $x(t)$ are "acceptably small" on the whole of the time interval $0 \leqslant t \leqslant t_f$;

(iii) ensuring that the elements of the control vector $u(t)$ do not take "excessively high" values in the interval $0 \leqslant t \leqslant t_f$.

The degree of compromise is represented by the $n \times n$ constant matrices F and Q and the $l \times l$ constant matrix R. These matrices cannot take arbitrary values! We will constrain them all to be *symmetric* (see Exercise 5.2.1) with F and Q *positive semi-definite* and R *positive definite* (see Exercise 5.2.2), i.e.

$$Q = Q^T \geqslant 0, \qquad F = F^T \geqslant 0, \qquad R = R^T > 0 \tag{5.25}$$

These assumptions guarantee the existence of an acceptable solution to the problem.

EXERCISE 5.2.1. If M is a constant $r \times r$ matrix then the function $f(x) = x^T M x$ is termed a *quadratic form* in the $r \times 1$ vector x. Verify that

$$x^T M x \equiv x^T \frac{(M + M^T)}{2} x \tag{5.26}$$

and that $M_0 = (M + M^T)/2$ is a symmetric matrix. Hence deduce that $f(x) = x^T M_0 x$. The moral of this story is that, given any quadratic form $x^T M x$, we can always replace it with a quadratic form $x^T M_0 x$ where M_0 is symmetric. Hence there is no loss of generality in the above analysis in supposing F, Q and R to be symmetric.

EXERCISE 5.2.2. A $r \times r$ matrix M is said to be positive definite (written $M > 0$) if the quadratic form

$$x^T M x > 0 \text{ for all non-zero } x \tag{5.27}$$

Using the result of Exercise 5.2.1, we can suppose, without loss of generality, that $M = M^T$. Using the properties of symmetric matrices, write

$$M = V \begin{bmatrix} m_1 & & & \\ & m_2 & & \\ & & \ddots & \\ & & & m_r \end{bmatrix} V^T \tag{5.28}$$

EXERCISE 5.2.2. contd.

where m_i, $1 \leqslant i \leqslant r$, are the r real eigenvalues of M, and V is an $r \times r$ *orthogonal* eigenvector matrix (i.e. $VV^T = V^TV = I_r$). Hence deduce that M is positive definite if, and only if, all its eigenvalues $m_i > 0$, $1 \leqslant i \leqslant m$. (Hint: write the vector $x = Vz$). Conclude that all positive-definite matrices are non-singular.

The matrix M is said to be positive semi-definite (written $M \geqslant 0$) if $x^TMx \geqslant 0$ for all vectors $x \neq 0$. Verify that the above theory still holds but that M is positive semi-definite if, and only if, all eigenvalues $m_i \geqslant 0$, $1 \leqslant i \leqslant m$. Note, however, that positive semi-definite matrices can be singular.

EXERCISE 5.2.3. An important variant on the performance index (5.24) requires only that the system outputs remain acceptably small on the time interval of interest, i.e. we choose

$$J(u) = \frac{1}{2} y^T(t_f)F_0 y(t_f) + \frac{1}{2} \int_0^{t_f} \{y^T(t)Q_0 y(t) + u^T(t)Ru(t)\}\mathrm{d}t \qquad (5.29)$$

where $y(t) = Cx(t) + Du(t)$ is the system output, $F_0 = F_0^T \geqslant 0$, $Q_0 = Q_0^T \geqslant 0$ and $R = R^T > 0$. In the case of $D = 0$, verify that (5.29) reduces to (5.24) with

$$Q = C^TQ_0C, \qquad F = C^TF_0C \qquad (5.30)$$

5.2.1 Optimal control on a finite time interval

Consider the solution of the linear quadratic optimal control problem in the case of t_f finite. There are a number of approaches to this problem varying in structure and complexity. The approach taken here is that of the inspired guess! More precisely, let us look for a solution to our problem of the linear time-varying state feedback form

$$u^*(t) = -R^{-1}B^TK(t)x^*(t) \qquad (5.31)$$

where $x^*(t)$ is the state trajectory resulting from the use of $u^*(t)$ and $K(t)$ is an $n \times n$ symmetric matrix of time-varying elements. Substituting into (5.23) it is clear that $x^*(t)$ is the solution of the time-varying linear equation

$$\frac{\mathrm{d}x^*(t)}{\mathrm{d}t} = (A - BR^{-1}B^TK(t))x^*(t), \qquad x^*(0) = x_0 \qquad (5.32)$$

The problem now, of course, is to find $K(t)$! (Note: those readers who have no interest in the proof may wish simply to note that the correct choice is obtained by solving (5.40) and (5.41).)

Let $u(t)$ be any other control input generating the state trajectory $x(t)$. Then we can express it in the form

$$u(t) = -R^{-1}B^TK(t)x(t) + v(t) \tag{5.33}$$

It is clear that $u = u^*$ if, and only if, $v(t) \equiv 0$, and substituting (5.33) into (5.23) indicates that $x(t)$ is the solution of the equation

$$\frac{\mathrm{d}x(t)}{\mathrm{d}t} = \{A - BR^{-1}B^TK(t)\}x(t) + Bv(t)$$

$$x(0) = x_0 \tag{5.34}$$

The analysis proceeds by the evaluation of the performance index corresponding to u. First, note that

$$\begin{aligned} u^T(t)Ru(t) &= (-R^{-1}B^TK(t)x(t) + v(t))^TR(-R^{-1}B^TK(t)x(t) + v(t)) \\ &= x^T(t)K(t)BR^{-1}B^TK(t)x(t) + v^T(t)Rv(t) \\ &\quad - (Bv(t))^TK(t)x(t) - x^T(t)K(t)(Bv(t)) \end{aligned} \tag{5.35}$$

where we have used the assumed symmetry of $K(t)$ and the relation $(R^{-1})^T = (R^T)^{-1}$. Substituting for $Bv(t)$ from (5.34) yields the relation, after a little manipulation,

$$\begin{aligned} u^T(t)Ru(t) &= x^T(t)\{-K(t)BR^{-1}B^TK(t) + A^TK(t) + K(t)A\}x(t) \\ &\quad - \left(\frac{\mathrm{d}x(t)}{\mathrm{d}t}\right)^T K(t)x(t) - x^T(t)K(t)\frac{\mathrm{d}x(t)}{\mathrm{d}t} + v^T(t)Rv(t) \end{aligned} \tag{5.36}$$

Next, note, from elementary calculus, that

$$\begin{aligned} x^T(t_f)K(t_f)x(t_f) - x^T(0)K(0)x(0) &= \int_0^{t_f} \frac{\mathrm{d}}{\mathrm{d}t}(x^T(t)K(t)x(t))\mathrm{d}t \\ = \int_0^{t_f} &\left(\left(\frac{\mathrm{d}x(t)}{\mathrm{d}t}\right)^T K(t)x(t) + x^T(t)\left(\frac{\mathrm{d}K(t)}{\mathrm{d}t}\right)x(t) + x^T(t)K(t)\frac{\mathrm{d}x(t)}{\mathrm{d}t}\right)\mathrm{d}t \end{aligned} \tag{5.37}$$

and finally, write

$$\begin{aligned} x^T(t_f)Fx(t_f) &= x^T(t_f)(F - K(t_f))x(t_f) + (x^T(t_f)K(t_f)x(t_f) \\ &\quad - x^T(0)K(0)x(0)) + x^T(0)K(0)x(0) \end{aligned} \tag{5.38}$$

Using (5.36)–(5.38) in (5.24) yields the following expression for the value of the performance index $J(u)$,

$$J(u) = \frac{1}{2}x^T(0)K(0)x(0) + \frac{1}{2}\int_0^{t_f} v^T(t)Rv(t)\mathrm{d}t + \frac{1}{2}\int_0^{t_f} x^T(t) \times$$

$$\left\{Q - K(t)BR^{-1}B^{T}K(t) + A^{T}K(t) + K(t)A + \frac{\mathrm{d}K(t)}{\mathrm{d}t}\right\}x(t)\mathrm{d}t$$

$$+\frac{1}{2}x^{T}(t_f)\{F - K(t_f)\}x(t_f) \tag{5.39}$$

This expression is valid for any choice of symmetric $K(t)$. We will choose $K(t)$ to eliminate the dependence of $J(u)$ on the state trajectory $x(t)$ for $t > 0$. More precisely, choose $K(t)$ to be the solution of the *matrix Riccati equation*

$$Q - K(t)BR^{-1}B^{T}K(t) + A^{T}K(t) + K(t)A + \frac{\mathrm{d}K(t)}{\mathrm{d}t} = 0 \tag{5.40}$$

with the terminal boundary condition

$$K(t_f) = F \tag{5.41}$$

In this case (5.39) reduces to

$$J(u) = \frac{1}{2}x^{T}(0)K(0)x(0) + \frac{1}{2}\int_0^{t_f} v^{T}(t)Rv(t)\mathrm{d}t \tag{5.42}$$

and, taking $v(t) \equiv 0$,

$$J(u^*) = \tfrac{1}{2}x^{T}(0)K(0)x(0) \tag{5.43}$$

It remains only to check that the feedback law (5.31) with $K(t)$ given as the solution of (5.40) and (5.41) is an optimal controller. This follows directly from the assumption that R is positive-definite as this implies that $v^{T}(t)Rv(t) \geqslant 0$ for all function vectors $v(t)$ and hence that $\frac{1}{2}\int_0^{t_f} v^{T}(t)Rv(t)\mathrm{d}t \geqslant 0$, i.e.

$$J(u) \geqslant \tfrac{1}{2}x^{T}(0)K(0)x(0) = J(u^*) \tag{5.44}$$

The optimality of u^* follows from this relation as u was chosen arbitrarily.

The above analysis can be summarized by the statement that the optimal controller for the finite-time linear quadratic optimal control problem is a time-varying linear state feedback control law of the form given in (5.31), where the $n \times n$ symmetric Riccati matrix $K(t)$ is the solution of the matrix Riccati equation (5.40) with boundary condition (5.41). We make the following observations.

(a) The solution $K(t)$ is independent of the initial state $x(0)$ and hence the optimal control law (5.31) can be implemented as the feedback configuration illustrated in Fig. 61, the resulting system response being optimal for all initial conditions. Implementation does, however, require measurement of all system states.

(b) The solution $K(t)$ is symmetric for all t in the range $0 \leqslant t \leqslant t_f$, i.e.

$$K(t) \equiv K^{T}(t), \qquad 0 \leqslant t \leqslant t_f \tag{5.45}$$

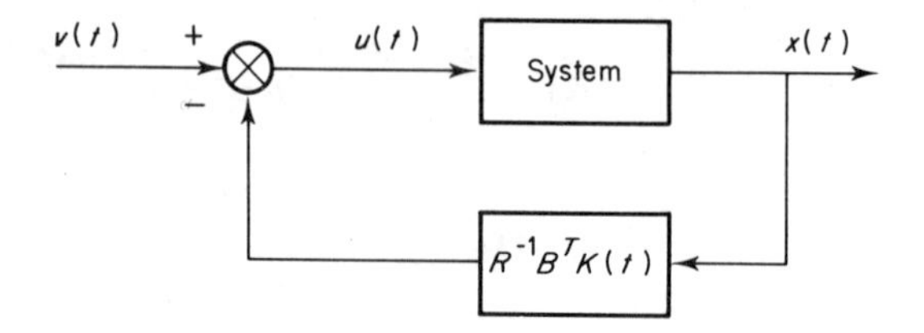

Fig. 61. Optimal feedback system.

(Note: to prove this, take the transpose of (5.40) and (5.41) and verify that $K^T(t)$ is also a solution of these equations. Equation (5.45) then follows from the uniqueness of the solution.)

(c) The symmetric matrix $K(0)$ is positive semi-definite. To prove this note that relations (5.25) ensure that $J(u) \geqslant 0$ for all inputs $u(t)$ and hence (equation (5.43)) that $J(u^*) = \frac{1}{2}x^T(0)K(0)x(0) = \min J(u) \geqslant 0$ for all initial conditions $x(0)$.

(d) The Riccati equation is a non-linear matrix differential equation with a terminal boundary condition. The nonlinear structure precludes analytic solution except in one special case (see Example 5.2.1) and hence it normally must be solved numerically. There is no problem in principle here – one simply has to use standard numerical integration procedures, starting at the final time $t = t_f$ and integrating *backwards* in time by taking negative time steps. The problem lies in the number of equations to be solved. As $K(t)$ is an $n \times n$ matrix, (5.40) is obviously equivalent to n^2 nonlinear ordinary differential equations. The symmetry property of $K(t)$ does reduce this number to $n(n+1)/2$ but the problem is obvious: if $n = 20$, say, it is necessary to solve simultaneously $20 \times 21/2 = 210$ differential equations. Not an easy problem!

EXAMPLE 5.2.1. To illustrate the application of the above theory, consider the system of Example 5.1.3:

$$\frac{dx(t)}{dt} = -x(t) + u(t), \qquad x(0) = 1 \tag{5.46}$$

with performance index (5.22),

$$J(u) = \frac{1}{2}\int_0^{t_f} (\lambda x^2(t) + (1-\lambda)u^2(t))dt \tag{5.47}$$

where $0 \leqslant \lambda < 1$ and the control input is subject to no constraints. It is clear that $n = l = 1$, $A = -1$, $B = 1$, $Q = \lambda \geqslant 0$, $R = 1 - \lambda > 0$ and

$F = 0$. The optimal controller for the system is the time-varying feedback controller (see (5.31))

$$u(t) = -\frac{1}{(1-\lambda)} K(t)x(t) \tag{5.48}$$

where the scalar gain function $K(t)$ is the solution of the Riccati equation (see (5.40) and (5.41))

$$\lambda - \frac{K^2(t)}{1-\lambda} - 2K(t) + \frac{\mathrm{d}K(t)}{\mathrm{d}t} = 0, \qquad K(t_f) = 0 \tag{5.49}$$

(Note: for notational convenience we have dropped the * notation for optimal control and state trajectories.) In this particular case, (5.49) can be solved analytically by defining $\psi(t) = K(t) + 1 - \lambda$ when (5.49) reduces to

$$\frac{\mathrm{d}\psi(t)}{\mathrm{d}t} = \frac{1}{(1-\lambda)} \psi^2(t) - 1, \qquad \psi(t_f) = 1 - \lambda \tag{5.50}$$

Setting $\epsilon^2 = 1 - \lambda$, $\epsilon > 0$, it is clear that (5.50) can be written in the form

$$\frac{\mathrm{d}\psi(t)}{\mathrm{d}t} = \frac{1}{\epsilon^2}(\psi^2(t) - \epsilon^2) = \frac{1}{\epsilon^2}(\psi(t) + \epsilon)(\psi(t) - \epsilon) \tag{5.51}$$

leading to the relation

$$\frac{\mathrm{d}}{\mathrm{d}t} \log_e \frac{(\psi(t) - \epsilon)}{(\psi(t) + \epsilon)} = \frac{\mathrm{d}\psi(t)}{\mathrm{d}t} \left\{ \frac{1}{(\psi(t) - \epsilon)} - \frac{1}{(\psi(t) + \epsilon)} \right\} = \frac{2}{\epsilon} \tag{5.52}$$

and hence

$$\frac{\psi(t) - \epsilon}{\psi(t) + \epsilon} = \gamma \mathrm{e}^{2t/\epsilon} \tag{5.53}$$

when γ is a real constant. Using the terminal boundary condition $\psi(t_f) = 1 - \lambda = \epsilon^2$, it follows that $\gamma = (\epsilon - 1)/(\epsilon + 1) \exp\{-2t_f/\epsilon\}$ and hence, solving (5.53) for $\psi(t) = K(t) + 1 - \lambda = K(t) + \epsilon^2$,

$$K(t) = \epsilon \frac{(1 + \epsilon + (\epsilon - 1)\mathrm{e}^{2(t-t_f)/\epsilon})}{(1 + \epsilon - (\epsilon - 1)\mathrm{e}^{2(t-t_f)/\epsilon})} - \epsilon^2 \tag{5.54}$$

For the purpose of implementation, the gain $K(t)$ could be computed off-line as above and the result stored digitally for use in the closed-loop system of Fig. 62(a). Alternatively, the initial condition $K(0)$ could be deduced from off-line calculations and the solution to the Riccati equation recomputed on-line in real-time as illustrated in Fig. 62(b). In both cases the action of the control system is only optimal in the time interval $0 \leqslant t \leqslant t_f$.

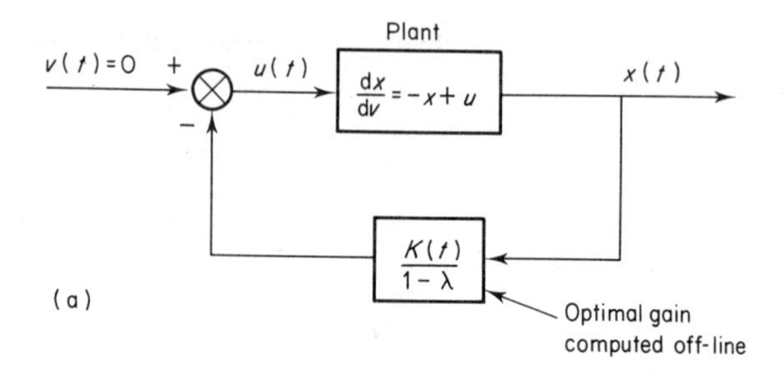

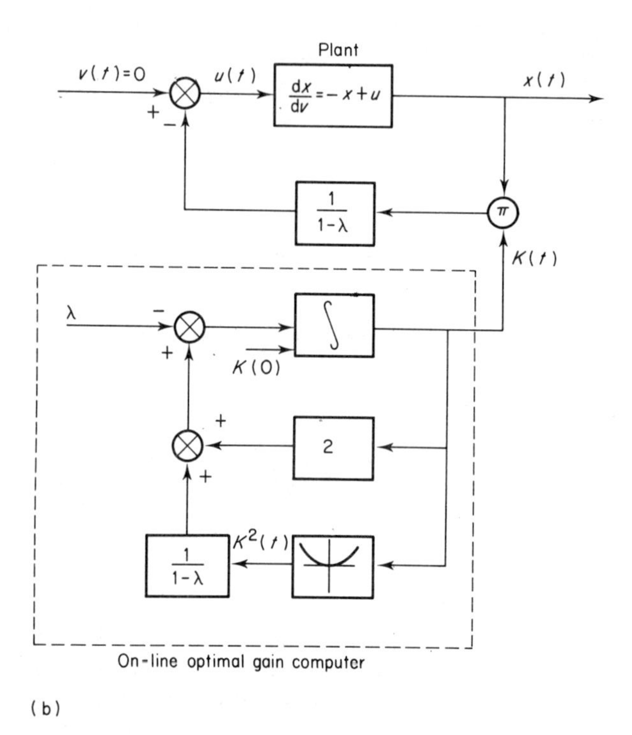

Fig. 62. Implementation of an optimal control scheme.

EXERCISE 5.2.4. In the above example, the overall time-varying gain is simply $(1-\lambda)^{-1}K(t) = \epsilon^{-2}K(t)$. Show that

$$\lim_{\epsilon \to 0} \epsilon(\epsilon^{-2}K(t)) = 1 \tag{5.55}$$

for any point in time $t < t_f$. Deduce that $(1-\lambda)^{-1}K(t) \simeq (1-\lambda)^{-1/2}, t < t_f$ if λ is very close to unity and hence that the optimal controller gain is high if

EXERCISE 5.2.4. contd.

the control term weighting in the performance index (5.47) is small. As a contrast, verify also that $\lim_{\epsilon \to 1} \epsilon^{-2} K(t) = 0$ for each time $0 \leqslant t \leqslant t_f$ and hence that control gains are small if the state weighting in (5.47) is small.

EXERCISE 5.2.5. For the system (5.46) with performance index

$$J(u) = \frac{1}{2} x(t_f)^2 + \frac{\lambda}{2} \int_0^{t_f} u^2(t)\mathrm{d}t, \qquad \lambda > 0 \tag{5.56}$$

and unconstrained input, show that the optimal controller is a feedback controller with time-varying gain $\lambda^{-1} K(t)$ where $K(t) = \psi(t) - \lambda$ and $\psi(t)$ is the solution of the nonlinear differential equation

$$\frac{\mathrm{d}\psi(t)}{\mathrm{d}t} = \frac{1}{\lambda}(\psi^2(t) - \lambda^2), \qquad \psi(t_f) = 1 + \lambda \tag{5.57}$$

Hence find the solution for $K(t)$.

As a final point in this section, it is emphasized that the approach taken above to the construction of the optimal controller is useful only in the linear quadratic problem. In other optimal control problems, the optimal controller does not have a "nice" explicit form. The structure that arises in general can be introduced by defining the $n \times 1$ "costate vector"

$$p(t) = K(t)x^*(t) \tag{5.58}$$

It is clear that the optimal controller (5.31) takes the form

$$u^*(t) = -R^{-1}B^T p(t) \tag{5.59}$$

and, using (5.41), that $p(t)$ is subject to the terminal boundary condition

$$p(t_f) = Fx^*(t_f) \tag{5.60}$$

We can obtain a differential equation representation of the costate by using (5.32) and (5.40) to verify that

$$\frac{\mathrm{d}p(t)}{\mathrm{d}t} = \frac{\mathrm{d}K(t)}{\mathrm{d}t} x^*(t) + K(t)\frac{\mathrm{d}x^*(t)}{\mathrm{d}t} = -A^T p(t) - Qx^*(t) \tag{5.61}$$

In summary, the optimal controller is a solution of the coupled differential equations

$$\begin{aligned} \frac{\mathrm{d}x^*(t)}{\mathrm{d}t} &= Ax^*(t) + Bu^*(t), \qquad x^*(0) = x_0 \\ u^*(t) &= -R^{-1}B^T p(t) \\ \frac{\mathrm{d}p(t)}{\mathrm{d}t} &= -A^T p(t) - Qx^*(t), \qquad p(t_f) = Fx^*(t_f) \end{aligned} \tag{5.62}$$

The presence of boundary conditions on the state and costate at different times gives rise to the description of this type of problem as a "two-point boundary value problem" (or, for short, TPBVP). More of these later!

EXERCISE 5.2.6. Verify that the TPBVP generating the optimal controller in Example 5.2.1 is

$$\frac{dx^*(t)}{dt} = -x^*(t) + u^*(t), \qquad x^*(0) = 1$$

$$u^*(t) = -\frac{1}{1-\lambda}p(t)$$

$$\frac{dp(t)}{dt} = p(t) - \lambda x^*(t), \qquad p(t_f) = 0 \tag{5.63}$$

In the case of Exercise 5.2.5, show that the second and third equations are replaced by $u^*(t) = -\lambda^{-1}p(t)$ and $dp(t)/dt = p(t)$, $p(t_f) = x^*(t_f)$ respectively.

We note finally that the Riccati matrix $K(t)$ can, in principle, be computed from (5.62) (see problem (5)).

5.2.2 *Optimal regulation on an infinite time interval*

In this section we attack the following two problems associated with the optimal control solutions discussed in Section 5.2.1, namely

(a) the need to compute the $n(n+1)/2$ independent time functions needed to define the Riccati matrix $K(t)$ is a major difficulty, particularly if the system state dimension n is large;

(b) the fact that, in applications where the optimal control is to have the role of a regulator to limit state deviations without excessive control inputs, the design engineer is interested in control over a long time interval characterized by letting $t_f \to +\infty$.

It is a fortunate fact that these two problems can be eliminated simultaneously by considering the optimal control problem

$$\frac{dx(t)}{dt} = Ax(t) + Bu(t), \qquad x(0) = x_0$$

$$J(u) = \frac{1}{2}\int_0^\infty (x^T(t)Qx(t) + u^T(t)Ru(t))\,dt$$

$$u(t) \text{ unconstrained}, \; Q = Q^T \geqslant 0 \text{ and } R = R^T > 0 \tag{5.64}$$

For technical reasons associated with the existence of a solution to this problem, we will also suppose that the plant is time-invariant and controllable. There is also no term outside the integral in this case as there is no sense in introducing terms in $x(\infty)$ when for any $u(t)$ such that $J(u)$ is finite, it is expected intuitively that $\lim_{t \to +\infty} x(t) = 0$.

The common sense background to the solution of the above problem is to note that, as the solution to the finite time problem has a linear state feedback form for all finite values of t_f, it is natural to expect the solution with $t_f = +\infty$ to have a linear state feedback form. It is also natural to expect this feedback law to be constant in time, as, after the optimal control has been operating for some time t_1 (say), it still has an infinite time interval left to continue to operate in an optimal manner. There is no *a priori* reason, therefore, to change the feedback gains.

Perhaps the easiest way to illustrate the principle is to consider the solution (5.54) of Example 5.2.1 as t_f varies and, in particular, as $t_f \to +\infty$. Figure 63 illustrates the behaviour of the solutions as t_f increases. Note that, at any particular point t in time, the Riccati solution $K(t)$ appears to converge to a specific value K_∞ that is independent of t. The value of K_∞ is easily obtained from (5.54) to be

$$K_\infty = \lim_{t_f \to \infty} K(t) = \epsilon(1 - \epsilon) > 0$$

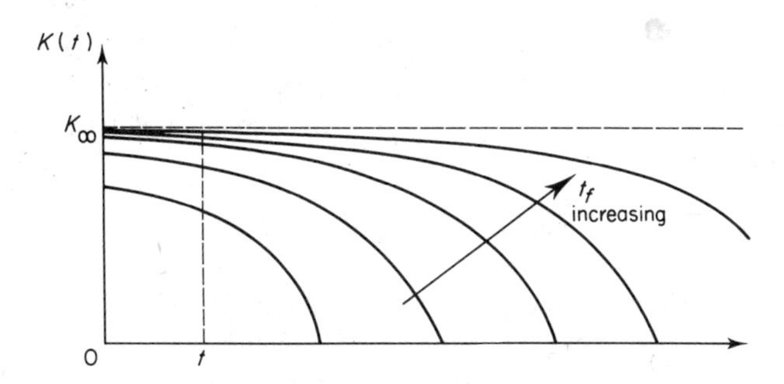

Fig. 63. Variation of $K(t)$ with final time t_f.

This result is only a special case of the general result that the solution of the infinite time linear quadratic optimal control problem takes the form of the linear, constant state-variable feedback controller

$$u^*(t) = -R^{-1}B^TK_\infty x^*(t) \tag{5.65}$$

where K_∞ is an $n \times n$, symmetric positive semi-definite matrix and $x^*(t)$ is the state trajectory resulting from the use of the optimal controller $u^*(t)$. As illustrated in Fig. 63, K_∞ can be obtained from the limiting operation

$$K_\infty = \lim_{t_f \to +\infty} K(t) \tag{5.66}$$

where $K(t)$ is the solution of (5.40) with the boundary condition $K(t_f) = 0$ and t is fixed but arbitrary. Alternatively, K_∞ can be obtained as illustrated in Fig. 64 by integrating the Riccati differential equation backwards in time from an arbitrary point t_f and noting that

$$K_\infty = \lim_{t \to -\infty} K(t) \tag{5.67}$$

is just the steady state value. Yet another approach is obtained by noting that the steady state of (5.40) can be characterized by setting $\mathrm{d}K(t)/\mathrm{d}t = 0$, i.e. K_∞ is the symmetric, positive-definite solution of the algebraic Riccati equation

$$Q - K_\infty BR^{-1}B^T K_\infty + A^T K_\infty + K_\infty A = 0 \tag{5.68}$$

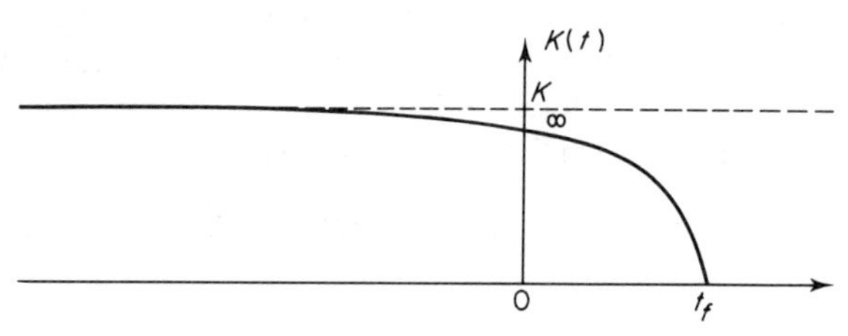

Fig. 64. Steady state solution of the Riccati equation by reverse-time integration.

The corresponding optimal value of the performance index is then obtained by using (5.66) with $t = 0$ in (5.43) to give

$$J(u^*) = \tfrac{1}{2} x^T(0) K_\infty x(0) \tag{5.69}$$

EXAMPLE 5.2.2. Consider the problem of the regulation of the linear system (5.46) on the time interval $t \geqslant 0$ in such a manner that acceptable transient state deviations from zero are maintained without excessive use of control input. We will represent this problem as an optimal control problem with performance index (c.f. (5.47))

$$J(u) = \frac{1}{2}\int_0^\infty (\lambda x^2(t) + (1-\lambda)u^2(t))\mathrm{d}t, \qquad 0 \leqslant \lambda < 1 \tag{5.70}$$

where λ is a parameter that is to be adjusted to obtain the required balance between state and control magnitudes.

Without any fuss we note the data $n = 1, Q = \lambda, R = 1 - \lambda, A = -1$, $B = 1$ and that the optimal controller is the feedback controller $u(t) =$

$-(1-\lambda)^{-1}K_\infty x(t)$ where the (scalar) gain K_∞ is a solution of the algebraic Riccati equation (5.68)

$$\lambda - \frac{K_\infty^2}{1-\lambda} - 2K_\infty = 0 \tag{5.71}$$

i.e. $K_\infty = -(1-\lambda) \pm \sqrt{1-\lambda}$. Remembering that K_∞ must be symmetric (trivial as, in this case, K_∞ is a scalar) and positive, the positive root is taken to yield the optimal feedback law

$$u(t) = -(-1 + (1-\lambda)^{-1/2})x(t) \tag{5.72}$$

Substituting into (5.46) it is easily verified that the optimal closed-loop response takes the form $x(t) = \exp\{-(1-\lambda)^{-1/2}t\}$ and hence, if λ is close to unity, we obtain a closed-loop system with high controller gain and rapid state responses.

EXAMPLE 5.2.3. Consider the regulation of the inertial system $d^2y(t)/dt^2 = u(t)$ represented by the state-variable model

$$\frac{dx(t)}{dt} = \begin{bmatrix} 0 & 1 \\ 0 & 0 \end{bmatrix} x(t) + \begin{bmatrix} 0 \\ 1 \end{bmatrix} u(t),$$

$$y(t) = x_1(t), \qquad x_2(t) = \frac{dy(t)}{dt} \tag{5.73}$$

to minimize the performance index

$$J(u) = \frac{1}{2}\int_0^\infty (x_1^2(t) + \alpha x_2^2(t) + \lambda u^2(t))\,dt \tag{5.74}$$

where $\alpha \geqslant 0$ and $\lambda > 0$. The performance index is aimed to reflect the objective of limiting the magnitude of output excursions without using excessive amounts of control action (represented by the "compromise" parameter λ) or allowing excessively large rates of change in the output (represented by the parameter α). The problem data is

$$A = \begin{bmatrix} 0 & 1 \\ 0 & 0 \end{bmatrix}, \qquad B = \begin{bmatrix} 0 \\ 1 \end{bmatrix}, \qquad Q = \begin{bmatrix} 1 & 0 \\ 0 & \alpha \end{bmatrix} \geqslant 0,$$

$$R = \lambda > 0 \tag{5.75}$$

We know that the optimal controller is a constant state feedback controller of the form $u(t) = -R^{-1}B^T K_\infty x(t)$ where

$$K_\infty = \begin{bmatrix} k_{11} & k_{12} \\ k_{12} & k_{22} \end{bmatrix} \tag{5.76}$$

is symmetric, positive semi-definite and a solution of the algebraic Riccati equation

$$\begin{bmatrix} 1 & 0 \\ 0 & \alpha \end{bmatrix} - \begin{bmatrix} k_{11} & k_{12} \\ k_{12} & k_{22} \end{bmatrix} \begin{bmatrix} 0 \\ 1 \end{bmatrix} \lambda^{-1} [0 \quad 1] \begin{bmatrix} k_{11} & k_{12} \\ k_{12} & k_{22} \end{bmatrix}$$

$$+ \begin{bmatrix} 0 & 0 \\ 1 & 0 \end{bmatrix} \begin{bmatrix} k_{11} & k_{12} \\ k_{12} & k_{22} \end{bmatrix} + \begin{bmatrix} k_{11} & k_{12} \\ k_{12} & k_{22} \end{bmatrix} \begin{bmatrix} 0 & 1 \\ 0 & 0 \end{bmatrix} = \begin{bmatrix} 0 & 0 \\ 0 & 0 \end{bmatrix} \tag{5.77}$$

That is,

$$u(t) = -R^{-1}B^TK_\infty x(t) = -\lambda^{-1}(k_{12}x_1(t) + k_{22}x_2(t)) \tag{5.78}$$

which, bearing in mind the state definitions in (5.73), is a *proportional plus derivative* feedback controller. Also, (5.77) reduces to the three algebraic equations

$$1 - \lambda^{-1}k_{12}^2 = 0 \tag{5.79}$$

$$k_{11} - \lambda^{-1}k_{22}k_{12} = 0 \tag{5.80}$$

$$\alpha - \lambda^{-1}k_{22}^2 + 2k_{12} = 0 \tag{5.81}$$

to be solved for k_{11}, k_{22} and k_{12} to ensure that K_∞ is positive semi-definite. Using the results of Exercise 5.2.2, this is equivalent to requiring that both eigenvalues of K_∞ are $\geqslant 0$. It is left as an exercise for the reader to show that this is the case if, and only if,

$$k_{11} + k_{22} \geqslant 0, \qquad k_{11}k_{22} - k_{12}^2 \geqslant 0 \tag{5.82}$$

(Hint: calculate the characteristic polynomial of K_∞). But the second inequality implies that $k_{11}k_{22} \geqslant 0$, which, combined with the first inequality, reduces (5.82) to

$$k_{11} \geqslant 0, \qquad k_{22} \geqslant 0, \qquad k_{11}k_{22} - k_{12}^2 \geqslant 0 \tag{5.83}$$

It follows immediately from (5.81) that

$$k_{22} = \sqrt{(\lambda(\alpha + 2k_{12}))} \tag{5.84}$$

Also, as $k_{11} = \lambda^{-1}k_{22}k_{12} \geqslant 0$, it is clear that we should look for $k_{12} \geqslant 0$, i.e. we must have

$$k_{12} = \lambda^{1/2}, \qquad k_{11} = \sqrt{(\alpha + 2\sqrt{\lambda})} \tag{5.85}$$

Substituting back into (5.78), the optimal controller is

$$u(t) = -\lambda^{-1/2}(x_1(t) + \sqrt{(\alpha + 2\sqrt{\lambda})}x_2(t)) \tag{5.86}$$

or, in Laplace transform terms, assuming zero initial conditions

$$u(s) = -(\lambda^{-1/2} + s\lambda^{-1/2}\sqrt{(\alpha + 2\sqrt{\lambda})})y(s) \tag{5.87}$$

Taking, for simplicity, the case of $\lambda = 1$, the resulting feedback system can be represented in either of the two forms indicated in Fig. 65.

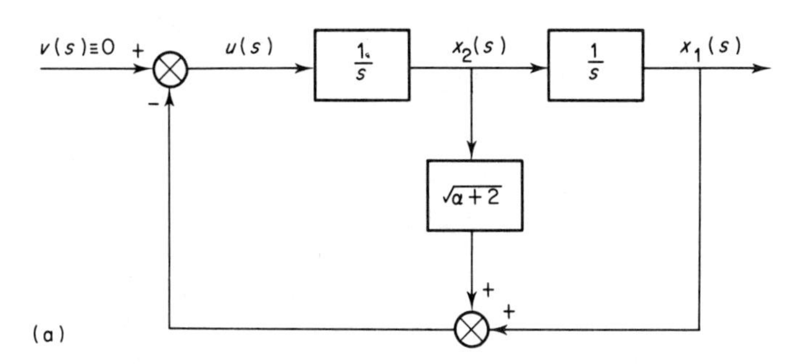

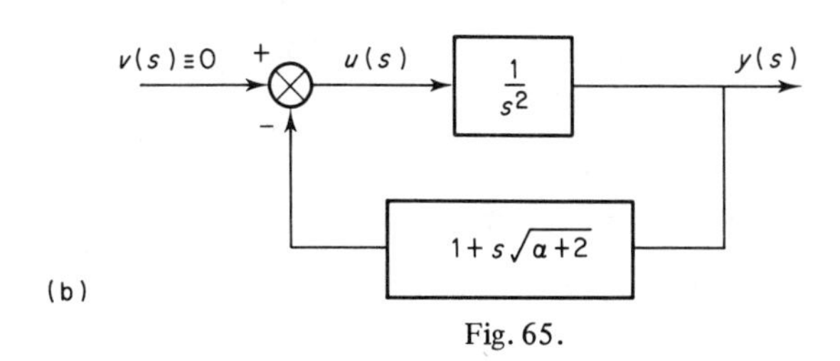

Fig. 65.

EXAMPLE 5.2.4. Verify that the closed-loop system of Example 5.2.3 has the form

$$\frac{dx(t)}{dt} = \begin{bmatrix} 0 & 1 \\ -1 & -\sqrt{\alpha+2} \end{bmatrix} x(t) \tag{5.88}$$

when $\lambda = 1$. Hence deduce that the optimal closed-loop system characteristic polynomial takes the form $\rho_c(s) = s^2 + s\sqrt{(\alpha+2)} + 1$, and verify that the "root-locus" obtained by varying α in the range $0 \leqslant \alpha < +\infty$ takes the form illustrated in Fig. 66. In particular prove that

(i) the closed-loop system is stable for all choices of $\alpha \geqslant 0$;
(ii) the system response is increasingly overdamped as α increases;
(iii) the system response is more oscillatory (underdamped) as α decreases;
(iv) the system is critically damped when $\alpha = 2$.

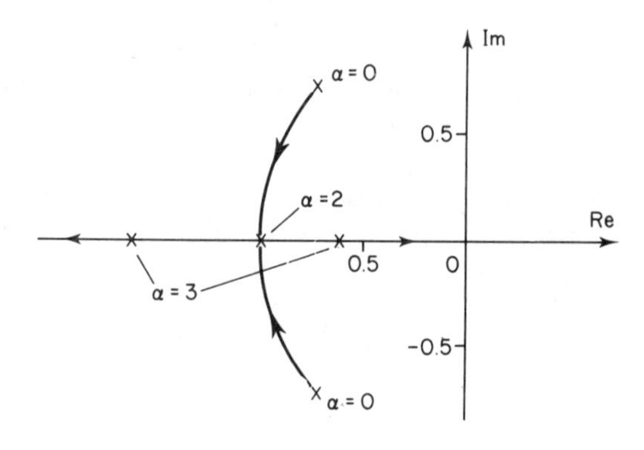

Fig. 66.

Finally, the example illustrates the obvious point that the transient response characteristics of the optimally controlled system depend critically on the choice of parameter matrices, F, Q and R in the performance index. In general, it is not possible to obtain analytic expressions for the optimal controller or to predict, *a priori*, what choice of performance index will produce the most satisfactory response. The attainment of satisfactory dynamic characteristics is normally approached, therefore, by an intelligent trial and error procedure based on varying the parameters in F, Q, R.

5.3 Minimum Energy Control: the Unconstrained Case

In this section we consider the optimal control of the linear, time-invariant system

$$\frac{dx(t)}{dt} = Ax(t) + Bu(t), \qquad x(0) = x_0 \tag{5.89}$$

on a fixed, finite time interval $0 \leqslant t \leqslant t_f$. The system is assumed to be subject to the terminal boundary condition

$$x(t_f) = x_f \text{ (specified)} \tag{5.90}$$

and the object of the optimal controller is to minimize the performance index

$$J(u) = \frac{1}{2} \int_0^{t_f} u^T(t) R u(t) dt \tag{5.91}$$

where R is a constant, symmetric, positive-definite $l \times l$ matrix. It is assumed that the elements of the input vector $u(t)$ are unconstrained.

In physical terms, there are two interpretations to the problem: either

(a) the introduction of the performance index (5.91) is an attempt to achieve the required terminal condition (5.90) using a control of "acceptable" magnitude, or

(b) the integrand $\frac{1}{2}u^T Ru$ is interpreted as a measure of the power provided by the controller and hence $J(u)$ is the total energy injected by the control inputs $u(t)$ on the interval $0 \leqslant t \leqslant t_f$. The objective of the optimal controller is then to minimize the energy required to satisfy the terminal conditions (5.90) (see Example 5.1.2).

The first intuitive step in the solution of the above "minimum-energy" problem is to note that it has the structure of the linear quadratic problem with $F = Q = 0$ but with the inclusion of the terminal boundary condition (5.90). Unfortunately this similarity is only superficial but it can be used to motivate the following construction. It is clear that the optimal controller cannot be obtained from (5.31), (5.40) and (5.41) as the data $F = Q = 0$ yields $K(t) \equiv 0$ and hence $u^*(t) \equiv 0$. This cannot be the case, in general, as a zero input will not normally produce a state trajectory satisfying (5.90). It is tempting, however, to conjecture that the optimal controller u^* and the resulting optimal state trajectory x^* are the solution of a two-point boundary value problem of the form of (5.62). This is indeed the case, the TPBVP being

$$
\begin{aligned}
\frac{dx^*(t)}{dt} &= Ax^*(t) + Bu^*(t) \\
x^*(0) &= x_0, \; x^*(t_f) \; = x_f \\
u^*(t) &= -R^{-1}B^T p(t) \\
\frac{dp(t)}{dt} &= -A^T p(t)
\end{aligned}
\tag{5.92}
$$

The only real difference between (5.62) and (5.92) is that the terminal boundary condition on the costate is replaced by the terminal boundary condition on the state. (Note: those readers who have no interest in the proof of (5.92) could now proceed directly to Exercise 5.3.1 and the material immediately following it.)

The proof that the optimal control and states, u^* and x^*, are the solutions of the TPBVP (5.92) is rather involved. We show, therefore, that if u^* and x^* satisfy (5.92) they are optimal (in effect, we will prove that (5.92) is a sufficient rather than a necessary condition of optimality). Let $u(t)$ and $x(t)$ be any input and state vectors satisfying (5.89) and (5.90) and write

$$
u(t) = -R^{-1}B^T p(t) + v(t) = u^*(t) + v(t) \tag{5.93}
$$

where $p(t)$ is the costate vector appearing in (5.92). Substituting into (5.91) it is easily shown that

$$J(u) = \frac{1}{2}\int_0^{t_f} p^T(t)BR^{-1}B^T p(t)\mathrm{d}t + \frac{1}{2}\int_0^{t_f} v^T(t)Rv(t)\mathrm{d}t$$

$$- \int_0^{t_f} p^T(t)Bv(t)\mathrm{d}t \tag{5.94}$$

and, hence, taking $v = 0$,

$$J(u^*) = \frac{1}{2}\int_0^{t_f} p^T(t)BR^{-1}B^T p(t)\mathrm{d}t \tag{5.95}$$

The proof of the optimality of u^* will be complete if we can show that

$$\int_0^{t_f} p^T(t)Bv(t)\mathrm{d}t = 0 \tag{5.96}$$

for all $v(t)$ of the form considered, for then (5.94) becomes

$$J(u) = J(u^*) + \frac{1}{2}\int_0^{t_f} v^T(t)Rv(t)\mathrm{d}t$$

$$\geqslant J(u^*) \tag{5.97}$$

as $R > 0$.

The proof of (5.96) follows by noting from (5.92) that

$$p(t) = \mathrm{e}^{-A^T t} p(0) \tag{5.98}$$

where $p(0)$ is the (unknown) boundary condition on the costate. Also, using (5.93) in (5.89) and representing the solution in the form of (1.159), we obtain

$$\mathrm{e}^{-At_f}x(t_f) = x(0) + \int_0^{t_f} \mathrm{e}^{-At}B(u^*(t) + v(t))\mathrm{d}t \tag{5.99}$$

But this must be satisfied for all $v(t)$ satisfying (5.90), including $v(t) \equiv 0$, i.e. we must have

$$\int_0^{t_f} \mathrm{e}^{-At}Bv(t)\mathrm{d}t = 0 \tag{5.100}$$

Multiplying from the left by $p^T(0)$ and using (5.98),

$$p^T(0)\int_0^{t_f} \mathrm{e}^{-At}Bv(t)\mathrm{d}t = \int_0^{t_f} p^T(0)\mathrm{e}^{-At}Bv(t)\mathrm{d}t$$

$$= \int_0^{t_f} (\mathrm{e}^{-A^T t}p(0))^T Bv(t)\mathrm{d}t = \int_0^{t_f} p^T(t)Bv(t)\mathrm{d}t = 0 \tag{5.101}$$

as required. This completes the proof of the statement that any solution u^* of the TPBVP (5.92) is a solution of the optimal control problem.

EXERCISE 5.3.1. Verify that the initial costate $p(0)$ is a solution of the matrix equation

$$W(t_f)p(0) = x_0 - e^{-At_f}x_f \tag{5.102}$$

where the $n \times n$ matrix $W(t_f)$ (a so-called "controllability Gramian") is given by

$$W(t_f) = \int_0^{t_f} e^{-At}BR^{-1}B^Te^{-A^Tt}\,dt \tag{5.103}$$

and, in the case where $W(t_f)$ is nonsingular, the optimal controller is given explicitly by the formula

$$u^*(t) = -R^{-1}B^Te^{-A^Tt}(W(t_f))^{-1}(x_0 - e^{-At_f}x_f) \tag{5.104}$$

(Hint: use (5.98) in (5.99) with $v(t) \equiv 0$.)

EXAMPLE 5.3.1. If, in Example 5.1.2, the input to the system is unconstrained and the rate of energy dissipation in the winding gear is assumed to take the form $L_e(x, u_0, t) = \frac{1}{2}u_0^2$, the optimal control problem takes the form

$$\frac{dx(t)}{dt} = \begin{bmatrix} 0 & 1 \\ 0 & 0 \end{bmatrix} x(t) + \begin{bmatrix} 0 \\ m^{-1} \end{bmatrix} u(t), \qquad u(t) = u_0(t) - mg$$

$$x(0) = \begin{bmatrix} L \\ 0 \end{bmatrix}, \qquad x(T) = \begin{bmatrix} 0 \\ 0 \end{bmatrix} \tag{5.105}$$

$$J(u) = \frac{1}{2}\int_0^T (u(t) + mg)^2\,dt \tag{5.106}$$

which is almost (but not quite) in the standard form (5.89)–(5.91). This problem is easily resolved by writing

$$\begin{aligned} J(u) &= \frac{1}{2}\int_0^T u^2(t)\,dt + mg\int_0^T u(t)\,dt + \frac{1}{2}(mg)^2T \\ &= \frac{1}{2}\int_0^T u^2(t)\,dt + \frac{1}{2}(mg)^2T \end{aligned} \tag{5.107}$$

as, from the state equations, we have $dx_2(t)/dt = m^{-1}u(t)$, $x_2(0) = x_2(T) = 0$ and hence

$$\int_0^T u(t)\,dt = 0 \tag{5.108}$$

Noting that the constant term $\frac{1}{2}(mg)^2T$ in (5.107) has no effect on the solution of the minimization problem, we can ignore it and replace the performance index (5.106) with the performance index

$$J(u) = \frac{1}{2}\int_0^T u^2(t)\mathrm{d}t \tag{5.109}$$

The resulting optimal control problem is now defined by (5.105) and (5.109) and takes the form of a minimum energy problem with the data

$$A = \begin{bmatrix} 0 & 1 \\ 0 & 0 \end{bmatrix}, \quad B = \begin{bmatrix} 0 \\ m^{-1} \end{bmatrix}, \quad x_0 = \begin{bmatrix} L \\ 0 \end{bmatrix},$$

$$x_f = \begin{bmatrix} 0 \\ 0 \end{bmatrix}, \quad t_f = T, \quad R = 1 \tag{5.110}$$

The optimal controller is a solution of the TPBVP

$$\frac{\mathrm{d}x^*(t)}{\mathrm{d}t} = \begin{bmatrix} 0 & 1 \\ 0 & 0 \end{bmatrix} x^*(t) + \begin{bmatrix} 0 \\ m^{-1} \end{bmatrix} u^*(t), \quad x(0) = \begin{bmatrix} L \\ 0 \end{bmatrix},$$

$$x(T) = \begin{bmatrix} 0 \\ 0 \end{bmatrix}, \quad \frac{\mathrm{d}p(t)}{\mathrm{d}t} = \begin{bmatrix} 0 & 0 \\ -1 & 0 \end{bmatrix} p(t)$$

$$u^*(t) = -[0 \quad m^{-1}]p(t) = -m^{-1}p_2(t) \tag{5.111}$$

We will consider two methods of solution.

Method 1 (direct parametric solution). The problem with the solution of (5.111) is the mixed boundary conditions and, in particular, the absence of a (known) boundary condition for the costate vector $p(t)$. Yet, without an initial condition $p(0)$ for the costate equation, we cannot obtain $p(t)$ directly and hence we cannot obtain $u^*(t)$. The way out of this difficulty is to obtain a parametric solution to the costate equation and adjust these parameters to satisfy the state boundary conditions.

It is clear that the costate equations take the form $\mathrm{d}p_1(t)/\mathrm{d}t = 0$ and $\mathrm{d}p_2(t)/\mathrm{d}t = -p_1(t)$ and, hence, that

$$p_1(t) = p_1(0), \qquad p_2(t) = p_2(0) - p_1(0)t \tag{5.112}$$

where $p_1(0)$ and $p_2(0)$ are the unknown initial conditions. The resultant optimal controller takes the form

$$u^*(t) = \pi_2 - \pi_1 t \tag{5.113}$$

where $\pi_i = -p_i(0)/m$ $(i = 1, 2)$ are (as, yet) unknown parameters. It is interesting to note that, even though we have not solved our problem, the costate equation has already yielded the information that the optimal control is a linear function of time.

The parameters π_1 and π_2 must now be chosen to satisfy the boundary conditions for $x^*(t)$. Noting that $\mathrm{d}x_2^*(t)/\mathrm{d}t = m^{-1}u^*(t)$, $x_2^*(0) = x_2^*(T) = 0$, we obtain

$$x_2^*(t) = (\pi_2 t - \pi_1 t^2/2)m^{-1} \tag{5.114}$$

and hence

$$\pi_2 - \pi_1 T/2 = 0 \tag{5.115}$$

Also, $\mathrm{d}x_1^*(t)/\mathrm{d}t = x_2^*(t), x_1^*(0) = L, x_1^*(T) = 0$ yields the relation

$$0 = L + T^2(\pi_2/2 - \pi_1 T/6)m^{-1} \tag{5.116}$$

The required parameters π_1 and π_2 are the solutions of the linear simultaneous algebraic equations (5.115) and (5.116), i.e. $\pi_1 = -m\,12L/T^3$ and $\pi_2 = -6mL/T^2$ and the optimal control is just

$$u^*(t) = \frac{12Lm}{T^3}\left(t - \frac{T}{2}\right) \tag{5.117}$$

(Note that the solution of the problem does not have a feedback form. Note also that the physical input to the system is obtained from $u_0^*(t) = u(t) - mg$.)

Method 2 (direct analytic solution). Applying the results of Exercise 5.3.1, we calculate $W(T)$ by noting that (Example 1.7.1)

$$\mathrm{e}^{At} = \begin{bmatrix} 1 & t \\ 0 & 1 \end{bmatrix} \tag{5.118}$$

and hence

$$W(T) = \int_0^T \begin{bmatrix} 1 & -t \\ 0 & 1 \end{bmatrix} \begin{bmatrix} 0 \\ m^{-1} \end{bmatrix} [0 \quad m^{-1}] \begin{bmatrix} 1 & 0 \\ -t & 1 \end{bmatrix} \mathrm{d}t$$

$$= \int_0^T m^{-2} \begin{bmatrix} t^2 & -t \\ -t & 1 \end{bmatrix} \mathrm{d}t = \frac{T}{m^2} \begin{bmatrix} T^2/3 & -T/2 \\ -T/2 & 1 \end{bmatrix} \tag{5.119}$$

Finally, substituting into (5.104),

$$u^*(t) = -[0 \quad m^{-1}]\begin{bmatrix} 1 & 0 \\ -t & 1 \end{bmatrix} \frac{m^2}{T} \begin{bmatrix} T^2/3 & -T/2 \\ -T/2 & 1 \end{bmatrix}^{-1} \begin{bmatrix} L \\ 0 \end{bmatrix}$$

$$= \frac{12Lm}{T^3}(t - T/2) \tag{5.120}$$

in agreement with (5.117).

EXERCISE 5.3.2. Show that the unconstrained optimal controller solving the problem

$$\frac{dx(t)}{dt} = u(t), \quad x(0) = L, \quad x(1) = 0, \quad J(u) = \frac{1}{2}\int_0^{t_f} u^2(t)dt \tag{5.121}$$

is simply the constant controller $u^*(t) = -L/t_f$. Repeat the problem for the system $dx(t)/dt = -\lambda x(t) + u(t)$ $(\lambda \neq 0)$ and verify that the optimal controller takes the form $u^*(t) = 2\lambda L \exp\{\lambda t\}/(1 - \exp\{2\lambda t_f\})$.

As a final point in this section, we note that, just as in Section 1.5 where transformation of state variables simplified the solution of the state equations, a transformation of variables can simplify the solution of the TPBVP (5.92). More precisely, if we introduce new state variables

$$z(t) = E^{-1}x(t) \tag{5.122}$$

where E is a $n \times n$ nonsingular matrix, then the unconstrained optimal control problem defined by (5.89)–(5.91) must be equivalent to the problem

$$\frac{dz(t)}{dt} = A^*z(t) + B^*u(t)$$

$$z(0) = z_0, \qquad z(t_f) = z_f$$

$$J(u) = \frac{1}{2}\int_0^{t_f} u^T(t)Ru(t)dt \tag{5.123}$$

where $A^* = E^{-1}AE$, $B^* = E^{-1}B$, $z_0 = E^{-1}x_0$ and $z_f = E^{-1}x_f$. In particular, we can now compute $u^*(t)$ by solving the TPBVP

$$\frac{dz^*(t)}{dt} = A^*z^*(t) + B^*u^*(t)$$

$$u^*(t) = -R^{-1}(B^*)^T p(t), \qquad z^*(0) = z_0, \qquad z^*(t_f) = z_f$$

$$\frac{dp(t)}{dt} = -(A^*)^T p(t) \tag{5.124}$$

The solution of the TPBVP will obviously be simplified if A^* has a simple form. For example, if E is a nonsingular eigenvector matrix for A, then A^* (and hence its transpose) is diagonal with diagonal elements equal to the eigenvalues λ_1, $\lambda_2, \ldots, \lambda_n$ of A. The reader should easily verify in this case that

$$p_i(t) = p_i(0)e^{-\lambda_i t}, \qquad 1 \leqslant i \leqslant n \tag{5.125}$$

and hence, defining

$$-R^{-1}(B^*)^T = [b_1^*, b_2^*, \ldots, b_n^*] \tag{5.126}$$

we obtain

$$u^*(t) = \sum_{i=1}^{n} b_i^* p_i(0) e^{-\lambda_i t} \tag{5.127}$$

The remaining problem is to find the unknown initial conditions $p_i(0)$, $1 \leqslant i \leqslant n$. These are chosen to satisfy the terminal boundary condition at $t = t_f$.

EXERCISE 5.3.3. Formulate and prove an equivalent to Exercise 5.3.1 for the TPBVP (5.124).

EXAMPLE 5.3.2. Consider the liquid level system (1.71) with data $a_1 = a_2 = \beta = 1$

$$\frac{dx(t)}{dt} = \begin{bmatrix} -1 & 1 \\ 1 & -1 \end{bmatrix} x(t) + \begin{bmatrix} 1 & 0 \\ 0 & 1 \end{bmatrix} u(t) \tag{5.128}$$

to be controlled using an unconstrained controller from the initial condition $x(0) = \begin{pmatrix} 1 \\ 0 \end{pmatrix}$ to the final condition $x(1) = \begin{pmatrix} 0 \\ 0 \end{pmatrix}$ whilst minimizing the performance index

$$J(u) = \frac{1}{2} \int_0^1 (u_1^2(t) + \alpha u_2^2(t))\,dt, \qquad \alpha > 0 \tag{5.129}$$

That is, we wish to return from the given initial condition back to the steady state in the specified time whilst not using too much control action. The data for the problem is

$$A = \begin{bmatrix} -1 & 1 \\ 1 & -1 \end{bmatrix}, \quad B = \begin{bmatrix} 1 & 0 \\ 0 & 1 \end{bmatrix}, \quad R = \begin{bmatrix} 1 & 0 \\ 0 & \alpha \end{bmatrix}, \quad t_f = 1 \tag{5.130}$$

and hence the optimal controller is the solution of the TPBVP

$$\frac{dx^*(t)}{dt} = \begin{bmatrix} -1 & 1 \\ 1 & -1 \end{bmatrix} x^*(t) + \begin{bmatrix} 1 & 0 \\ 0 & 1 \end{bmatrix} u^*(t)$$

$$u^*(t) = \begin{bmatrix} -p_1(t) \\ -\alpha^{-1} p_2(t) \end{bmatrix}, \qquad x(0) = \begin{bmatrix} 1 \\ 0 \end{bmatrix}, \qquad x(1) = \begin{bmatrix} 0 \\ 0 \end{bmatrix}$$

$$\frac{dp(t)}{dt} = \begin{bmatrix} 1 & -1 \\ -1 & 1 \end{bmatrix} p(t) \tag{5.131}$$

It is clear that the off-diagonal terms in A and A^T will complicate the solution of this problem. We will therefore diagonalize them by calculating the eigenvector matrix $E = \begin{bmatrix} 1 & -1 \\ 1 & 1 \end{bmatrix}$ of A and transforming the original problem to the variables $z(t) = E^{-1}x(t)$. The problem (5.128)–(5.129) is hence equivalent to the problem

$$\frac{dz(t)}{dt} = \begin{bmatrix} 0 & 0 \\ 0 & -2 \end{bmatrix} z(t) + \begin{bmatrix} \frac{1}{2} & \frac{1}{2} \\ -\frac{1}{2} & \frac{1}{2} \end{bmatrix} u(t),$$

$$z(0) = \begin{bmatrix} \frac{1}{2} \\ -\frac{1}{2} \end{bmatrix}, \qquad z(1) = \begin{bmatrix} 0 \\ 0 \end{bmatrix} \tag{5.132}$$

with performance index (5.129). The TPBVP to be solved in this case is simply

$$\frac{dz^*(t)}{dt} = \begin{bmatrix} 0 & 0 \\ 0 & -2 \end{bmatrix} z^*(t) + \begin{bmatrix} \frac{1}{2} & \frac{1}{2} \\ -\frac{1}{2} & \frac{1}{2} \end{bmatrix} u^*(t),$$

$$z^*(0) = \begin{bmatrix} \frac{1}{2} \\ -\frac{1}{2} \end{bmatrix}, \qquad z^*(1) = \begin{bmatrix} 0 \\ 0 \end{bmatrix}$$

$$u^*(t) = -\begin{bmatrix} \dfrac{1}{2} & -\dfrac{1}{2} \\ \dfrac{1}{2\alpha} & \dfrac{1}{2\alpha} \end{bmatrix} p(t), \qquad \frac{dp(t)}{dt} = \begin{bmatrix} 0 & 0 \\ 0 & 2 \end{bmatrix} p(t) \tag{5.133}$$

Solving the costate equations is a simple exercise yielding $p_1(t) \equiv p_1(0)$ and $p_2(t) = p_2(0)e^{2t}$, and hence

$$u_1^*(t) = \tfrac{1}{2}(p_2(0)e^{2t} - p_1(0)),$$

$$u_2^*(t) = -\frac{1}{2\alpha}(p_1(0) + p_2(0)e^{2t}) \qquad (5.134)$$

We obtain the required values of $p_1(0)$ and $p_2(0)$ from the state boundary conditions, the calculations being simplified as A^* is diagonal, i.e.

$$0 = \frac{1}{2} + \int_0^1 \frac{1}{2}(u_1^*(t) + u_2^*(t))\,dt$$

$$0 = \frac{-e^{-2}}{2} + e^{-2}\int_0^1 e^{2t}\frac{(u_2^*(t) - u_1^*(t))}{2}\,dt \qquad (5.135)$$

Performing the integrations and rearranging yields the two simultaneous algebraic equations

$$0 = 1 - \tfrac{1}{2}(1+\alpha^{-1})p_1(0) + \tfrac{1}{4}(1-\alpha^{-1})(e^2-1)p_2(0)$$

$$0 = 1 - \tfrac{1}{4}(1-\alpha^{-1})(e^2-1)p_1(0) + \tfrac{1}{8}(1+\alpha^{-1})(e^4-1)p_2(0) \qquad (5.136)$$

to be solved for $p_1(0)$ and $p_2(0)$. Taking for simplicity the case of $\alpha = 1$, we obtain $p_1(0) = 1, p_2(0) = -4/(e^4-1)$ and hence

$$u_1^*(t) = -0.037e^{2t} - 0.5, \qquad u_2^*(t) = +0.037e^{2t} - 0.5 \qquad (5.137)$$

EXERCISE 5.3.4. In the above example, show that $u_2^* \to 0, 0 \leqslant t \leqslant 1$, as the weighting parameter $\alpha \to +\infty$, and hence deduce that an increase in the relative weightings of u_2 and u_1 in the performance index will tend to reduce the magnitude of u_2^*. (Hint: let $\alpha \to +\infty$ in (5.136).) Formulate and prove a similar result for the case of $\alpha \to 0+$.

5.4 Introduction to Variational Calculus

Sections 5.2 and 5.3 have restricted our attention to two cases of unconstrained optimal control of particular interest. It will not be surprising to note that these results in the main are special cases of a more general theoretical result obtained from the general methodology of variational calculus. The purpose of this short section is to obtain (i) the basic structure of variational calculus in the form of the Euler-Lagrange equations and (ii) its relation to the results of Sections 5.2 and 5.3. There will be no attempt at mathematical rigour and, for convenience, we will restrict attention to a number of special cases of practical interest.

Case One (fixed-time, unconstrained optimal control with no terminal boundary conditions). Consider the problem of optimal control of the non-linear system

$$\frac{dx(t)}{dt} = f(x(t), u(t), t), \qquad x(0) = x_0 \tag{5.138}$$

to minimize the performance index

$$J(u) = \phi(x(t_f)) + \int_0^{t_f} L(x(t), u(t), t)dt \tag{5.139}$$

There are no terminal boundary conditions on x, t_f is assumed to be finite and fixed and the elements of the $l \times 1$ input vector are assumed to be unconstrained.

Suppose that $u^*(t)$ and $x^*(t), 0 \leqslant t \leqslant t_f$, are optimal control inputs and state trajectories satisfying

$$\frac{dx^*(t)}{dt} = f(x^*(t), u^*(t), t), \qquad x^*(0) = x_0 \tag{5.140}$$

then for any other input $u(t)$ we have

$$J(u) \geqslant J(u^*) \tag{5.141}$$

The mathematical problem that variational calculus addresses is the construction of a two-point boundary value problem for which u^* and x^* are solutions.

Adjoin the differential equation constraints (5.138) to the performance index using an $n \times 1$ Lagrange multiplier (or costate) function vector $p(t)$ to obtain

$$J(u) = \phi(x(t_f)) + \int_0^{t_f} \left\{ H(x(t), p(t), u(t), t) - p^T(t)\frac{dx(t)}{dt} \right\} dt \tag{5.142}$$

where

$$H(x, p, u, t) = L(x, u, t) + p^T f(x, u, t) \tag{5.143}$$

is the *Hamiltonian* function for the problem. Using integration by parts on the last term in (5.142) and expressing the inner products as summations yields

$$J(u) = \phi(x(t_f)) + \sum_{k=1}^{n} (p_k(0)x_k(0) - p_k(t_f)x_k(t_f))$$

$$+ \int_0^{t_f} \left\{ H(x(t), p(t), u(t), t) + \sum_{k=1}^{n} x_k(t)\frac{dp_k(t)}{dt} \right\} dt \tag{5.144}$$

A simple calculation using Taylor series now yields the relation

$$J(u) - J(u^*) = \phi(x(t_f)) - \phi(x^*(t_f)) - \sum_{k=1}^{n} p_k(t_f)(x_k(t_f) - x_k^*(t_f))$$

$$+\int_0^{t_f}\{H(x(t),p(t),u(t),t)-H(x^*(t),p(t),u^*(t),t)\}\mathrm{d}t$$

$$+\int_0^{t_f}\sum_{k=1}^{n}(x_k(t)-x_k^*(t))\frac{\mathrm{d}p_k(t)}{\mathrm{d}t}\mathrm{d}t$$

$$=\sum_{k=1}^{n}\left(\left.\frac{\partial\phi}{\partial x_k}\right|_*-p_k(t_f)\right)\delta x_k(t_f)$$

$$+\int_0^{t_f}\sum_{k=1}^{n}\left(\left.\frac{\partial H}{\partial x_k}\right|_*+\frac{\mathrm{d}p_k(t)}{\mathrm{d}t}\right)\delta x_k(t)\mathrm{d}t$$

$$+\int_0^{t_f}\sum_{k=1}^{l}\left.\frac{\partial H}{\partial u_k}\right|_*\delta u_k(t)\mathrm{d}t+\text{higher order terms}\tag{5.145}$$

where $\delta x = x - x^*$ and $\delta u = u - u^*$ and the $*$-notation on the partial derivatives indicates that they are to be evaluated at the data points $x^*(t), p(t), u^*(t), t$.

Although the rigorous justification is rather involved, we will now choose the costates to eliminate the terms in δx in (5.145), i.e.

$$\frac{\mathrm{d}p_k(t)}{\mathrm{d}t}=-\left.\frac{\partial H}{\partial x_k}\right|_*,\qquad p_k(t_f)=\left.\frac{\partial\phi}{\partial x_k}\right|_*,\qquad 1\leqslant k\leqslant n\tag{5.146}$$

when (5.145) reduces to

$$J(u)-J(u^*)=\int_0^{t_f}\sum_{k=1}^{l}\left.\frac{\partial H}{\partial u_k}\right|_*\delta u_k(t)\mathrm{d}t+\text{higher order terms}\tag{5.147}$$

If we now consider inputs where $\max\limits_{0\leqslant t\leqslant t_f}|u(t)-u^*(t)|$ is small, it is possible to neglect the higher order terms. But, apart from the smallness requirement, the functional form of $u(t)$ and hence $\delta u(t)$ is unconstrained. Hence, if (5.141) is to hold, we must have

$$\left.\frac{\partial H}{\partial u_k}\right|_*=0,\qquad 1\leqslant k\leqslant l\tag{5.148}$$

In summary the above analysis indicates that the optimal control is a solution of the TPBVP (5.140), (5.146) and (5.148), i.e.

$$\frac{\mathrm{d}x^*(t)}{\mathrm{d}t}=f(x^*(t),u^*(t),t),\qquad x^*(0)=x_0$$

$$\left.\frac{\partial H}{\partial u_k}\right|_*=0,\qquad 1\leqslant k\leqslant l$$

$$\frac{dp_k(t)}{dt} = -\left.\frac{\partial H}{\partial x_k}\right|_*, \qquad p_k(t_f) = \left.\frac{\partial \phi}{\partial x_k}\right|_*, \qquad 1 \leqslant k \leqslant n$$

$$0 \leqslant t \leqslant t_f \tag{5.149}$$

These equations are examples of "Euler-Lagrange Equations", the costate vector appearing in the role of a Lagrange multiplier.

EXERCISE 5.4.1. Verify that the Euler-Lagrange TPBVP for the unconstrained optimal control problems (5.23)–(5.25) reduce to equations (5.62). As an example, show that equations (5.63) in Exercise 5.2.6 can be obtained from (5.149).

Case Two (fixed-time, unconstrained optimal control with terminal state boundary conditions). Consider now the problem of optimal control of the nonlinear system (5.138) to minimize the performance index

$$J(u) = \int_0^{t_f} L(x(t), u(t), t)\mathrm{d}t \tag{5.150}$$

whilst satisfying the terminal boundary condition $x(t_f) = x_f$. The terminal time t_f is assumed fixed and the control input unconstrained.

The analysis closely follows that of Case One but with $\phi = 0$ and $\delta x(t_f) = 0$ to yield the result that the optimal controller is a solution of the TPBVP

$$\frac{dx^*(t)}{dt} = f(x^*(t), u^*(t), t), \quad x^*(0) = x_0, \quad x^*(t_f) = x_f$$

$$\left.\frac{\partial H}{\partial u_k}\right|_* = 0, \quad 1 \leqslant k \leqslant l, \quad \frac{dp_k(t)}{dt} = -\left.\frac{\partial H}{\partial x_k}\right|_*, \quad 1 \leqslant k \leqslant n \tag{5.151}$$

which is identical to (5.149) except that the presence of the terminal condition on the states removes our knowledge of the terminal costate boundary conditions.

EXERCISE 5.4.2. Verify that the Euler-Lagrange equations for the minimum energy problem (5.89)–(5.91) reduce to the TPBVP (5.92).

Problems

(1) A general quadratic form in the n variables $x_1, x_2, \ldots, x_n$ has the form of a double summation $\sum_{i=1}^{n} \sum_{k=1}^{n} x_i \gamma_{ik} x_k$. Show that

$$\sum_{i=1}^{n}\sum_{k=1}^{n} x_i \gamma_{ik} x_k = \sum_{i=1}^{n}\sum_{k=1}^{n} x_i \frac{(\gamma_{ik}+\gamma_{ki})}{2} x_k$$

and hence that the quadratic form has the structure $\frac{1}{2}x^T M x$ where the $n \times n$ symmetric matrix M has elements $M_{ik} = \gamma_{ik} + \gamma_{ki}$. Hence deduce that

$$x_1^2 + x_1x_2 + x_2^2 + x_1x_3 + x_3^2 = \tfrac{1}{2}(x_1\ x_2\ x_3)\begin{bmatrix} 2 & 1 & 1 \\ 1 & 2 & 0 \\ 1 & 0 & 2 \end{bmatrix}\begin{bmatrix} x_1 \\ x_2 \\ x_3 \end{bmatrix}$$

and check this result by multiplying out the triple matrix product.

(2) The linear time-invariant system $\mathrm{d}x(t)/\mathrm{d}t = Ax(t) + Bu(t)$ is to be controlled from the initial condition $x(0) = x_0$ using an unconstrained controller, to minimize the index

$$J(u) = \frac{1}{2}\int_0^{t_f} \mathrm{e}^{2\alpha t}\{x^T(t)Qx(t) + u^T(t)Ru(t)\}\mathrm{d}t$$

Show that the optimal controller has the feedback form

$$u(t) = -R^{-1}B^T K_\alpha(t)x(t)$$

where $K_\alpha(t)$ is the solution of the Riccati equation

$$Q - K_\alpha(t)BR^{-1}B^T K_\alpha(t) + (A+\alpha I)^T K_\alpha(t) + K_\alpha(t)(A+\alpha I) + \frac{\mathrm{d}K_\alpha(t)}{\mathrm{d}t} = 0,$$

$$K_\alpha(t_f) = 0$$

(Hint: write $z(t) = \mathrm{e}^{\alpha t}x(t)$ and $v(t) = \mathrm{e}^{\alpha t}u(t)$ and rewrite the system model and performance index in terms of these new variables.) (Note: the use of $\alpha > 0$ in the above performance index can be interpreted as an attempt to put more emphasis on reducing state and control magnitudes in the vicinity of $t = t_f$ rather than in the vicinity of $t = 0$.) What do you think happens if $t_f = +\infty$?

(3) Let x_∞ be a steady state solution of the linear, time-invariant system $\mathrm{d}x(t)/\mathrm{d}t = Ax(t) + Bu(t)$ corresponding to the constant input vector u_∞. Prove that the optimal unconstrained controller minimizing the performance index

$$J(u) = \frac{1}{2}\int_0^{t_f} \{(x(t)-x_\infty)^T Q(x(t)-x_\infty) + (u(t)-u_\infty)^T R(u(t)-u_\infty)\}\mathrm{d}t$$

(where $Q = Q^T \geqslant 0$ and $R = R^T > 0$) has the feedback form

$$u(t) = u_\infty - R^{-1}B^T K(t)(x(t)-x_\infty)$$

where $K(t)$ is the solution of the Riccati equation

$$Q - K(t)BR^{-1}B^TK(t) + A^TK(t) + K(t)A + \frac{dK(t)}{dt} = 0, \qquad K(t_f) = 0$$

(Hint: note that $Ax_\infty + Bu_\infty = 0$ and change variables to $z(t) = x(t) - x_\infty$, $v(t) = u(t) - u_\infty$.)

Formulate a similar result for the case of $t_f = +\infty$ and hence show that the optimal controller for the problem

$$\frac{dx(t)}{dt} = -x(t) + u(t), \qquad J(u) = \frac{1}{2}\int_0^\infty ((x(t)-1)^2 + (u(t)-1)^2)dt$$

is the feedback controller $u(t) = 1 - 0.414\,(x(t) - 1)$.

(4) Prove that, if $K(t)$ is nonsingular for $t < t_f$, the inverse $P(t) = K^{-1}(t)$ of the Riccati gain matrix solving (5.40) and (5.41) is a solution of the matrix differential equation

$$\frac{dP(t)}{dt} = P(t)QP(t) - BR^{-1}B^T + P(t)A^T + AP(t)$$

If F is positive and semi-definite, it is singular (Exercise 5.2.2) and hence $P(t)$ is infinite at $t = t_f$. However, if F is positive definite, it is nonsingular and the above "inverse Riccati equation" has the terminal boundary condition $P(t_f) = F^{-1}$.

(5) Show that the TPBVP (5.62) characterizing the solution of the finite time linear quadratic optimal control problem can be written in the $2n \times 2n$ form

$$\frac{d}{dt}\begin{bmatrix} x^*(t) \\ p(t) \end{bmatrix} = \begin{bmatrix} A & -BR^{-1}B^T \\ -Q & -A^T \end{bmatrix}\begin{bmatrix} x^*(t) \\ p(t) \end{bmatrix} = \tilde{A}\begin{bmatrix} x^*(t) \\ p(t) \end{bmatrix}$$

with the boundary conditions $x^*(0) = x_0$ and $p(t_f) = Fx^*(t_f)$. Hence verify that

$$\begin{bmatrix} x^*(t_f) \\ p(t_f) \end{bmatrix} = e^{\tilde{A}(t_f - t)}\begin{bmatrix} x^*(t) \\ p(t) \end{bmatrix}$$

and, writing

$$e^{\tilde{A}(t_f - t)} = \begin{bmatrix} \Phi_{11}(t_f, t) & \Phi_{12}(t_f, t) \\ \Phi_{21}(t_f, t) & \Phi_{22}(t_f, t) \end{bmatrix}$$

(where each block is $n \times n$), deduce that

$$(F\Phi_{11}(t_f, t) - \Phi_{21}(t_f, t))x^*(t) + (F\Phi_{12}(t_f, t) - \Phi_{22}(t_f, t))p(t) \equiv 0$$

Conclude (assuming invertibility where necessary) that the Riccati gain matrix

$$K(t) = (\Phi_{22}(t_f, t) - F\Phi_{12}(t_f, t))^{-1}(F\Phi_{11}(t_f, t) - \Phi_{21}(t_f, t))$$

Apply these results to the problem

$$\frac{dx(t)}{dt} = -x(t) + u(t), \qquad J(u) = \frac{1}{2}\int_0^1 (x^2(t) + u^2(t)dt$$

by verifying that the TPBVP takes the form

$$\frac{d}{dt}\begin{bmatrix} x^*(t) \\ p(t) \end{bmatrix} = \begin{bmatrix} -1 & -1 \\ -1 & 1 \end{bmatrix}\begin{bmatrix} x^*(t) \\ p(t) \end{bmatrix}$$

with $p(1) = 0$, that

$$e^{\tilde{A}(1-t)} = \frac{1}{2\sqrt{2}}\begin{bmatrix} (\sqrt{2}-1)e^{\sqrt{2}(1-t)} + (1+\sqrt{2})e^{-\sqrt{2}(1-t)}, & -e^{\sqrt{2}(1-t)} + e^{-\sqrt{2}(1-t)} \\ -e^{\sqrt{2}(1-t)} + e^{-\sqrt{2}(1-t)}, & (1+\sqrt{2})e^{\sqrt{2}(1-t)} + (\sqrt{2}-1)e^{-\sqrt{2}(1-t)} \end{bmatrix}$$

and hence that $K(t) = ((1+\sqrt{2})e^{\sqrt{2}(1-t)} + (\sqrt{2}-1)e^{-\sqrt{2}(1-t)})^{-1}(e^{\sqrt{2}(1-t)} - e^{-\sqrt{2}(1-t)})$. Check this result by solving the relevant Riccati equation.

(6) Occasionally, transformation of state variables can ease the solution of optimal control problems. Using the transformation $z(t) = E^{-1}x(t)$, show that the linear quadratic problem (t_f finite or infinite)

$$\frac{dx(t)}{dt} = Ax(t) + Bu(t), \qquad J(u) = \frac{1}{2}\int_0^{t_f} (x^T(t)Qx(t) + u^T(t)Ru(t))dt$$

with $Q = Q^T \geqslant 0$ and $R = R^T > 0$ becomes the linear quadratic problem

$$\frac{dz(t)}{dt} = A^*z(t) + B^*u(t), \qquad J(u) = \frac{1}{2}\int_0^{t_f} (z^T(t)Q^*z(t) + u^T(t)Ru(t))dt$$

where

$$A^* = E^{-1}AE, \qquad B^* = E^{-1}B, \qquad Q^* = E^TQE.$$

(7) The liquid level system (1.71) with data $a_1 = a_2 = \beta = 1$ is to be controlled to minimize the performance index

$$J(u) = \int_0^\infty \left(\frac{1}{2}x_1^2(t) + \frac{1}{2}x_2^2(t) + \frac{1}{2}u_1^2(t) + \frac{1}{2}u_2^2(t)\right)dt$$

use problem (1) to show that $Q = R = \begin{bmatrix} 1 & 0 \\ 0 & 1 \end{bmatrix}$ and hence that the optimal controller takes the form $u^*(t) = -K_\infty x^*(t)$ where K_∞ (equation (5.76)) is a positive definite solution of the algebraic Riccati equation. Verify that the elements of K_∞ are the solutions of the equations

$$1-k_{11}^2-k_{12}^2+2k_{12}-2k_{11} = 0$$

$$1-k_{12}^2-k_{22}^2+2k_{12}-2k_{22} = 0$$

$$-k_{11}k_{12}-k_{12}k_{22}+k_{22}-2k_{12}+k_{11} = 0$$

with

$$k_{11}>0, \qquad k_{22}>0 \quad \text{and} \quad k_{11}k_{22}-k_{12}^2>0$$

Use the ideas of problem (7) to simplify the above problem. More precisely, if we choose $E=\frac{1}{\sqrt{2}}\begin{bmatrix}1 & -1\\ 1 & 1\end{bmatrix}$ to be an eigenvector matrix of A, the optimization problem becomes the problem

$$\frac{\mathrm{d}z(t)}{\mathrm{d}t} = \begin{bmatrix}0 & 0\\ 0 & -2\end{bmatrix}z(t)+\frac{1}{\sqrt{2}}\begin{bmatrix}1 & 1\\ -1 & 1\end{bmatrix}u(t)$$

$$J(u) = \frac{1}{2}\int_0^{\infty}(z^T(t)z(t)+u^T(t)u(t))\mathrm{d}t$$

and hence that the solution has the feedback form

$$u^*(t) = -\frac{1}{\sqrt{2}}\begin{bmatrix}1 & -1\\ 1 & 1\end{bmatrix}K_\infty^* z^*(t) = -\frac{1}{2}\begin{bmatrix}1 & -1\\ 1 & 1\end{bmatrix}K_\infty\begin{bmatrix}1 & 1\\ -1 & 1\end{bmatrix}x^*(t)$$

where K_∞^* is the solution of an algebraic Riccati equation. Show that K_∞^* is diagonal of the form

$$\begin{bmatrix}k_{11} & 0\\ 0 & k_{22}\end{bmatrix}, \quad k_{11}>0, \;\; k_{22}>0, \;\; 1-k_{11}^2 = 0, \;\; 1-k_{22}^2-4k_2 = 0$$

Hence solve the original problem.

(8) An nth order integrating process is described by the differential equation

$$\frac{\mathrm{d}^n y(t)}{\mathrm{d}t^n} = u(t) \qquad (u(t) \text{ unconstrained})$$

If this system is to be controlled from the initial condition $y(0)=(x_0)_1$, $\mathrm{d}y(t)/\mathrm{d}t|_{t=0}=(x_0)_2, \ldots, \mathrm{d}^{n-1}y(t)/\mathrm{d}t^{n-1}|_{t=0}=(x_0)_n$ to the final condition $y(t_f)=(x_f)_1$, $\mathrm{d}y(t)/\mathrm{d}t|_{t=t_f}=(x_f)_2, \ldots, \mathrm{d}^{n-1}y(t)/\mathrm{d}t^{n-1}|_{t=t_f}=(x_f)_n$ whilst minimizing the performance index $J(u)=\frac{1}{2}\int_0^{t_f}u^2(t)\mathrm{d}t$, verify that the problem can be formulated in the state-variable form

$$\frac{dx(t)}{dt} = \begin{bmatrix} 0 & 1 & 0 & \cdots & 0 \\ 0 & 0 & & & \vdots \\ \vdots & & & & 0 \\ \vdots & & & & 1 \\ 0 & \cdots & \cdots & 0 & 0 \end{bmatrix} x(t) + \begin{bmatrix} 0 \\ 0 \\ \vdots \\ 0 \\ 1 \end{bmatrix} u(t)$$

with initial and terminal conditions $x(0) = x_0, x(t_f) = x_f$. Show that the costate variables for this problem are solutions of the equations

$$\frac{dp_1(t)}{dt} = 0, \qquad \frac{dp_2(t)}{dt} = -p_1(t), \ldots, \qquad \frac{dp_n(t)}{dt} = -p_{n-1}(t)$$

and that the optimal controller $u^*(t) = -p_n(t)$. Hence verify that

$$u^*(t) = -\sum_{k=1}^{n} (-1)^{n-k} p_k(0) \frac{t^{n-k}}{(n-k)!}$$

where the initial conditions $p_k(0)$, $1 \leqslant k \leqslant n$, are chosen to satisfy the state boundary conditions. In the case of $n = 3$, $x_0 = (0, 0, 0)^T$, $x_f = (1, 0, 0)^T$, $t_f = 1$, show that $u^*(t) = 60 - 360t + 360t^2$.

(9) Show that the unconstrained minimum energy problem

$$\frac{dx(t)}{dt} = \begin{bmatrix} -1 & 1 \\ 0 & -2 \end{bmatrix} x(t) + \begin{bmatrix} 0 \\ 1 \end{bmatrix} u(t), \qquad x(0) = \begin{bmatrix} 0 \\ 0 \end{bmatrix}, \qquad x(1) = \begin{bmatrix} 1 \\ 0 \end{bmatrix}$$

$J(u) = \frac{1}{2} \int_0^1 u^2(t) dt$ is equivalent to the (simpler!) minimum energy problem

$$\frac{dz(t)}{dt} = \begin{bmatrix} -1 & 0 \\ 0 & -2 \end{bmatrix} z(t) + \begin{bmatrix} \alpha^{-1} \\ \beta^{-1} \end{bmatrix} u(t), \qquad z(0) = \begin{bmatrix} 0 \\ 0 \end{bmatrix}, \qquad z(1) = \begin{bmatrix} \alpha^{-1} \\ 0 \end{bmatrix}$$

with the same performance index and where α and β are non-zero scalars. (Hint: use an eigenvector transformation. The actual value of α and β will depend upon the transformation that you choose.) Hence verify that the optimal controller is $u^*(t) = -\alpha^{-1} p_1(0) e^t - \beta^{-1} p_2(0) e^{2t}$, where $p_1(0)$ and $p_2(0)$ are the costate boundary conditions satisfying the algebraic equations

$$e = \alpha^{-1} p_1(0) \frac{(1 - e^2)}{2} + \beta^{-1} p_2(0) \frac{(1 - e^3)}{3}$$

$$0 = \alpha^{-1} p_1(0) \frac{(1 - e^3)}{3} + \beta^{-1} p_2(0) \frac{(1 - e^4)}{4}$$

(10) Apply variational calculus to the unconstrained, nonlinear optimal control problem

$$\frac{dx(t)}{dt} = \begin{bmatrix} x_1(t) - x_1(t)x_2(t) + u(t) \\ -x_2(t) + x_1(t) \end{bmatrix}, \qquad x(0) = \begin{bmatrix} 1 \\ 1 \end{bmatrix}$$

$$J(u) = \frac{1}{2}(x_1(1))^2 + \frac{1}{2}\int_0^1 ((x_2(t))^4 + u^2(t))dt$$

More precisely, define the terminal function $\phi = \frac{1}{2}(x_1(1))^2$ and the Hamiltonian

$$H = \tfrac{1}{2}x_2^4 + \tfrac{1}{2}u^2 + p_1(x_1 - x_1x_2 + u) + p_2(-x_2 + x_1)$$

and compute the derivatives

$$\frac{\partial H}{\partial x_1} = p_1(1 - x_2) + p_2, \qquad \frac{\partial H}{\partial x_2} = 2x_2^3 - p_1x_1 - p_2$$

$$\frac{\partial \phi}{\partial x_1} = x_1, \qquad \frac{\partial \phi}{\partial x_2} = 0, \qquad \frac{\partial H}{\partial u} = u + p_1$$

Finally show that the optimal controller $u^*(t)$ is a solution of the TPBVP

$$\frac{dx^*(t)}{dt} = \begin{bmatrix} x_1^*(t)(1 - x_2^*(t)) + u^*(t) \\ -x_2^*(t) + x_1^*(t) \end{bmatrix}, \qquad x^*(0) = \begin{bmatrix} 1 \\ 1 \end{bmatrix}$$

$$u^*(t) = -p_2(t)$$

$$\frac{dp(t)}{dt} = \begin{bmatrix} -p_1(t)(1 - x_2^*(t)) - p_2(t) \\ -2(x_2^*(t))^3 + p_1(t)x_1^*(t) + p_2(t) \end{bmatrix}, \qquad p(1) = \begin{bmatrix} x_1^*(1) \\ 0 \end{bmatrix}$$

(Note: this TPBVP cannot be solved analytically.)

Remarks and Further Reading

Optimal control is a big subject in its own right with a scope that extends from the simple beginnings outlined in this chapter to the control of differential delay systems and systems described by partial differential equations (Curtain and Pritchard, 1979) and beyond that into the general mathematical discipline of optimization theory (see, for example, Burley, 1974; Luenberger, 1969). There are many texts that are easily accessible to the engineering reader, e.g. Layton (1976), Takahashi *et al.* (1972), Ogata (1967), Prime (1969), Noton (1965), Lawden (1975), Greensite (1970) and Burley (1974). There are also many specialist texts, e.g. Anderson and Moore (1971), Athans and Falb (1966),

Bryson and Ho (1969), Kwakernaak and Sivan (1972), Citron (1969), Sage (1968) and Kirk (1970). Detailed examples and descriptions of applications of optimal control can be found in these texts and others including the text edited by Munro (1969a). The reader interested in more advanced mathematical treatments should see, for example, Craven (1978), Adby and Dempster (1974), Sagan (1969) or Luenberger (1969).

The following individual points may be of particular interest to the reader:

(1) In almost all of the material in this chapter (the exception being Section 5.2.2), the plant model and Q and R could be time-varying with no change in the obtained results.

(2) In the case of optimal regulation on an infinite time interval, it is generally true that the closed-loop system is asymptotically stable. This can be deduced by noting that $J(u^*) = \frac{1}{2}x^T(0)K_\infty x(0)$ is a Liapunov function for the closed-loop system (see Anderson and Moore, 1971).

(3) Some information on the effects of non-linearities introduced at the stage of implementation of linear optimal regulators can be found in Anderson and Moore (1971).

(4) Interpretations of optimality in the frequency domain can be found in Kalman (1964) and MacFarlane (1970c).

(5) The choice of Q, R and F are design decisions that are not completely straightforward. Trial and error choice is an obvious way out of this difficulty. Alternatively systematic techniques (e.g. Stein, 1979; Owens, 1979) based on the theory of optimal system root-loci (see, for example, Owens, 1980; Postlethwaite, 1979) can be attempted, although it is not yet clear how successful these techniques are in practice.

(6) The implementation of the optimal feedback laws derived in Section 5.2.2 suffers from the normal problems of state feedback, i.e. the need to measure all of the system states. The obvious way out of this difficulty is to use an observer but it must be recognized that the resulting system is then not optimal. This is not necessarily a problem and, indeed, in the case of large scale systems control (Singh, 1977), suboptimal designs are of great importance.

(7) The mathematical methods of functional analysis can be of great help in unifying, extending and providing insight into optimization problems. Luenberger (1969), Leigh (1980) and Porter (1966) provide useful introductions to these techniques for control theorists.

(8) A description of feedback solutions to linear quadratic and minimum energy problems with terminal constraints can be found, for example, in Bryson and Ho (1969).

(9) Introductory accounts of variational calculus can be found in Layton (1976) and Burley (1974). More detailed treatments can be obtained in Luenberger (1969) and Sagon (1969).

6. Optimal Control in the Presence of Control Constraints

The material described in Chapter five provides an introduction to almost all of the essential concepts and problems arising in the non-classical discipline of optimal control. Most of the remaining new features arise in optimal control problems where the plant inputs $u_1(t), \ldots, u_l(t)$ are subjected to *constraints*. In such cases the methods of variational calculus do not apply (basically because the presence of constraints means that $\delta u(t)$ in (5.147) cannot necessarily be chosen arbitrarily and hence we cannot deduce that (5.148) holds), although, as we shall see, a modified TPBVP will still describe the optimal solution. It is the aim of this chapter to introduce this material at a level that highlights the source and structure of several of the main results in this area without the need for complex mathematics and prepares the reader for more advanced theoretical or applications studies.

The reader will have noted that the solution of the TPBVP can be a nontrivial matter. He will not be reassured or surprised to find out that the introduction of control constraints into the problem simply makes things much worse! For this reason we devote the whole of the first section (Section 6.1) to the detailed analysis of a class of optimal control problems that are of little relevance to applications but have the property that the structure of the TPBVP problem is so simple that we can concentrate entirely on the new features introduced by the control constraints. In this way we obtain a painless introduction to the effect of constraints that forms a useful foundation for the "more relevant" problems introduced in Sections 6.2 and 6.3.

6.1 Optimal Control with Performance Index Linear in the States

As a simple vehicle for the analysis of the effect of constraints on optimal control problems, consider the following linear, time-invariant system

$$\frac{\mathrm{d}x(t)}{\mathrm{d}t} = Ax(t) + Bu(t), \qquad x(0) = x_0 \tag{6.1}$$

with inputs subjected to constraints (see Section 5.1)

$$u(t) \in \Omega(t), \qquad t \geqslant 0 \tag{6.2}$$

and suppose that the system is to be controlled in such a manner that the performance index

$$J(u) = \sum_{k=1}^{n} \alpha_k x_k(t_f) + \int_0^{t_f} \left\{ \sum_{k=1}^{n} \beta_k(t) x_k(t) + g(u(t), t) \right\} \mathrm{d}t \tag{6.3}$$

is minimized. The terminal time t_f is assumed to be prespecified and the final state $x(t_f)$ is unspecified. An alternative description of the performance index uses the definitions

$$\alpha = \begin{bmatrix} \alpha_1 \\ \alpha_2 \\ \cdot \\ \cdot \\ \alpha_n \end{bmatrix}, \qquad \beta(t) = \begin{bmatrix} \beta_1(t) \\ \beta_2(t) \\ \cdot \\ \cdot \\ \beta_n(t) \end{bmatrix} \tag{6.4}$$

when it is clear that

$$J(u) = \alpha^T x(t_f) + \int_0^{t_f} \{\beta^T(t) x(t) + g(u(t), t)\} \mathrm{d}t \tag{6.5}$$

The performance index is linear in the state variables and hence, for simplicity, the optimal control problem defined by (6.1), (6.2) and (6.5) will be termed the "linear cost problem".

6.1.1 *Derivation of the TPBVP*

We produce the required relations defining the optimal controller by analysis of the variation of J with u. Let $u^{(1)}(t)$ and $u^{(2)}(t)$, $0 \leqslant t \leqslant t_f$, be two admissible controllers generating state trajectories $x^{(1)}(t)$ and $x^{(2)}(t)$ originating at the same initial state x_0 at $t = 0$, and let $p(t)$ be an arbitrary (for the moment that is) $n \times 1$ matrix function of time. Then integration by parts yields the *identity*

$$\int_0^{t_f} \left\{ p^T(t) \left(\frac{\mathrm{d}x^{(2)}(t)}{\mathrm{d}t} - \frac{\mathrm{d}x^{(1)}(t)}{\mathrm{d}t} \right) + \left(\frac{\mathrm{d}p(t)}{\mathrm{d}t} \right)^T (x^{(2)}(t) - x^{(1)}(t)) \right\} \mathrm{d}t$$

$$- p^T(t_f)(x^{(2)}(t_f) - x^{(1)}(t_f)) + p^T(0)(x^{(2)}(0) - x^{(1)}(0)) = 0 \tag{6.6}$$

But $x^{(2)}(0) = x^{(1)}(0)$ and

$$\frac{\mathrm{d}x^{(2)}(t)}{\mathrm{d}t} - \frac{\mathrm{d}x^{(1)}(t)}{\mathrm{d}t} = A(x^{(2)}(t) - x^{(1)}(t)) + B(u^{(2)}(t) - u^{(1)}(t)) \tag{6.7}$$

so that (6.6) takes the form, after a little manipulation,

$$\int_0^{t_f}\left\{\left(\frac{dp(t)}{dt}+A^Tp(t)\right)^T(x^{(2)}(t)-x^{(1)}(t))+p^T(t)B(u^{(2)}(t)-u^{(1)}(t))\right\}dt$$

$$-p^T(t_f)(x^{(2)}(t_f)-x^{(1)}(t_f)) = 0 \tag{6.8}$$

Adding the left-hand side of this expression to the expression for $J(u^{(2)})-J(u^{(1)})$ yields the following formula for the difference between the values of the performance index

$$J(u^{(2)})-J(u^{(1)}) = (\alpha-p(t_f))^T(x^{(2)}(t_f)-x^{(1)}(t_f))$$

$$+\int_0^{t_f}\{p^T(t)B(u^{(2)}(t)-u^{(1)}(t))+g(u^{(2)}(t),t)-g(u^{(1)}(t),t)\}dt$$

$$+\int_0^{t_f}\left(\frac{dp(t)}{dt}+A^Tp(t)+\beta(t)\right)^T(x^{(2)}(t)-x^{(1)}(t))dt \tag{6.9}$$

We will now choose our costate vector $p(t)$ to eliminate the dependence of $J(u^{(2)})-J(u^{(1)})$ on $x^{(2)}$ (we wish to assess the dependence of J on the controller, not the state). More precisely, choose $p(t)$ to be the solution of the differential equations

$$\frac{dp(t)}{dt} = -A^Tp(t)-\beta(t) \tag{6.10}$$

subject to the terminal boundary condition

$$p(t_f) = \alpha \tag{6.11}$$

With this choice (6.9) reduces to

$$J(u^{(2)})-J(u^{(1)}) =$$

$$\int_0^{t_f}\{p^T(t)B(u^{(2)}(t)-u^{(1)}(t))$$

$$+g(u^{(2)}(t),t)-g(u^{(1)}(t),t)\}dt \tag{6.12}$$

or, following the development of Section 5.4, introducing the Hamiltonian (c.f. (5.143))

$$H(x,p,u,t) = \beta^T(t)x+g(u,t)+p^T(Ax+Bu) \tag{6.13}$$

it is easily verified that (6.12) can be written in the form

$$J(u^{(2)}) - J(u^{(1)}) =$$

$$\int_0^{t_f} \{H(x^{(1)}(t), p(t), u^{(2)}(t), t) - H(x^{(1)}(t), p(t), u^{(1)}(t), t)\}\mathrm{d}t \quad (6.14)$$

(Note: the same state vector $x^{(1)}(t)$ occurs in both Hamiltonians in (6.14).)

Suppose now that $u^{(1)}$ is an optimal controller u^*, then we must have

$$J(u^{(2)}) \geqslant J(u^{(1)}) \quad (6.15)$$

for every choice of admissible controller $u^{(2)}$. Equation (6.14) then implies that

$$\int_0^{t_f} \{H(x^{(1)}(t), p(t), u^{(2)}(t), t) - H(x^{(1)}(t), p(t), u^{(1)}(t), t)\}\mathrm{d}t \geqslant 0 \quad (6.16)$$

for every choice of admissible controller $u^{(2)}$. But, for this to be the case, it is both necessary and sufficient that the integrand is positive at each point in the interval $0 \leqslant t \leqslant t_f$, i.e.

$$H(x^{(1)}(t), p(t), u^{(2)}(t), t) \geqslant H(x^{(1)}(t), p(t), u^{(1)}(t), t) \quad (6.17)$$

for all $u^{(2)}(t) \in \Omega(t)$ and $0 \leqslant t \leqslant t_f$

which simply means that, at each point of time t in the interval $0 \leqslant t \leqslant t_f$, the Hamiltonian function assumes its minimum value with respect to all controls satisfying the constraints when evaluated at the optimal control. We will use the compact notation

$$H(x^{(1)}(t), p(t), u^{(1)}(t), t) = \min_{u \in \Omega(t)} H(x^{(1)}(t), p(t), u, t)\,, \quad 0 \leqslant t \leqslant t_f \quad (6.18)$$

to signify this result.

In summary the above analysis has proved that a necessary and sufficient condition for an admissible controller $u^*(t)$ to be an optimal controller for the linear cost problem is that it is a solution of the TPBVP

$$\frac{\mathrm{d}x^*(t)}{\mathrm{d}t} = Ax^*(t) + Bu^*(t), \qquad x^*(0) = x_0$$

$$u^*(t) \in \Omega(t)$$

$$\frac{\mathrm{d}p(t)}{\mathrm{d}t} = -A^T p(t) - \beta(t), \qquad p(t_f) = \alpha \quad (6.19)$$

$$H(x^*(t), p(t), u^*(t), t) = \min_{u \in \Omega(t)} H(x^*(t), p(t), u, t)\,, \quad 0 \leqslant t \leqslant t_f \quad (6.20)$$

This TPBVP has an obvious structural similarity to the TPBVP (5.149) obtained from the use of variational calculus. The main difference is that the relations

$$\left.\frac{\partial H}{\partial u_k}\right|_* = 0, \qquad 1 \leqslant k \leqslant l$$

are replaced by the algebraic Hamiltonian minimization problem (6.20). In general the solution of (6.20) does not occur at a stationary point of the Hamiltonian (see Fig. 67(a)) although, if the problem is unconstrained, the solution does reduce to the calculation of the relevant solution of the relations

$$\left.\frac{\partial H}{\partial u_k}\right|_* = 0, \qquad 1 \leqslant k \leqslant l \tag{6.21}$$

(see Fig. 67(b)).

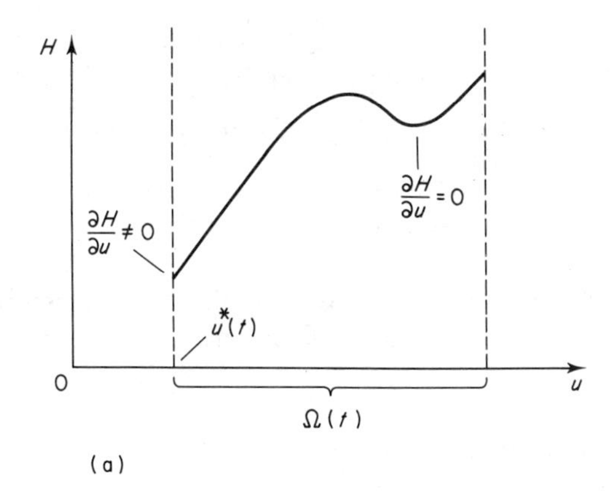

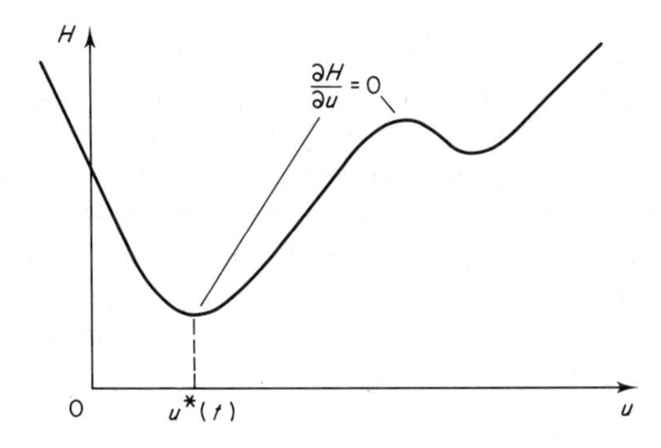

Fig. 67. (a) Hamiltonian minimization in the presence of constraints. (b) Hamiltonian minimization in the absence of constraints.

Finally, before moving on to look at some typical problems and, in particular, the effect of control weighting $g(u, t)$ on optimal control structure, it is useful and instructive to outline the procedure for solving the TPBVP in the following step by step manner. Note that steps 1 and 3 are straightforward but that step 2 requires careful computational and conceptual consideration.

Step 1. Noting that the costate equation and its boundary condition are independent of both x^* and u^*, solve for $p(t)$ directly. (Note: it is this property that simplifies this particular TPBVP.)

Step 2. Solve for the optimal controller $u^*(t)$ at each point of time $0 \leqslant t \leqslant t_f$ by minimizing the Hamiltonian

$$\begin{aligned} H(x^*(t), p(t), u, t) &= \beta^T(t)x^*(t) + g(u, t) + p^T(t)(Ax^*(t) + Bu) \\ &= g(u, t) + p^T(t)Bu + \text{terms independent of } u \quad (6.22) \end{aligned}$$

with respect to all controllers satisfying the constraints $\Omega(t)$. It is clear that this problem is equivalent to the minimization of the "reduced Hamiltonian"

$$H_r(p(t), u, t) = g(u, t) + p^T(t)Bu \quad (6.23)$$

with respect to all controllers satisfying the constraints $\Omega(t)$ and hence that we do not need to know $x^*(t)$ in order to compute $u^*(t)$.

Step 3. Integrate the state equations with the optimal controller $u^*(t)$ and known initial condition to obtain the optimal state trajectory $x^*(t)$. Hence compute $J(u^*)$.

6.1.2 *Energy weighting of control signals*

Consider the linear cost problem with

$$g(u, t) = \tfrac{1}{2}u^T Ru \quad (6.24)$$

where R is an $l \times l$ symmetric, positive-definite matrix. The solution to the optimal control problem can be approached in the step by step manner outlined above. Suppose that the costate trajectory $p(t)$, $0 \leqslant t \leqslant t_f$, has been computed and concentrate on the calculation of the optimal controller $u^*(t)$ by solving the Hamiltonian minimization problem (6.20) or, equivalently, the minimization of the reduced Hamiltonian with respect to all controllers satisfying the constraints $\Omega(t)$. For simplicity, assume that the constraints take the form of amplitude limitations,

$$-M_k^{(1)} \leqslant u_k(t) \leqslant M_k^{(2)}, \qquad 1 \leqslant k \leqslant l, \qquad 0 \leqslant t \leqslant t_f \quad (6.25)$$

There are two separate cases to consider.

The case of unconstrained control. In this case (represented by taking $M_k^{(1)} = M_k^{(2)} = +\infty$, $1 \leqslant k \leqslant l$) the problem of minimizing H_r with respect to u reduces to the solution of the equalities

$$\left.\frac{\partial H_r}{\partial u_k}\right|_* = 0, \qquad 1 \leqslant k \leqslant l \tag{6.26}$$

The reader should verify that these equations can be written in the matrix form

$$Ru^*(t) + B^T p(t) = 0, \qquad 0 \leqslant t \leqslant t_f \tag{6.27}$$

and hence that the optimal controller is given explicitly by the formula (c.f. (5.62) and (5.92))

$$u^*(t) = -R^{-1} B^T p(t), \qquad 0 \leqslant t \leqslant t_f \tag{6.28}$$

The case of constrained control. Even in this relatively simple case, there is no analytical way of representing the solution. We will therefore restrict our attention to the special case when

$$g(u, t) = \tfrac{1}{2} \sum_{k=1}^{l} r_k u_k^2 \qquad \left(\text{i.e. } R = \begin{bmatrix} r_1 & 0 & 0 \\ 0 & r_2 & \\ 0 & & \ddots \; r_l \end{bmatrix}\right) \tag{6.29}$$

where r_k, $1 \leqslant k \leqslant l$, are real, strictly positive scalars. It is convenient to define the $l \times 1$ matrix

$$z(t) = B^T p(t), \qquad 0 \leqslant t \leqslant t_f \tag{6.30}$$

when our problem reduces to the minimization of

$$\begin{aligned} H_r(p(t), u, t) &= \sum_{k=1}^{l} \{\tfrac{1}{2} r_k u_k^2 + z_k(t) u_k\} \\ &= \sum_{k=1}^{l} r_k \left\{\tfrac{1}{2} u_k^2 + \frac{z_k(t)}{r_k} u_k\right\} \end{aligned} \tag{6.31}$$

with respect to controls satisfying (6.25). Finally, as there are no "coupling terms" in (6.31) and (6.25), the minimization of H_r reduces yet again to the solution of l distinct minimization problems of the form:

$$\text{Minimize } H_r^{(k)} = \tfrac{1}{2} u_k^2 + \frac{z_k(t)}{r_k} u_k \text{ with respect to}$$

$$\text{all controllers } u_k \text{ satisfying the constraints } -M_k^{(1)} \leqslant u_k \leqslant M_k^{(2)} \tag{6.32}$$

The presence of constraints means that the solution of these problems cannot, in general, be obtained by solving (6.26). A more involved approach is required.

The simplest approach to the solution of (6.32) uses a graphical argument to verify that

$$u_k^*(t) = \begin{cases} M_k^{(2)} & \text{if} \quad -\dfrac{z_k(t)}{r_k} > M_k^{(2)} \\ -\dfrac{z_k(t)}{r_k} & \text{if} \quad -M_k^{(1)} \leqslant -\dfrac{z_k(t)}{r_k} \leqslant M_k^{(2)} \\ -M_k^{(1)} & \text{if} \quad -\dfrac{z_k(t)}{r_k} < -M_k^{(1)} \end{cases}$$

$$0 \leqslant t \leqslant t_f, \qquad 1 \leqslant k \leqslant l \tag{6.33}$$

i.e. the optimal control $u_k^*(t)$ is obtained by "clipping-off" the function $-z_k(t)/r_k$ as illustrated in Fig. 68. The graphical proof is outlined below.

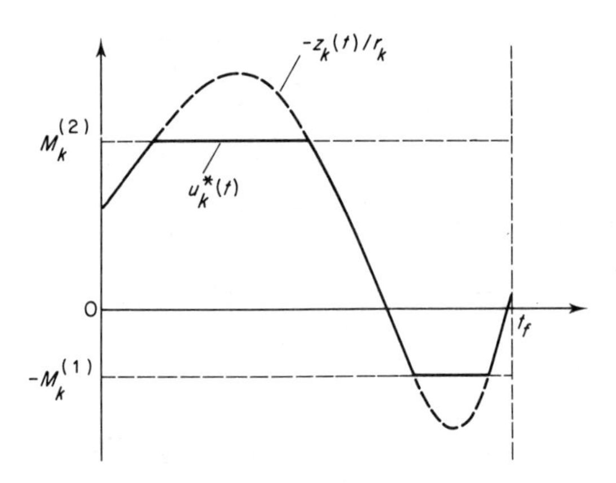

Fig. 68. Constrained control and the "clipping-off" procedure.

A simple differentiation argument easily shows that the parabola $\frac{1}{2}u_k^2 + (z_k(t)/r_k)u_k$ has an absolute minimum (in the absence of constraints) when $u_k = -z_k(t)/r_k$. Suppose initially that $-M_k^{(1)} \leqslant -(z_k(t)/r_k) \leqslant M_k^{(2)}$ then, using Fig. 69(a), it is clear that $H_r^{(k)}(u_k)$ has a minimum with respect to all controllers satisfying the constraints at the point $-(z_k(t)/r_k)$. Suppose now that $-(z_k(t)/r_k) > M_k^{(2)}$, then it is not an admissible control value, but using Fig. 69(b) it can easily be deduced that $H_r^{(k)}(u_k)$ has a minimum with respect to all controller values satisfying the constraints at the point $M_k^{(2)}$. If $-(z_k(t)/r_k) < -M_k^{(1)}$ then a similar graphical argument indicates that $H_r^{(k)}$ is minimized with respect to admissible control values at the point $-M_k^{(1)}$, which completes the verification of (6.33).

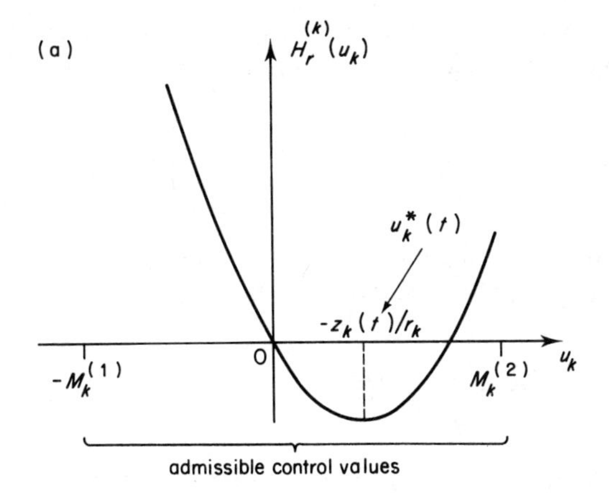

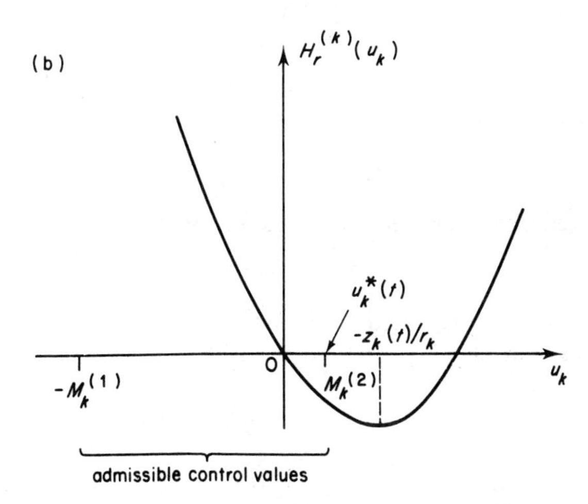

Fig. 69. Hamiltonian minimization. (a) $-M_k^{(1)} \leqslant -z_k(t)/r_k \leqslant M_k^{(2)}$. (b) $-z_k(t)/r_k > M_k^{(2)}$.

EXAMPLE 6.1.1. Consider the solution of the linear cost problem

$$\frac{dx(t)}{dt} = \begin{bmatrix} 0 & 1 \\ 0 & 0 \end{bmatrix} x(t) + \begin{bmatrix} 0 \\ 1 \end{bmatrix} u(t), \qquad x(0) = \begin{bmatrix} 0 \\ 0 \end{bmatrix}$$

$$J(u) = x_1(t_f) + \int_0^{t_f} \tfrac{1}{2} \lambda u^2(t) dt \tag{6.34}$$

where $\lambda > 0$ and we initially suppose that the input $u(t)$ is unconstrained, i.e. the amplitude limits $M_k^{(1)} = M_k^{(2)} = +\infty$. The system data is

$$A = \begin{bmatrix} 0 & 1 \\ 0 & 0 \end{bmatrix}, \qquad B = \begin{bmatrix} 0 \\ 1 \end{bmatrix}, \qquad x_0 = \begin{bmatrix} 0 \\ 0 \end{bmatrix} \tag{6.35}$$

and, by comparing the performance index with (6.3), we have $\alpha_1 = 1$, $\alpha_2 = 0$, $\beta_1(t) \equiv \beta_2(t) \equiv 0$ or, in vector form,

$$\alpha = \begin{bmatrix} 1 \\ 0 \end{bmatrix}, \qquad \beta(t) = \begin{bmatrix} 0 \\ 0 \end{bmatrix} \quad \text{and} \quad g(u, t) = \tfrac{1}{2}\lambda u^2 \tag{6.36}$$

It follows immediately that the costate equation (6.10) and its boundary condition (6.11) take the form

$$\frac{dp(t)}{dt} = \begin{bmatrix} 0 & 0 \\ -1 & 0 \end{bmatrix} p(t), \qquad p(t_f) = \begin{bmatrix} 1 \\ 0 \end{bmatrix} \tag{6.37}$$

yielding the solutions

$$p_1(t) = 1 \qquad p_2(t) = t_f - t \tag{6.38}$$

The "reduced Hamiltonian" H_r has the form

$$H_r = \tfrac{1}{2}\lambda u^2 + p_2(t)u \tag{6.39}$$

which is minimized at the solution of the relation $\partial H/\partial u = 0$, i.e. the optimal controller is given by the relation

$$u^*(t) = -\lambda^{-1} p_2(t) = \frac{t - t_f}{\lambda} \tag{6.40}$$

which is illustrated in Fig. 70(a). It is left as an exercise for the reader to show that

$$x_1^*(t) = \lambda^{-1} t^2 \left(\frac{t}{6} - \frac{t_f}{2}\right), \qquad x_2^*(t) = \lambda^{-1} t \left(\frac{t}{2} - t_f\right)$$

$$J(u^*) = -\tfrac{1}{6}\lambda^{-1} t_f^3 \tag{6.41}$$

Consider now the problem (6.34) with the control amplitude constraint $|u(t)| \leqslant M$, $0 \leqslant t \leqslant t_f$, or, in the form of (6.25)

$$-M \leqslant u(t) \leqslant M \tag{6.42}$$

The solution for the costate (6.38) is unchanged by the introduction

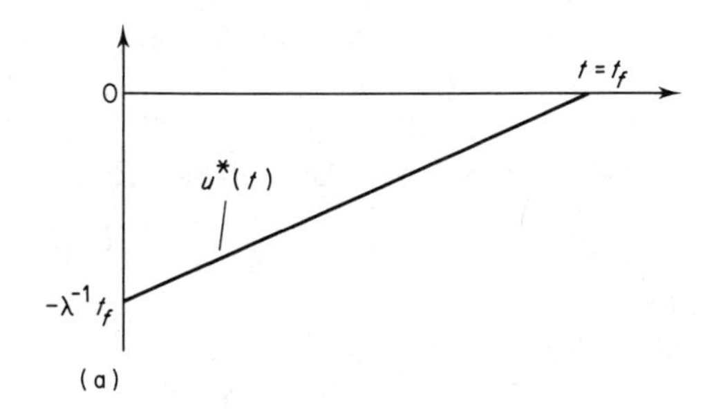

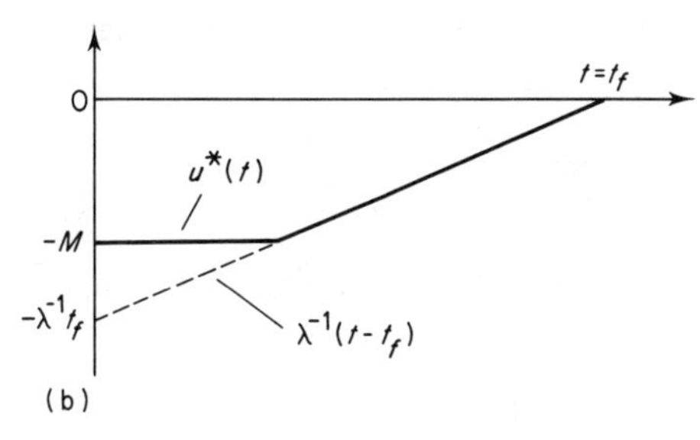

Fig. 70.

of these constraints. All that changes is the expression for $u^*(t)$ obtained from minimization of the reduced Hamiltonian, i.e. using a similar argument to the proof of (6.33)

$$u^*(t) = \begin{cases} M & \text{if} \quad -\lambda^{-1}p_2(t) > M \\ -\lambda^{-1}p_2(t) & \text{if} \quad |\lambda^{-1}p_2(t)| \leqslant M \\ -M & \text{if} \quad -\lambda^{-1}p_2(t) < -M \end{cases} \tag{6.43}$$

or, introducing the "sign function"

$$\operatorname{sgn} \eta = \begin{cases} 1 & \text{if} \quad \eta > 0 \\ \text{not defined} & \text{if} \quad \eta = 0, \\ -1 & \text{if} \quad \eta < 0 \end{cases} \tag{6.44}$$

we obtain

$$u^*(t) = \begin{cases} -\lambda^{-1}p_2(t) & \text{if} \quad |\lambda^{-1}p_2(t)| \leqslant M \\ -M \operatorname{sgn}(\lambda^{-1}p_2(t)) & \text{if} \quad |\lambda^{-1}p_2(t)| > M \end{cases} \tag{6.45}$$

There are now two possibilities, either (a) $|\lambda^{-1}p_2(t)| \leqslant M$ for all $0 \leqslant t \leqslant t_f$ or (b) $|\lambda^{-1}p_2(t)| > M$ in some time interval. It is trivially verified that case (a) holds if, and only if, $t_f \leqslant \lambda M$ and that in this case, the optimal controller is simply that obtained above in the case of unconstrained control (Fig. 70(a)). Case (b) holds if, and only if, $t_f > \lambda M$, the optimal controller taking the form illustrated in Fig. 70(b). (Note: in this case the optimal controller in the presence of constraints can be obtained from the unconstrained optimal controller by the "clipping-off" procedure. It is emphasized that this is not, in general, the case!)

EXERCISE 6.1.1. Verify that the solution of the problem

$$\frac{dx(t)}{dt} = -x(t) + u(t), \qquad x(0) = 0$$

$$J(u) = 2x(1) + \int_0^1 (x(t) + \tfrac{1}{2}u^2(t))\,dt \tag{6.46}$$

in the presence of the constraint $|u(t)| \leqslant M$ is the optimal control

$$u^*(t) = \begin{cases} -p(t) & \text{if} \quad |p(t)| \leqslant M \\ -M \operatorname{sgn} p(t) & \text{if} \quad |p(t)| > M \end{cases} \tag{6.47}$$

where $p(t) = 1 + e^{t-1}$ is the solution of the costate equation

$$\frac{dp(t)}{dt} = p(t) - 1, \qquad p(1) = 2 \tag{6.48}$$

Hence verify that $u^*(t) \equiv -M$ if $M \leqslant 1 + e^{-1}$, that $u^*(t) = -1 - e^{t-1}$ if $M \geqslant 2$ and that

$$u^*(t) = \begin{cases} -1 - e^{t-1}, & 0 \leqslant t \leqslant 1 + \log_e(M-1) \\ -M & t \geqslant 1 + \log_e(M-1) \end{cases} \tag{6.49}$$

if $1 + e^{-1} \leqslant M \leqslant 2$. (Hint: consider the graphical form of $-p_2(t)$ and the clipping off procedure).

6.1.3 *Bang-bang control*

The linear cost problem with no control weighting

$$g(u, t) \equiv 0 \tag{6.50}$$

and the control amplitude constraints

$$-M_k^{(1)} \leqslant u_k(t) \leqslant M_k^{(2)}, \qquad 1 \leqslant k \leqslant l,\ 0 \leqslant t \leqslant t_f \tag{6.51}$$

give rise to a distinctive optimal control structure that is seen in many other more complex problems. In this case the Hamiltonian takes the form

$$H(x, p, u, t) = p^T Bu + \text{terms in state and costate only} \tag{6.52}$$

and hence the optimal controller can be obtained by minimizing the reduced Hamiltonian

$$H_r(p(t), u, t) = p(t)^T Bu \tag{6.53}$$

with respect to the constraints (6.51).

Using the definition (6.30) it is easily verified that

$$H_r = (B^T p(t))^T u = z^T(t)u(t) = \sum_{k=1}^{l} z_k(t)u_k \tag{6.54}$$

and hence that

(a) if $z_k(t) > 0$, H_r is minimized by choosing $u_k^*(t) = -M_k^{(1)}$

(b) if $z_k(t) < 0$, H_r is minimized by choosing $u_k^*(t) = M_k^{(2)}$

(c) if $z_k(t) = 0$ then H_r can be minimized by choosing any value of control in the range $-M_k^{(1)}$ to $M_k^{(2)}$.

For simplicity, we will restrict our attention to the case of constraints of the form

$$|u_k(t)| \leqslant M_k, \qquad 1 \leqslant k \leqslant l, \quad 0 \leqslant t \leqslant t_f \tag{6.55}$$

(and hence, $-M_k \leqslant u_k(t) \leqslant M_k$, $1 \leqslant k \leqslant l$, $0 \leqslant t \leqslant t_f$) when the above results can be expressed in terms of the sign function (6.44) in the form

$$u_k^*(t) = -M_k \operatorname{sgn}(z_k(t)), \qquad 1 \leqslant k \leqslant l, \quad 0 \leqslant t \leqslant t_f \tag{6.56}$$

The form of $u_k^*(t)$ can be deduced graphically by plotting $z_k(t)$ and applying the sign function definition as illustrated in Fig. 71. This form of control is called "bang-bang" or "switching control" as, except at times when $z_k(t) = 0$, the controller only takes its boundary values $\pm M_k$ and switches between these boundaries (Fig. 71) at the so-called "switching times" obtained from the solution of the equation

$$z_k(t) = 0 \tag{6.57}$$

In classical terms, the optimal control is just the response of an ideal relay to the input $-z_k(t)$.

EXAMPLE 6.1.2. Consider the linear cost problem

$$\frac{dx(t)}{dt} = \begin{bmatrix} 0 & 1 \\ 0 & 0 \end{bmatrix} x(t) + \begin{bmatrix} 0 \\ 1 \end{bmatrix} u(t), \qquad x(0) = \begin{bmatrix} 0 \\ 0 \end{bmatrix}$$

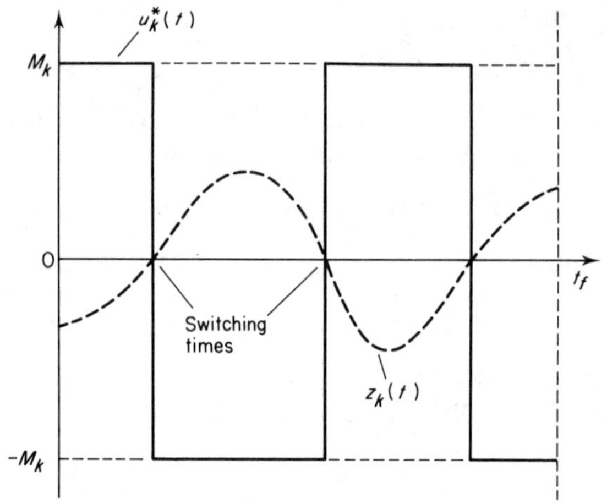

Fig. 71. Bang-bang optimal control.

$$|u(t)| \leqslant 1, \qquad J(u) = -x_1(1) + \frac{3}{16}x_2(1) + \int_0^1 2x_1(t)\mathrm{d}t \tag{6.58}$$

with the obvious data

$$A = \begin{bmatrix} 0 & 1 \\ 0 & 0 \end{bmatrix}, \quad B = \begin{bmatrix} 0 \\ 1 \end{bmatrix}, \quad \alpha = \begin{bmatrix} -1 \\ 3/16 \end{bmatrix}, \quad \beta(t) = \begin{bmatrix} 2 \\ 0 \end{bmatrix}, \quad t_f = 1 \tag{6.59}$$

The costate equations to be solved take the form

$$\frac{\mathrm{d}p(t)}{\mathrm{d}t} = \begin{bmatrix} 0 & 0 \\ -1 & 0 \end{bmatrix} p(t) - \begin{bmatrix} 2 \\ 0 \end{bmatrix}, \qquad p(1) = \begin{bmatrix} -1 \\ 3/16 \end{bmatrix} \tag{6.60}$$

or, in element form, $\mathrm{d}p_1(t)/\mathrm{d}t = -2$, $p_1(1) = -1$, $\mathrm{d}p_2(t)/\mathrm{d}(t) = -p_1(t), p_2(1) = 3/16$, i.e.

$$p_1(t) = 1 - 2t, \qquad p_2(t) = t^2 - t + 3/16 \tag{6.61}$$

The reduced Hamiltonian for the problem is $H_r = p^T Bu = p_2 u$ and hence, applying the above ideas, we deduce that

$$\begin{aligned} u^*(t) &= -\operatorname{sgn} p_2(t) \\ &= -\operatorname{sgn}\,(t - \tfrac{3}{4})(t - \tfrac{1}{4}) \\ &= \begin{cases} -1, & 0 \leqslant t < \tfrac{1}{4} \\ +1, & \tfrac{1}{4} < t < \tfrac{3}{4} \\ -1, & \tfrac{3}{4} < t \leqslant 1 \end{cases} \end{aligned} \tag{6.62}$$

as illustrated in Fig. 72. The switching times for the problem are clearly $t = \frac{1}{4}$ and $t = \frac{3}{4}$.

EXERCISE 6.1.2. Repeat the problem discussed in Example 6.1.2 with the performance index

$$J(u) = -\tfrac{3}{2}x_1(1) + \tfrac{1}{2}x_2(1) + 2\int_0^1 x_1(t)\,\mathrm{d}t \tag{6.63}$$

Verify that the resultant optimal controller takes the form

$$u^*(t) = -\operatorname{sgn}(t(t-\tfrac{1}{2})) = \begin{cases} 1, & 0 < t < \frac{1}{2} \\ -1, & \frac{1}{2} < t \leqslant 1 \end{cases} \tag{6.64}$$

with switching times at $t = 0$ and $t = \frac{1}{2}$.

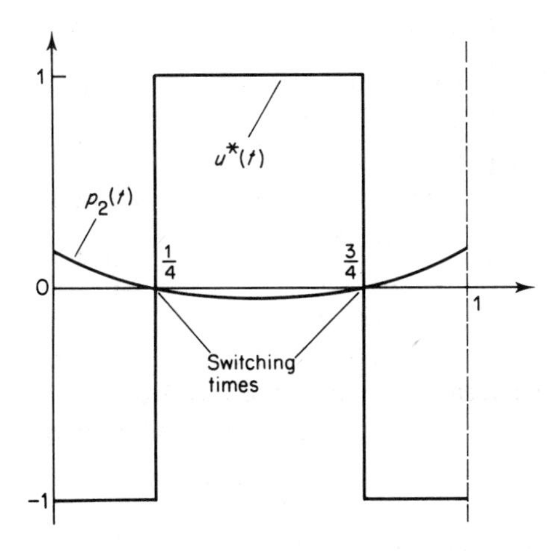

Fig. 72.

6.1.4 *Fuel weighting of control signals*

The linear cost problem with fuel weighting of control signals

$$g(u, t) = \sum_{k=1}^{l} r_k |u_k|, \qquad r_k > 0\,(1 \leqslant k \leqslant l) \tag{6.65}$$

and the amplitude constraints (6.51) on the control signals give rise to an optimal control with a distinct but, in many ways, similar structure to bang-bang control. The relevant Hamiltonian is simply

$$H(x, p, u, t) = \sum_{k=1}^{l} r_k |u_k| + p^T Bu + \text{terms in } x \text{ and } p \text{ only}$$

$$= \sum_{k=1}^{l} r_k \left\{ |u_k| + \frac{z_k}{r_k} u_k \right\} + \text{terms in } x \text{ and } p \text{ only} \tag{6.66}$$

It follows directly that the optimal controller $u_k^*(t)$ is a solution of the problem:

Minimize $|u_k| + r_k^{-1} z_k(t) u_k$ with respect to controllers satisfying the amplitude constraints (6.51) (6.67)

The solution to this takes the form (the proof is left until later in the section)

$$u_k^*(t) = \begin{cases} M_k^{(2)} & \text{if} \quad r_k^{-1} z_k(t) < -1 \\ \text{undefined} & \text{if} \quad r_k^{-1} z_k(t) = -1 \\ 0 & \text{if} \quad -1 < r_k^{-1} z_k(t) < 1 \\ \text{undefined} & \text{if} \quad r_k^{-1} z_k(t) = 1 \\ -M_k^{(1)} & \text{if} \quad r_k^{-1} z_k(t) > 1 \end{cases} \tag{6.68}$$

The graphical form of $u_k^*(t)$ can be obtained by consideration of a plot of $r_k^{-1} z_k(t)$ against time t as illustrated in Fig. 73. As can be seen the optimal control switches between its boundary values $M_k^{(2)}$ and $-M_k^{(1)}$ and the value of zero, the switching times being obtained from solution of the equations

$$r_k^{-1} z_k(t) = \pm 1 \tag{6.69}$$

Taking, for simplicity, the case of $M_k^{(1)} = M_k^{(2)} = M_k$, the control constraints take the form $|u_k(t)| \leq M$, $0 \leq t \leq t_f$, and Fig. 73 indicates that the optimal controller $u_k^*(t)$ is simply the output from an ideal relay with deadzone responding to the input $r_k^{-1} z_k(t)$.

Returning now to the proof of (6.68), consider initially the minimization of $|u_k| + r_k^{-1} z_k(t) u_k$ in the case of $r_k^{-1} z_k(t) < -1$. The easiest approach is to regard this expression as the summation of the two functions $|u_k|$ and $r_k^{-1} z_k(t) u_k$ as illustrated in Fig. 74(a) where it is self evident that the minimum occurs at the point $u_k^* = M_k^{(2)}$. Taking the case now of $r_k^{-1} z_k(t) = -1$ leads to the graphical form indicated in Fig. 74(b) where it is seen that any point in the range $0 \leq u_k^* \leq M_k^{(2)}$ achieves the minimum. The case of $-1 < r_k^{-1} z_k(t) \leq 0$ generates

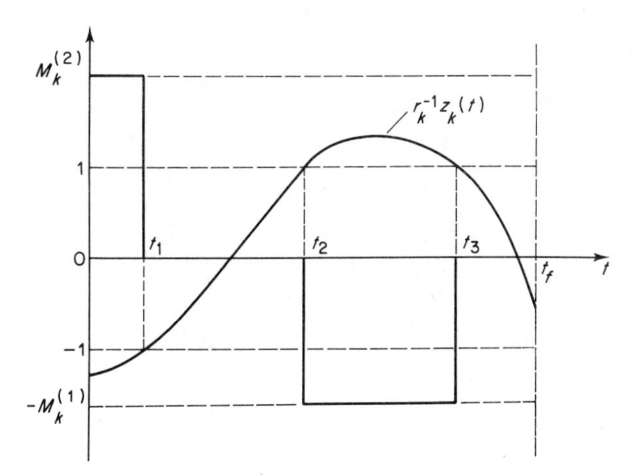

Fig. 73. Optimal control structure with fuel weighting.

Fig. 74(c) and hence the minimization at the point $u_k^* = 0$. It is left as an exercise for the reader to use similar arguments in the cases of $0 \leqslant r_k^{-1} z_k(t) < 1$, $r_k^{-1} z_k(t) = 1$ and $r_k^{-1} z_k(t) > 1$ to complete the verification argument.

EXAMPLE 6.1.3. Consider the linear cost problem

$$\frac{dx(t)}{dt} = \begin{bmatrix} 0 & 1 \\ 0 & 0 \end{bmatrix} x(t) + \begin{bmatrix} 0 \\ 1 \end{bmatrix} u(t), \qquad x(0) = \begin{bmatrix} 0 \\ 0 \end{bmatrix}$$

$$J(u) = -x_1(1) + \tfrac{3}{16} x_2(1) + \int_0^1 \{2x_1(t) + \tfrac{3}{32}|u(t)|\} dt \quad (6.70)$$

subjected to the control amplitude constraints $|u(t)| \leqslant 1$. This problem is identical to that of Example 6.1.2 except for the addition of the fuel control weighting in the performance index. The reader will be interested to note the significant effect on the optimal control strategy due to this simple change.

The data for this problem is given by (6.59) with $g(u, t) = \frac{3}{32}|u(t)|$ and the costate equations are given by (6.60) with solution (6.61). The Hamiltonian for the problem is simply

$$H(x, p, u, t) = \tfrac{3}{32}|u| + p_2 u + \text{terms in } x \text{ and } p \text{ only} \quad (6.71)$$

which indicates that the optimal controller $u^*(t)$ is generated by minimizing the reduced Hamiltonian $H_r = |u| + \frac{32}{3} p_2(t) u$ subject to the constraints $-1 \leqslant u \leqslant 1$. Using (6.68)

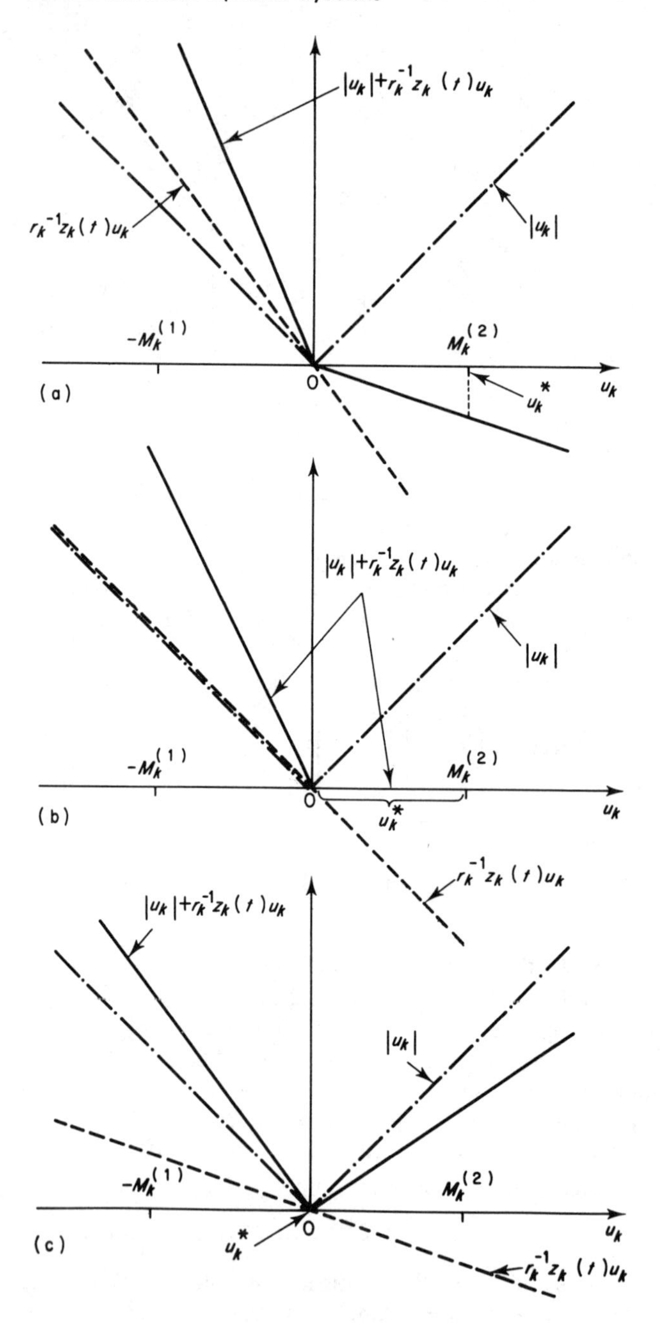

Fig. 74. Minimization of $|u_k| + r_k^{-1} z_k(t)u_k$.

$$u^*(t) = \begin{cases} 1 & \text{if} \quad \frac{32}{3}p_2(t) < -1 \\ \text{undefined} & \text{if} \quad \frac{32}{3}p_2(t) = -1 \\ 0 & \text{if} \quad -1 < \frac{32}{3}p_2(t) < 1 \\ \text{undefined} & \text{if} \quad \frac{32}{3}p_2(t) = 1 \\ -1 & \text{if} \quad \frac{32}{3}p_2(t) > 1 \end{cases} \tag{6.72}$$

or, using the construction of Fig. 75,

$$u^*(t) = \begin{cases} -1 & 0 \leqslant t < t_1 \\ 0 & t_1 < t < t_2 \\ -1 & t_2 < t \leqslant 1 \end{cases} \tag{6.73}$$

where the switching times $t_1 < t_2$ are the solutions of the equation $p_2(t) = 3/32$.

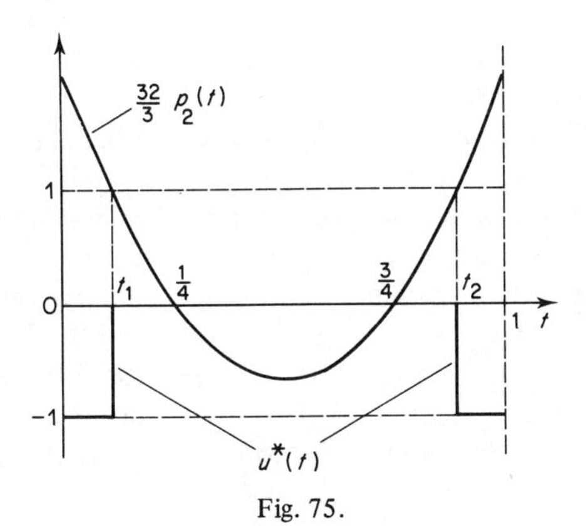

Fig. 75.

EXERCISE 6.1.3. With the problem described in Example 6.1.3 suppose that the control weighting in the performance index is decreased to give the new performance index

$$J(u) = -x_1(1) + \tfrac{3}{16}x_2(1) + \int_0^1 \{2x_1(t) + \tfrac{3}{64}|u(t)|\}\,dt \tag{6.74}$$

then verify that the optimal controller takes the form

EXERCISE 6.1.3. contd.

$$u^*(t) = \begin{cases} -1 & 0 \leqslant t < t_1 \\ 0 & t_1 < t < t_2 \\ +1 & t_2 < t < t_3 \\ 0 & t_3 < t < t_4 \\ -1 & t_4 < t < 1 \end{cases} \tag{6.75}$$

where t_1 and t_4 are the solutions of the equation $p_2(t) = 3/64$ and t_2 and t_3 are the solutions of the equation $p_2(t) = -3/64$. Also calculate the total amount of fuel used by (6.73) and (6.75) and conclude that a decrease in control weighting in the performance index tends to increase the total fuel usage.

6.1.5 *Physical interpretation of the costate and Hamiltonian*

It is clear from the examples described above that the structure of the optimal controller depends primarily upon the form of the costate and upon the nature of the Hamiltonian function and its dependence on the control vector u. It is natural, therefore, to ask whether or not the costate and Hamiltonian have an interpretation in physical terms that can perhaps explain their arrival in optimal control problems. The answer is very involved but we can provide a simple interpretation in the context of the linear cost problem that perhaps gives a hint to the reason for their general importance.

The clue lies in expression (6.14) which relates the difference between the performance indices, corresponding to two distinct controls $u^{(2)}$ and $u^{(1)}$, and the difference between two Hamiltonian functions. Consider the choice of $u^{(2)}$ of the form

$$u^{(2)}(t) = \begin{cases} u^{(1)}(t) & \text{if} \quad 0 \leqslant t < t' \quad \text{or} \quad t' + \delta < t \leqslant t_f \\ u^{(1)}(t) + \Delta u & \text{if} \quad t' \leqslant t \leqslant t' + \delta \end{cases} \tag{6.76}$$

where Δu is a constant perturbation and δ is a small positive real number. This control is represented schematically in Fig. 76, together with a schematic representation of the effect on the resulting response of the state vector. It is important to note that the state response in the interval $t < t'$ is unchanged by the control perturbation but that the state response can be changed significantly at "future" times $t > t'$. The resulting change in the performance index is given by (6.14)

$$J(u^{(2)}) - J(u^{(1)})$$

$$= \int_0^{t_f} \{H(x^{(1)}(t), p(t), u^{(2)}(t), t) - H(x^{(1)}(t), p(t), u^{(1)}(t), t)\}\,\mathrm{d}t$$

$$= \int_{t'}^{t'+\delta} \{H(x^{(1)}(t), p(t), u^{(2)}(t), t) - H(x^{(1)}(t), p(t), u^{(1)}(t), t)\}\,\mathrm{d}t$$

(by (6.76))

$$\simeq \delta\{H(x^{(1)}(t'), p(t'), u^{(1)}(t') + \Delta u, t') - H(x^{(1)}(t'), p(t'), u^{(1)}(t'), t')\} \tag{6.77}$$

if δ is small. That is, the change in the value of the performance index can be expressed entirely in terms of values of the states, costates and controls at the time $t = t'$. In particular it is seen that we do not need to know the future ($t > t'$) effect of the control perturbation to evaluate the change in J. We must therefore conclude that all the necessary information about this future is contained in the costate $p(t')$ and the structure of the Hamiltonian! More precisely the costate $p(t')$(which is obtained by integrating (6.10) subject to (6.11) over future times $t' < t \leqslant t_f$) contains all the relevant information about future effects of control perturbations at $t = t'$ and the Hamiltonian describes the way that this information can be used to describe the change in the value of the performance index J.

EXERCISE 6.1.4. Show that

$$x^{(2)}(t) - x^{(1)}(t) = \begin{cases} 0 & 0 \leqslant t \leqslant t' \\ \int_{t'}^{t'+\delta} \mathrm{e}^{A(t-t'')} B\,\Delta u\,\mathrm{d}t'', & t' + \delta \leqslant t \leqslant t_f \end{cases} \tag{6.78}$$

and hence that $x^{(2)}(t) - x^{(1)}(t) \simeq \mathrm{e}^{A(t-t')} B\,\Delta u\,\delta$ if δ is small enough and $t \geqslant t' + \delta$.

EXERCISE 6.1.5. Assuming, for simplicity, the case of a single-input system, prove that

$$\frac{\partial J}{\partial \Delta u} = \int_{t'}^{t'+\delta} \frac{\partial}{\partial u} H(x^{(1)}(t), p(t), u, t)\Big|_{u = u^{(1)}(t)} \mathrm{d}t \tag{6.79}$$

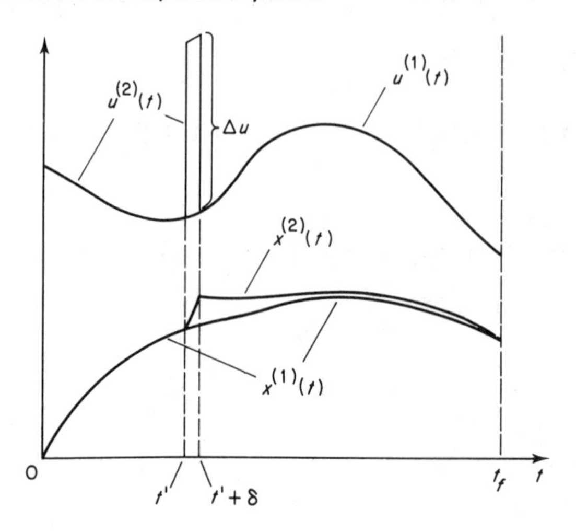

Fig. 76. Effect of a control perturbation at $t = t'$.

EXERCISE 6.1.5 contd.

and hence that $\partial J/\partial \Delta u \simeq \delta (\partial/\partial u)H(x^{(1)}(t'), p(t'), u, t')|_{u=u^{(1)}(t')}$. Deduce that the partial derivative of the Hamiltonian with respect to the control input is a measure of the *sensitivity* of the performance index to control variations of the type considered.

6.2 Minimum Energy and Minimum Fuel Problems with Control Constraints

In this section we consider the control of the linear, time-invariant system

$$\frac{dx(t)}{dt} = Ax(t) + Bu(t), \qquad x(0) = x_0 \tag{6.80}$$

on a specified time interval $0 \leqslant t \leqslant t_f$. The system is assumed to be subjected to the terminal boundary conditions

$$x(t_f) = x_f \qquad \text{(specified)} \tag{6.81}$$

and the control constraints

$$u(t) \in \Omega(t), \qquad 0 \leqslant t \leqslant t_f \tag{6.82}$$

The control problem is to choose an *optimal* controller $u^*(t)$ to minimize the performance index

$$J(u) = \int_0^{t_f} g(u(t), t)\mathrm{d}t \tag{6.83}$$

where $g(u, t)$ is interpreted as a measure of the power (or fuel flow) required to generate the input u at time t. The performance index is hence a measure of the total energy (or fuel) consumption in the time interval $0 \leqslant t \leqslant t_f$ and an optimal controller is an admissible controller that minimizes the total energy (or fuel) consumption. In this sense the material of this section is a generalization of the problem considered in Section 5.4.

6.2.1 *Derivation of the TPBVP*

Following the procedure of Section 6.1.1 let $u^{(1)}(t)$ and $u^{(2)}(t)$ be admissible controllers generating state trajectories $x^{(1)}(t)$ and $x^{(2)}(t)$ originating at the initial state x_0 at $t = 0$ and terminating at the final state x_f at $t = t_f$. Let $p(t)$ be an arbitrary (for the moment that is) $n \times 1$ matrix function of time and note that relations (6.6) and (6.7) still hold but that, as $x^{(1)}(t_f) = x^{(2)}(t_f) = x_f$, (6.8) reduces to the identity

$$\int_0^{t_f} \left\{ \left(\frac{\mathrm{d}p(t)}{\mathrm{d}t} + A^T p(t) \right)^T (x^{(2)}(t) - x^{(1)}(t)) + p^T(t)B\left(u^{(2)}(t) - u^{(1)}(t)\right) \right\} \mathrm{d}t = 0 \tag{6.84}$$

which can be regarded as a relationship between the controllers $u^{(1)}$, $u^{(2)}$ and their state trajectories. We can eliminate the state trajectories from the relationship by choosing our *costate vector* $p(t)$ to be a solution of the differential equations

$$\frac{\mathrm{d}p(t)}{\mathrm{d}t} = -A^T p(t) \tag{6.85}$$

when (6.84) reduces to the compact relation

$$\int_0^{t_f} p^T(t)B(u^{(2)}(t) - u^{(1)}(t))\mathrm{d}t = 0 \tag{6.86}$$

(Note: in contrast to the linear cost problem discussed in Section 6.1.1 the elimination of the states does not require us to specify the costate boundary conditions.)

The difference between the values of the performance index is as follows

$$
\begin{aligned}
J(u^{(2)})-J(u^{(1)}) &= \int_0^{t_f} \{g(u^{(2)}(t),t)-g(u^{(1)}(t),t)\}\mathrm{d}t \\
&= \int_0^{t_f} \{g(u^{(2)}(t),t)-g(u^{(1)}(t),t)\}\mathrm{d}t \\
&\quad + \int_0^{t_f} p^T(t)B(u^{(2)}(t)-u^{(1)}(t))\mathrm{d}t
\end{aligned} \tag{6.87}
$$

or, introducing the Hamiltonian function

$$
H(x,p,u,t) = g(u,t)+p^T(Ax+Bu), \tag{6.88}
$$

we obtain

$$
\begin{aligned}
J(u^{(2)})-J(u^{(1)}) &= \int_0^{t_f} \{H(x^{(1)}(t),p(t),u^{(2)}(t),t) \\
&\quad - H(x^{(1)}(t),p(t),u^{(1)}(t),t)\}\mathrm{d}t
\end{aligned} \tag{6.89}
$$

which is identical to (6.14). An argument similar to that following (6.14) soon yields the fact that a *sufficient* (and, using more advanced reasoning than is possible in this text, necessary) condition for the admissible controller $u^*(t)$ to be an optimal controller for the problem considered is that it is a solution of the TPBVP

$$
\begin{aligned}
\frac{\mathrm{d}x^*(t)}{\mathrm{d}t} &= Ax^*(t)+Bu^*(t) \\
x^*(0) &= x_0, \qquad x^*(t_f) = x_f \\
u^*(t) &\in \Omega(t) \\
\frac{\mathrm{d}p(t)}{\mathrm{d}t} &= -A^Tp(t) \qquad \text{(boundary condition not known)} \\
H(x^*(t),p(t),u^*(t),t) &= \min_{u\in\Omega(t)} H(x^*(t),p(t),u,t) \\
0 &\leqslant t \leqslant t_f
\end{aligned} \tag{6.90}
$$

In many ways this TPBVP is identical to that obtained in Section 5.4 with the exception that the stationary point condition

$$
\left.\frac{\partial H}{\partial u}\right|_* \equiv 0
$$

is replaced by the Hamiltonian minimization. In this sense we can expect to have computational problems due to the unknown costate boundary conditions. In other ways the TPBVP (6.90) is similar to that obtained for the linear cost problem (see (6.19) and (6.20)) if we set $\beta(t) \equiv 0$ and remove the (known) terminal boundary condition on the costate. We can expect, therefore, to have similar difficulties in Hamiltonian minimization to those observed in Sections 6.1.2–6.1.4. In fact, as the following examples will perhaps indicate, these two problems will interact to produce a TPBVP that is, in general, extremely difficult to solve.

6.2.2 A minimum energy example

Consider the problem of mine-winder optimization outlined in Example 5.3.1 with the data $T = m = L = 1$ and introducing the amplitude constraint $|u(t)| \leqslant M$ to represent the limits on generated torque in the winding engine. More precisely consider the constrained minimum energy optimal control problem

$$\frac{dx(t)}{dt} = \begin{bmatrix} 0 & 1 \\ 0 & 0 \end{bmatrix} x(t) + \begin{bmatrix} 0 \\ 1 \end{bmatrix} u(t), \qquad x(0) = \begin{bmatrix} 1 \\ 0 \end{bmatrix}, \qquad x(1) = \begin{bmatrix} 0 \\ 0 \end{bmatrix}$$

$$-M \leqslant u(t) \leqslant M$$

$$J(u) = \tfrac{1}{2} \int_0^1 u^2(t) dt \tag{6.91}$$

This problem fits into the general type discussed above with the data

$$A = \begin{bmatrix} 0 & 1 \\ 0 & 0 \end{bmatrix}, \qquad B = \begin{bmatrix} 0 \\ 1 \end{bmatrix}, \qquad x_0 = \begin{bmatrix} 1 \\ 0 \end{bmatrix}$$

$$x_f = \begin{bmatrix} 0 \\ 0 \end{bmatrix}, \qquad t_f = 1, \qquad g(u, t) = \tfrac{1}{2} u^2 \tag{6.92}$$

and can be solved by finding the solution to the TPBVP (6.90).

To begin the solution of the TPBVP, consider the costate equations

$$\frac{dp(t)}{dt} = -A^T p(t) = \begin{bmatrix} 0 & 0 \\ -1 & 0 \end{bmatrix} p(t) \tag{6.93}$$

written down in element by element form $dp_1(t)/dt = 0$, $dp_2(t)/dt = -p_1(t)$. As we do not know the boundary conditions we will attempt a parametric solution by writing $p_1(0)=\pi_1$ and $p_2(0)=\pi_2$ where π_1 and π_2 are unknown parameters. It is easily verified that

$$p_1(t) \equiv \pi_1, \qquad p_2(t) = \pi_2 - \pi_1 t \tag{6.94}$$

and hence, in particular, that $p_2(t)$ is a linear function of time.

The next step in the solution is to minimize the Hamiltonian

$$\begin{aligned} H(x,p,u,t) &= \tfrac{1}{2}u^2 + (p_1 \quad p_2)\left\{\begin{bmatrix}0 & 1\\0 & 0\end{bmatrix}\begin{bmatrix}x_1\\x_2\end{bmatrix}+\begin{bmatrix}0\\1\end{bmatrix}u\right\} \\ &= \tfrac{1}{2}u^2 + p_2 u + \text{terms in } x \text{ and } p \text{ only} \end{aligned} \tag{6.95}$$

with respect to controllers satisfying the constraint $-M \leqslant u \leqslant M$. This problem has been considered in Section 6.1.2 and leads to the following characterization of the optimal controller

$$u^*(t) = \begin{cases} -p_2(t) & \text{if} \quad |p_2(t)| \leqslant M \\ -M \operatorname{sgn} p_2(t) & \text{if} \quad |p_2(t)| > M \end{cases} \tag{6.96}$$

or, using (6.94),

$$u^*(t) = \begin{cases} \pi_1 t - \pi_2 & \text{if} \quad |\pi_1 t - \pi_2| \leqslant M \\ -M \operatorname{sgn}(\pi_2 - \pi_1 t) & \text{if} \quad |\pi_1 t - \pi_2| > M \end{cases} \tag{6.97}$$

The optimal control problem will now be solved if we choose the parameters π_1 and π_2 to ensure that $u^*(t)$ transfers the specified initial state to the specified final state in the specified time. This was easily achieved in Example 5.3.1, where the control was unconstrained, as the linear dependence of $u^*(t)$ on π_1 and π_2 made possible the construction of a set of simultaneous linear equations describing the required values. The presence of constraints, however, can mean that $u^*(t)$ is a nonlinear function of π_1 and π_2 due to the existence of the "sign function" in (6.97). To get round this problem we need to use a bit of common sense and guesswork!

To begin the analysis it is intuitively obvious that, if M is large enough, the required choice of π_1 and π_2 will be such that $|\pi_1 t - \pi_2| \leqslant M$ for all times in the interval $0 \leqslant t \leqslant 1$ and, hence, that $u^*(t) = \pi_1 t - \pi_2$, $0 \leqslant t \leqslant 1$. But this is simply the unconstrained controller (5.113) and the appropriate values of π_1 and π_2 have been found to give (5.117), i.e.

$$u^*(t) = 12t - 6 \tag{6.98}$$

which has the form sketched in Fig. 77(a). A simple calculation soon yields the fact that this controller satisfies the constraint $|u^*(t)| \leqslant M$, $0 \leqslant t \leqslant 1$, if and only if $M \geqslant 6$. Hence, $u^*(t)$(as given in (6.98)) solves the problem if $M \geqslant 6$.

Consider now the case of $M < 6$. It follows from the above that $|\pi_1 t - \pi_2|$

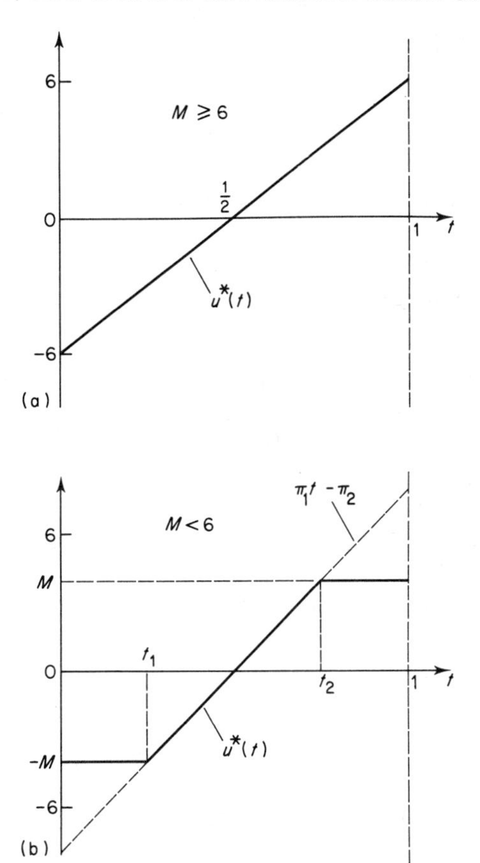

Fig. 77.

cannot be less than M at all points in the interval $0 \leqslant t \leqslant 1$ and hence that $u^*(t)$ cannot be the simple straight line $\pi_1 t - \pi_2$. It is at this stage that we need a little guesswork. Bearing in mind the form of the solution when $M \geqslant 6$ (Fig. 76(a)) it is natural to consider the possibility that the optimal control when $M < 6$ has the "clipped-off" structure illustrated in Fig. 77(b) and parameterized by the times t_1 and t_2. If this is correct, then we will have solved the optimal control problem if we find those values of the (new) parameters t_1 and t_2 that ensure that the initial and final state boundary conditions are satisfied. This problem is considered below.

The state x_2^* satisfies the differential equation $\mathrm{d}x_2^*(t)/\mathrm{d}t = u(t)$ with the boundary conditions $x_2^*(0) = x_2^*(1) = 0$, i.e.

$$x_2^*(t) = \int_0^t u(t')\mathrm{d}t' \tag{6.99}$$

and hence, from Fig. 77(b),

$$x_2^*(1) = 0 = \int_0^1 u(t)\mathrm{d}t = -Mt_1 + M(1-t_2) \tag{6.100}$$

so that, for $M \neq 0$, we obtain the simple relation

$$t_2 = 1 - t_1 \tag{6.101}$$

The resulting state trajectory is obtained by evaluating the integral in (6.99) to be

$$x_2^*(t) = \begin{cases} -Mt & \text{if} \quad 0 \leqslant t \leqslant t_1 \\ \dfrac{-Mt(t-1)}{(2t_1-1)} - \dfrac{Mt_1^2}{2t_1-1} & \text{if} \quad t_1 \leqslant t \leqslant t_2 \\ -Mt_1 + M(t-t_2) & \text{if} \quad t_2 \leqslant t \leqslant 1 \end{cases} \tag{6.102}$$

and is illustrated in Fig. 78. Note that $x_2(t)$ lies entirely within the "bounding triangle" indicated.

The state $x_1^*(t)$ must satisfy the differential equation $\mathrm{d}x_1^*(t)\mathrm{d}t = x_2^*(t)$ with the boundary conditions $x_1^*(0) = 1, x_1^*(1) = 0$, i.e.

$$x_1^*(1) = 0 = x_1^*(0) + \int_0^1 x_2^*(t)\mathrm{d}t = 1 + \int_0^1 x_2^*(t)\mathrm{d}t$$

$$= 1 + \int_0^{t_1} x_2^*(t)\mathrm{d}t + \int_{t_1}^{t_2} x_2^*(t)\mathrm{d}t + \int_{t_2}^1 x_2^*(t)\mathrm{d}t$$

$$= 1 - \frac{M}{2}t_1^2 - \frac{M}{(2t_1-1)}\{\tfrac{1}{3}(t_2^3 - t_1^3) - \tfrac{1}{2}(t_2^2 - t_1^2) + t_1^2(t_2 - t_1)\}$$

$$-Mt_1(1-t_2) + \frac{M}{2}(1-t_2^2) - Mt_2(1-t_2) \tag{6.103}$$

Substitution for t_2 from (6.101) leads, after a little manipulation, to a cubic equation for t_1 that must be solved numerically for each choice of M of interest. The time $t_2 = 1 - t_1$ is then trivially found and the solution of the optimal control problem is completed.
(Note: the cubic equation will have *three* solutions in general so that it is necessary to find the solution t_1 that lies in the range $0 \leqslant t_1 \leqslant t_2 = 1 - t_1$, i.e. $0 \leqslant t_1 \leqslant \frac{1}{2}$.)

Finally, it is possible to obtain some important information concerning the dependence of the solution on the constraint parameter M. More precisely, for (6.103) to hold we need that

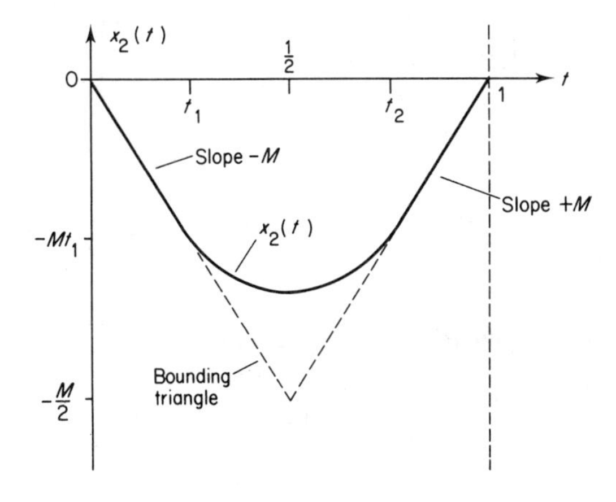

Fig. 78.

$$0 = 1 + \int_0^1 x_2^*(t)\,\mathrm{d}t$$

but

$$1 + \int_0^1 x_2^*(t)\,\mathrm{d}t \geqslant 1 - \text{area of the bounding triangle in Fig. 78}$$

$$= 1 - \frac{M}{4} \tag{6.104}$$

and hence we cannot satisfy the boundary condition on $x_1^*(t)$ at $t = 1$ if $M < 4$. There are two possible explanations of this phenomenon – *either* the guess for $u^*(t)$ given in Fig. 77(b) is not correct, *or* a solution to our optimal control problem does not exist for this range of M. In this particular example a solution does not exist for $M < 4$ because M is then so small that no admissible controller can be constructed to transfer the specified initial state to the specified final state in the specified time, i.e. we do not have enough control available to actually complete the task!

6.2.3 *A minimum fuel example*

Consider the problem of minimum fuel control of a land vehicle as introduced in Example 5.1.1 and for simplicity assume the data $m = L = t_f = 1$ and $M_1 = M_2 = M$. Also, by assuming that the rate of fuel consumption $L_f(x, u, t)$ is proportional to the absolute magnitude of the input u, it follows that the problem of finding the driving strategy that produces minimum fuel consumption reduces to the solution of the constrained optimal control problem

$$\frac{dx(t)}{dt} = \begin{bmatrix} 0 & 1 \\ 0 & 0 \end{bmatrix} x(t) + \begin{bmatrix} 0 \\ 1 \end{bmatrix} u(t), \qquad x(0) = \begin{bmatrix} 0 \\ 0 \end{bmatrix}, \quad x(1) = \begin{bmatrix} 1 \\ 0 \end{bmatrix}$$

$$-M \leqslant u(t) \leqslant M$$

$$J(u) = \int_0^1 |u(t)| \, dt \tag{6.105}$$

This problem is of the general type considered with data

$$A = \begin{bmatrix} 0 & 1 \\ 0 & 0 \end{bmatrix}, \quad B = \begin{bmatrix} 0 \\ 1 \end{bmatrix}, \quad x_0 = \begin{bmatrix} 0 \\ 0 \end{bmatrix}, \quad x_f = \begin{bmatrix} 1 \\ 0 \end{bmatrix},$$

$$g(u, t) = |u|, \qquad t_f = 1 \tag{6.106}$$

and can be solved by computing the solution of the TPBVP (6.90).

The costate equations for this problem take the form of (6.93) with solutions (6.94) characterized by unknown parameters π_1 and π_2. The Hamiltonian takes the form

$$H(x, p, u, t) = |u| + p_2 u + \text{terms in } x \text{ and } p \text{ only} \tag{6.107}$$

The minimization of Hamiltonians of this form in the presence of amplitude constraints has been considered in Section 6.1.4 and leads to the following expression for the optimal controller

$$u^*(t) = \begin{cases} -M & \text{if} \quad p_2(t) > 1 \\ \text{undefined} & \text{if} \quad p_2(t) = 1 \\ 0 & \text{if} \quad -1 < p_2(t) < 1 \\ \text{undefined} & \text{if} \quad p_2(t) = -1 \\ M & \text{if} \quad p_2(t) < -1 \end{cases} \tag{6.108}$$

The solution to the problem would now be complete if we knew the correct values of the parameters π_1 and π_2 and hence $p_2(t) = \pi_2 - \pi_1 t$. Unfortunately we do not know the correct values. We are therefore reduced (yet again!) to the use of common sense and guesswork in choosing appropriate values such that the controller (6.108) generates a state trajectory originating and terminating at the required states at the required times.

The simplest approach to the problem is to consider the general structure of all possible solutions (6.108) generated by different choices of π_1 and π_2 and to use a process of elimination based on physical and/or mathematical reasoning. Assuming for simplicity that $\pi_1 \neq 0$, there are only seven possible general structures for $u^*(t)$ due to the fact that $p_2(t)$ is a straight line and hence can intersect each of the lines ± 1 once only. These possibilities are illustrated in Fig. 79(a)–(g). (Note: the assumption that $\pi_1 \neq 0$ ensures that $p(t) \not\equiv 1$ and $p(t) \not\equiv -1$ and hence removes the need to consider the undefined values of $u^*(t)$ in (6.108). This assumption is fully justified in this problem!) We can immediately eliminate (a), (b), (c) and (d) as, in these cases,

$$x_2^*(1) = x_2^*(0) + \int_0^1 u^*(t)\mathrm{d}t = \int_0^1 u^*(t)\mathrm{d}t \neq 0 \tag{6.109}$$

and we cannot therefore satisfy the terminal boundary condition $x_2(1) = 0$. We can eliminate (g) also as, although $x_2^*(1) = 0$, it is also true that $x_2^*(t) \equiv 0$, $0 \leqslant t \leqslant 1$, and hence

$$x_1^*(1) = x_1^*(0) + \int_0^1 x_2^*(t)\mathrm{d}t = 0 \tag{6.110}$$

so that we cannot satisfy the terminal boundary condition $x_1(1) = 1$. This leaves us with (e) and (f) as the only two possibilities. Although we can eliminate (e) by a mathematical argument, we prefer to use a physical argument. More precisely, glancing at Fig. 59 it is clear that the candidate optimal controller shown in Fig. 79(e) consists of a period of acceleration from rest at $y = 0$ in the negative direction followed by a period of cruising followed by positive acceleration to the final condition. It is clearly a waste of fuel to reverse away from your destination and retread your way! We must conclude that Fig. 79(f) is the only possible candidate for the optimal control. All we need do now is calculate suitable values for the switching times t_1 and t_2.

The first step is to use the boundary conditions on $x_2^*(t)$ to verify that

$$x_2^*(t) = \int_0^t u^*(t)\mathrm{d}t$$

$$= \begin{cases} Mt & \text{if} \quad 0 \leqslant t \leqslant t_1 \\ Mt_1 & \text{if} \quad t_1 \leqslant t \leqslant t_2 \\ Mt_1 - M(t-t_2) & \text{if} \quad t_2 \leqslant t \leqslant 1 \end{cases} \tag{6.111}$$

and hence that we need

$$x_2^*(1) = Mt_1 - M(1-t_2) = 0 \tag{6.112}$$

or

$$t_2 = 1 - t_1 \tag{6.113}$$

We can now satisfy the boundary conditions on $x_1^*(t)$ by using (6.111) to verify that

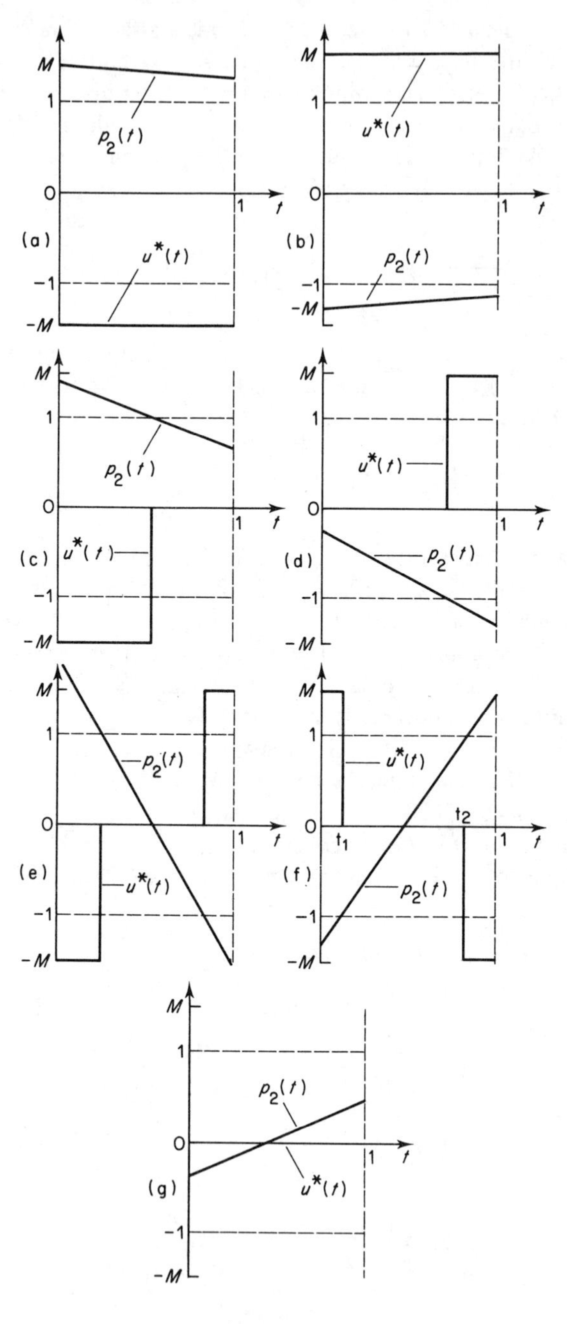

Fig. 79.

$$1 = x_1^*(1) = x_1^*(0) + \int_0^1 x_2^*(t)\mathrm{d}t = \int_0^1 x_2^*(t)\mathrm{d}t$$

$$= \int_0^{t_1} x_2^*(t)\mathrm{d}t + \int_{t_1}^{t_2} x_2^*(t)\mathrm{d}t + \int_{t_2}^{1} x_2^*(t)\mathrm{d}t$$

$$= \frac{M}{2}t_1^2 + Mt_1(t_2 - t_1) + \frac{Mt_1}{2}(1 - t_2) \tag{6.114}$$

Eliminating t_2 using (6.113) and rearranging yields the quadratic equation

$$t_1^2 - t_1 + \frac{1}{M} = 0 \tag{6.115}$$

that is to be solved for t_1. The two solutions take the form $\frac{1}{2}(1 \pm \sqrt{1 - 4/M})$. However, as we require $0 \leqslant t_1 \leqslant t_2 = 1 - t_1$, we must choose the solution in the range $0 \leqslant t_1 \leqslant \frac{1}{2}$, i.e.

$$t_1 = \tfrac{1}{2}(1 - \sqrt{1 - 4/M}), \qquad t_2 = 1 - t_1 = \tfrac{1}{2}(1 + \sqrt{1 - 4/M}) \tag{6.116}$$

which completes our computation of $u^*(t)$. Finally note that t_1 and t_2 are complex if $M < 4$, which suggests that a solution to the specified minimum fuel problem exists if, and only if, $M \geqslant 4$. This is in accord with the results of Section 6.2.2.

EXERCISE 6.2.1. Use a graphical argument similar to that based on Fig. 78 for the case of minimum energy control to verify that a solution to the minimum fuel problem exists if, and only if, $M \geqslant 4$. In the case of $M = 4$, show that

$$u^*(t) = \begin{cases} M & 0 \leqslant t < \frac{1}{2} \\ -M & \frac{1}{2} < t \leqslant 1 \end{cases} \tag{6.117}$$

and that this is the *only* controller that will transfer the initial state to the final state in the required time.

6.3 Time Optimal Control to a Specified Final State

In all of the examples discussed in previous sections and Chapter five, the time t_f taken to accomplish the control objective was taken to be constant and the objective of the optimal controller was to optimize other aspects of system performance (e.g. fuel consumption). In this section we introduce a class of problems where the objective of the optimal controller is simply to accomplish a specified dynamic task *as quickly as possible* and with no regard to fuel or energy consumption or intermediate transient characteristics of the state vector. Problems of this type are termed *time optimal* or *minimum time control problems.* The specific problem discussed here is the choice of an optimal controller $u^*(t)$ satisfying the amplitude constraints

$$|u_k(t)| \leqslant M_k, \qquad 0 \leqslant t \leqslant t_f, \qquad 1 \leqslant k \leqslant l \tag{6.118}$$

that drives the state vector $x(t)$ satisfying the equation

$$\frac{\mathrm{d}x(t)}{\mathrm{d}t} = Ax(t) + Bu(t) \tag{6.119}$$

from the specified initial condition $x(0) = x_0$ to a specified final condition $x(t_f) = x_f$ in such a manner that the time t_f required to accomplish this task is minimized, i.e. the performance index

$$J(u) = t_f = \int_0^{t_f} \mathrm{d}t \tag{6.120}$$

A simple example of a problem of this type can be constructed by consideration of the control of a land vehicle as introduced in Example 5.1.1. Suppose that the objective of minimizing fuel consumption is replaced by the transfer of the vehicle at rest at $y = 0$ at $t = 0$ to the point $y = L$ at rest in a minimum time t_f. Suppose also, for simplicity, that $M_1 = M_2 = M$ in (5.11) and hence that the maximum acceleration and deceleration capabilities of the vehicle are identical. The solution to this problem is easily deduced from physical reasoning, i.e. the driver will accelerate the car at maximum acceleration to the point $L/2$ and then instantly switch to maximum deceleration. The optimal control input has the *bang-bang* form

$$u^*(t) = \begin{cases} +M & \text{if} \quad 0 \leqslant t \leqslant t_f^*/2 \\ -M & \text{if} \quad t_f^*/2 < t \leqslant t_f^* \end{cases} \tag{6.121}$$

where $t_f^*/2$ is the time taken to reach the point $y = L/2$ at maximum acceleration,

$$\frac{t_f^*}{2} = \sqrt{\frac{mL}{M}} \tag{6.122}$$

The actual minimum time is, therefore,

$$J(u^*) = t_f^* = \sqrt{\frac{4mL}{M}} \tag{6.123}$$

The above example has many of the features seen in the solution of the general time optimal control problem but the proof of the technical details is a non-trivial problem. We therefore state the result without proof: a *necessary* condition for the admissible controller $u^*(t)$ to be an optimal controller for the minimum time problem with minimum time t_f^* is that it is a solution of the TPBVP

$$\frac{\mathrm{d}x^*(t)}{\mathrm{d}t} = Ax^*(t) + Bu^*(t), \qquad x^*(0) = x_0,\ x^*(t_f^*) = x_f$$

$$|u_k^*(t)| \leqslant M_k, \qquad 1 \leqslant k \leqslant l$$

$$\frac{\mathrm{d}p(t)}{\mathrm{d}t} = -A^T p(t), \qquad 0 \leqslant t \leqslant t_f^* \tag{6.124}$$

where, if we define the Hamiltonian function,

$$H(x, p, u, t) = 1 + p^T(Ax + Bu) \tag{6.125}$$

then $u^*(t)$ also solves the Hamiltonian minimization problem

$$H(x^*(t), p(t), u^*(t), t) = \min_{u \in \Omega(t)} H(x^*(t), p(t), u, t)$$

$$0 \leqslant t \leqslant t_f^* \tag{6.126}$$

and the transversality condition

$$H(x_f, p(t_f^*), u^*(t_f^*), t_f^*) = 0 \tag{6.127}$$

must be satisfied at $t = t_f^*$.

The above TPBVP is certainly not easy to solve in general. This will become clear when the reader works through the examples at the end of the section. Before this we can make a useful statement concerning the form of $u^*(t)$ as deduced from (6.126). More precisely, defining the $l \times 1$ vector $z = B^T p$ then

$$\begin{aligned} H(x, p, u, t) &= p^T Bu + p^T Ax + 1 \\ &= (B^T p)u + p^T Ax + 1 \\ &= z^T u + p^T Ax + 1 \\ &= \sum_{k=1}^{l} z_k u_k + \text{terms in } x \text{ and } p \text{ only} \end{aligned} \tag{6.128}$$

and hence, using a similar analysis to that of Section 6.1.3, it follows that the minimization of H with respect to the constraints (6.118) yields the optimal control form (see (6.56))

$$u_k^*(t) = -M_k \operatorname{sgn} z_k(t)$$

$$0 \leqslant t \leqslant t_f, \qquad 1 \leqslant k \leqslant l \tag{6.129}$$

i.e. the time-optimal control has a bang-bang form!

EXAMPLE 6.3.1. Consider the minimum time control of the land vehicle of Example 5.1.1 with the data $M = L = 1$. The optimal control problem then takes the form

$$\frac{dx(t)}{dt} = \begin{bmatrix} 0 & 1 \\ 0 & 0 \end{bmatrix} x(t) + \begin{bmatrix} 0 \\ 1 \end{bmatrix} u(t), \qquad x(0) = \begin{bmatrix} 0 \\ 0 \end{bmatrix}, \qquad x(t_f) = \begin{bmatrix} 1 \\ 0 \end{bmatrix}$$

$$|u(t)| \leqslant M, \qquad J(u) = t_f \tag{6.130}$$

which has the form of (6.118)–(6.120) with the data

$$A = \begin{bmatrix} 0 & 1 \\ 0 & 0 \end{bmatrix}, \qquad B = \begin{bmatrix} 0 \\ 1 \end{bmatrix}, \qquad x_0 = \begin{bmatrix} 0 \\ 0 \end{bmatrix}, \qquad x_f = \begin{bmatrix} 1 \\ 0 \end{bmatrix} \tag{6.131}$$

We begin the solution of this problem with the solution of the costate equations

$$\frac{dp(t)}{dt} = -A^T p(t) = \begin{bmatrix} 0 & 0 \\ -1 & 0 \end{bmatrix} p(t) \tag{6.132}$$

As in the case of minimum energy and fuel control, the boundary conditions are not known. We therefore attempt a parametric solution by setting $p_1(0) = \pi_1$ and $p_2(0) = \pi_2$ where π_1 and π_2 are unknown parameters. The reader should verify that

$$p_1(t) \equiv \pi_1, \qquad p_2(t) = \pi_2 - \pi_1 t \tag{6.133}$$

that $z(t) = B^T p(t) = p_2(t)$ and hence, minimizing the Hamiltonian, that

$$u^*(t) = -M \operatorname{sgn} p_2(t) \tag{6.134}$$

As $p_2(t)$ is a linear function of time then, assuming that $p_2(t) \not\equiv 0$, it is obvious that the optimal control has, at most, one switching time in any interval $0 \leqslant t \leqslant t_f^*$, i.e. $u^*(t)$ has the form

$$u^*(t) = \begin{cases} \lambda M, & 0 \leqslant t < t_1 \\ -\lambda M & t_1 < t \leqslant t_f^* \end{cases} \tag{6.135}$$

for some time t_1 in the range $0 \leqslant t_1 \leqslant t_f^*$ and $\lambda = +1$ or -1. The reader should verify that

$$x_2^*(t) = \begin{cases} \lambda M t & \text{if} \quad 0 \leqslant t < t_1 \\ \lambda M t_1 - \lambda M (t - t_1) & \text{if} \quad t_1 < t \leqslant t_f^* \end{cases} \tag{6.136}$$

and hence that the boundary condition $x_2^*(t_f^*) = 0$ yields the relation

$$\lambda M t_1 - \lambda M(t_f^* - t_1) = 0 \qquad \text{(i.e. } t_1 = t_f^*/2) \qquad (6.137)$$

Also, to satisfy the boundary conditions on $x_1^*(t)$ we need

$$x_1^*(t_f^*) = 1 = \int_0^{t_f^*} x_2^*(t)\,dt = \tfrac{1}{4}\lambda M t_f^{*2} \qquad (6.138)$$

which immediately tells us that we must choose $\lambda = +1$ and $t_f^* = \sqrt{4/M}$. Note that all these results are consistent with (6.121) and (6.123) as derived from physical considerations.

An alternative and instructive graphical approach to the solution of this problem can be obtained by regarding x_1 and x_2 as Cartesian coordinates in the phase plane and plotting the so-called *switching curve*, which consists of the two loci generated by the state equations when they are integrated backwards in time from the terminal condition (1,0) using controllers $u(t) \equiv M$ and $u(t) \equiv -M$. More precisely, if $u(t) \equiv M$ then $x_2(t) = M(t - t_f^*)$ and $x_1(t) = 1 + M(t_f^* - t)^2/2$ and hence the switching curve is defined in part by

$$M(x_1 - 1) = x_2^2/2, \qquad x_2 < 0 \qquad (6.139)$$

It is left as an exercise for the reader to verify that the rest of the switching curve has the form (take the case of $u(t) \equiv -M$)

$$M(x_1 - 1) = -x_2^2/2, \qquad x_2 > 0 \qquad (6.140)$$

and hence that the switching curve has the form shown in Fig. 80 for the case of $M = 1$. If, as shown, we superimpose the curves $x_1 = x_2^2/2$ ($x_2 > 0$) and $x_1 = -x_2^2/2$ ($x_2 < 0$) representing the response of the system from the initial condition to constant inputs $u(t) \equiv 1$ and $u(t) \equiv -1$ respectively, then obviously the only way that the system can be transferred from the initial condition to the final state using a bang-bang controller with, at most, one switching time is that it follows the route $a \to b \to c$. This route consists of an initial period $0 \leqslant t \leqslant t_1$ when $u(t) \equiv +1$. The controller then switches to the value -1 for the remaining period $t_1 < t \leqslant t_f^*$ at the time $t = t_1$. Given the coordinates of b then the minimum time can be calculated from $t_f^* = t_1 + (t_f^* - t_1)$ where t_1 is the time taken to travel from the origin to b using $u(t) = +1$ and $t_f^* - t_1$ is the time taken to travel from b to the final state using $u(t) = -1$.

EXERCISE 6.3.1. If, in example 6.3.1, we take $M = 1$ and replace the initial condition $x_0 = 0$ by an arbitrary initial condition $x_0 = \begin{bmatrix} x_{10} \\ x_{20} \end{bmatrix}$, use a graphical argument similar to that of Fig. 80 to prove that

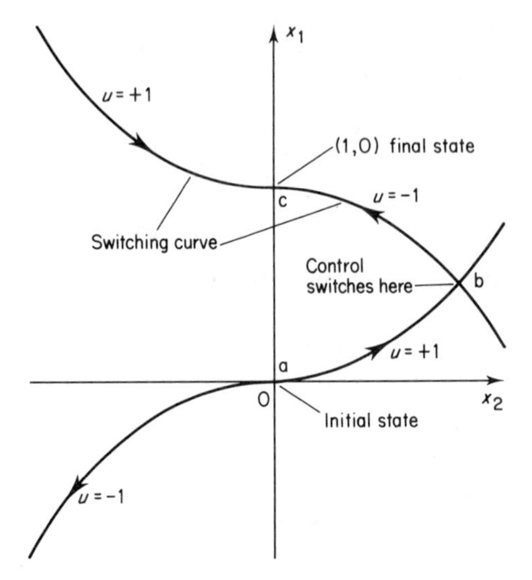

Fig. 80. Graphical calculation of a time-optimal control.

EXERCISE 6.3.1 contd.

(a) $u^*(t) \equiv +1$ if x_0 lies on the switching curve and $x_{20} < 0$
(b) $u^*(t) \equiv -1$ if x_0 lies on the switching curve and $x_{20} > 0$
(c) if x_0 lies *below* the switching curve, then $u^*(t) = +1, 0 \leqslant t \leqslant t_1$, where t_1 is the time when the state trajectory meets the switching curve, and $u^*(t) = -1, t_1 \leqslant t \leqslant t_f^*$
(d) if x_0 lies *above* the switching curve then $u^*(t) = -1, 0 \leqslant t \leqslant t_1$, where t_1 is the time when the state trajectory meets the switching curve and then $u^*(t) = +1, t_1 \leqslant t \leqslant t_f^*$.

Verify that this controller can be written in the (nonlinear) *feedback* form

$$u^*(t) = \begin{bmatrix} +1 \text{ if } x_1 - 1 < -\frac{1}{2}x_2^2 \operatorname{sgn} x_2 \text{ or } x_1 - 1 = -\frac{1}{2}x_2^2 \operatorname{sgn} x_2 \text{ and } x_2 < 0 \\ -1 \text{ if } x_1 - 1 > -\frac{1}{2}x_2^2 \operatorname{sgn} x_2 \text{ or } x_1 - 1 = -\frac{1}{2}x_2^2 \operatorname{sgn} x_2 \text{ and } x_2 > 0 \end{bmatrix} \tag{6.141}$$

and consider how this controller could be realized in terms of available state measurements using a digital computer.

6.4 Introduction to the Minimum Principle of Pontryagin

Just as the material of Sections 5.1.–5.3 represents a selection of important special cases of the general method of variational calculus, so the material of

Sections 6.1–6.3 represents a number of special cases of the (so-called) "Minimum Principle of Pontryagin" which, in general terms, is the generalization of the results of variational calculus (i.e. the Euler-Lagrange equations) to include control problems where the control input is subject to constraints. It is not within the scope of this text to give a general statement of the Minimum Principle, so we will content ourselves with a statement relevant to a large class of problems that includes those described in Sections 6.1 and 6.2. More precisely we restrict our attention to control problems of the form

$$\frac{dx(t)}{dt} = f(x(t), u(t), t), \qquad x(0) = x_0$$

$$u(t) \in \Omega(t), \qquad 0 \leqslant t \leqslant t_f$$

$$J(u) = \phi(x(t_f)) + \int_0^{t_f} L(x(t), u(t), t)\,dt \tag{6.142}$$

where t_f is finite and fixed and there are no terminal boundary conditions on the state at $t = t_f$. Note the assumed presence of control constraints.

Defining the Hamiltonian function

$$H(x, p, u, t) = L(x, u, t) + p^T f(x, u, t) \tag{6.143}$$

then the Minimum Principle states that a *necessary* condition for the admissible controller $u^*(t)$ to solve (6.142) is that it is a solution of the TPBVP

$$\frac{dx^*(t)}{dt} = f(x^*(t), u^*(t), t), \qquad x^*(0) = x_0$$

$$u^*(t) \in \Omega(t)$$

$$\frac{dp_k(t)}{dt} = -\left.\frac{\partial H}{\partial x_k}\right|_*, \qquad p_k(t_f) = \left.\frac{\partial \phi}{\partial x_k}\right|_*$$

$$1 \leqslant k \leqslant n$$

$$H(x^*(t), p(t), u^*(t), t) = \min_{u \in \Omega(t)} H(x^*(t), p(t), u, t)$$

$$0 \leqslant t \leqslant t_f \tag{6.144}$$

where $|_*$ means that the derivative is evaluated at $(x^*(t), p(t), u^*(t), t)$. These equations are of Euler-Lagrange type (see (5.149)) with the conditions $\partial H/\partial u_k|_* = 0$ replaced by the Hamiltonian minimization with respect to the constraints. In the case of unconstrained control, these conditions are, of course, equivalent!

EXERCISE 6.4.1. Verify that the linear cost problem of Section 6.1 takes the form of 6.142 with

$$\phi(x) = \alpha^T x, \qquad L(x, u, t) = \beta^T(t)x + g(u, t)$$

$$f(x, u, t) = Ax + Bu \tag{6.145}$$

Verify that the TPBVP (6.144) reduces to (6.19) and (6.20) in this case.

We can also easily state the minimum principle when

$$J(u) = \int_0^{t_f} L(x(t), u(t), t)\mathrm{d}t \tag{6.146}$$

and the state also must satisfy the terminal boundary condition $x(t_f) = x_f$. In fact $u^*(t)$ must still be a solution of the TPBVP (6.144) with the costate boundary conditions $p_k(t_f) = \partial\phi/\partial x_k|_*$, $1 \leqslant k \leqslant n$, replaced by the state boundary condition $x(t_f) = x_f$. It is left as an exercise for the reader to verify that the minimum fuel and minimum energy problems of Section 6.2 fall into this category with $L(x, u, t) = g(u, t)$ and $f(x, u, t) = Ax + Bu$ and that the TPBVP generated by the Minimum Principle reduces to the TPBVP (6.90) previously derived. (Note: for a statement of the Minimum Principle for problems where the terminal time is free to vary, the reader is referred to the reading list.)

Problems

(1) Verify that the scalar problem

$$\frac{\mathrm{d}x(t)}{\mathrm{d}t} = -x(t) + u(t), \qquad x(0) = 1, \qquad |u(t)| \leqslant M$$

$$J(u) = x(1) + \int_0^1 \lambda u^2(t)\mathrm{d}t, \qquad \lambda > 0$$

is a linear cost problem with $t_f = 1$, $A = -1$, $B = 1$, $\alpha = 1$, $\beta(t) \equiv 0$ and $g(u, t) = \lambda u^2$. For the case of $M = +\infty$ (the unconstrained control problem), verify that the optimal controller $u^*(t) = -\frac{1}{2}\lambda^{-1}p(t)$ where $p(t) = \mathrm{e}^{t-1}$ is the solution of the costate equation

$$\frac{dp(t)}{dt} = p(t), \qquad p(1) = 1$$

Sketch the form of $u^*(t)$ and use your results to verify that, for the case of M finite,

(a) if $2\lambda M \geqslant 1$ then $u^*(t) = -\frac{1}{2}\lambda^{-1}e^{t-1}$, or

(b) if $2\lambda M < 1$ then $u^*(t) = \begin{cases} -\frac{1}{2}\lambda^{-1}e^{t-1}, & 0 \leqslant t \leqslant 1 + \log_e 2\lambda M \\ -M, & 1 + \log_e 2\lambda M \leqslant t \leqslant 1 \end{cases}$

and hence that $u^*(t) \equiv -M$, $0 \leqslant t \leqslant 1$, if $M \leqslant e^{-1}/2\lambda$.

(2) Consider the problem in (1) with the alternative performance index

$$J(u) = x(1) + \int_0^1 \lambda|u(t)|\,dt, \qquad \lambda > 0$$

Verify that the optimal control takes the form

(a) if $\lambda \geqslant 1$ then $u^*(t) \equiv 0$

(b) if $\lambda e \leqslant 1$ then $u^*(t) \equiv -M$

(c) if $e^{-1} < \lambda < 1$ then

$$u^*(t) = \begin{cases} 0 & 0 \leqslant t \leqslant t_1 \\ -M & t_1 < t < 1 \end{cases}$$

where $t_1 = 1 + \log_e \lambda$ is the control switching time.

(3) Consider the bang-bang control law $u^*(t) = -M \operatorname{sgn} z(t)$ for a single-input system $dx(t)/dt = Ax(t) + Bu(t)$ where $z(t) = B^T p(t)$ and the costate $p(t)$ is a solution of the equation $dp(t)/dt = -A^T p(t)$. In the case that A has distinct eigenvalues $\lambda_1, \lambda_2, \ldots, \lambda_n$, prove that

$$z(t) = \sum_{k=1}^{n} c_k e^{-\lambda_k t}$$

for some constant scalars c_k, $1 \leqslant k \leqslant n$. The control switching times are the solutions of the equation

$$z(t_s) = \sum_{k=1}^{n} c_k e^{-\lambda_k t_s} = 0$$

Prove, in the case of all eigenvalues of A being real, that the bang-bang controller has, at most, $n-1$ switches. (Hint: use mathematical induction and a graphical argument.)

(4) The linear system

$$\mathrm{d}x_1(t)/\mathrm{d}t = x_2(t) + u_1(t), \qquad \mathrm{d}x_2(t)/\mathrm{d}t = u_2(t)$$

is to be controlled by controllers satisfying the constraints

$$|u_1(t)| \leqslant M, \qquad u_2(t) \text{ unconstrained}$$

in such a manner as to minimize the performance index

$$J(u) = 2x_1(1) + x_2(1) + \int_0^1 \{-2x_1(t) + |u_1(t)| + \tfrac{1}{2}u_2^2(t)\}\mathrm{d}t$$

Verify that the optimal control takes the form

$$u_1^*(t) = \begin{cases} -M, & p_1(t) > 1 \\ \text{undefined}, & p_1(t) = 1 \\ 0, & -1 < p_1(t) < 1 \\ \text{undefined}, & p_1(t) = -1 \\ M, & p_1(t) < -1 \end{cases}$$

and

$$u_2^*(t) = \begin{cases} -p_2(t), & |p_2(t)| \leqslant M \\ -M \operatorname{sgn} p_2(t), & |p_2(t)| > M \end{cases}$$

where the costates are the solutions of the equations

$$\mathrm{d}p_1(t)/\mathrm{d}t = 2, \qquad p_1(1) = 2, \qquad \mathrm{d}p_2(t)/\mathrm{d}t = -p_1(t), \qquad p_2(1) = 1$$

Hence find the precise form for the optimal controls.

(5) Using an analysis similar to that of Section 6.1.1, prove that a *sufficient* condition for an admissible controller $u^*(t)$ to be an optimal controller for the linear cost problem (6.1), (6.2) and (6.5) together with a *terminal* state constraint $x(t_f) = x_f$ and $\alpha = 0$ is that it is a solution of the TPBVP

$$\frac{\mathrm{d}x^*(t)}{\mathrm{d}t} = Ax^*(t) + Bu^*(t), \qquad x^*(0) = x_0, \qquad x^*(t_f) = x_f$$

$$u^*(t) \in \Omega(t)$$

$$\frac{\mathrm{d}p(t)}{\mathrm{d}t} = -A^T p(t) - \beta(t) \quad \text{(boundary conditions unspecified)}$$

$$H(x^*(t), p(t), u^*(t), t) = \min_{u \in \Omega(t)} H(x^*(t), p(t), u, t)$$

$$0 \leqslant t \leqslant t_f$$

Show also that this condition is necessary by verifying that this TPBVP is identical to that generated from Pontryagin's Minimum Principle.

Using the model introduced in Example 5.1.1 with the data $m = 1$ and $M_1 = M_2 = M$, show that the problem of maximizing the distance travelled from the point $y = 0$ at rest to rest in the specified time $t_f = 1$ can be formulated as the optimal control problem

$$\frac{dx(t)}{dt} = u(t), \qquad x(0) = x(1) = 0, \qquad |u(t)| \leqslant M$$

$$J(u) = -\int_0^1 x(t)dt$$

where the state $x(t) = dy(t)/dt$. Note that this problem takes the above form with $A = 0$, $B = 1$, $x_0 = x_f = 0$, $\beta(t) \equiv -1$. Solve this TPBVP to find the bang-bang optimal controller

$$u^*(t) = \begin{cases} M & 0 \leqslant t < \frac{1}{2} \\ -M & \frac{1}{2} < t \leqslant 1 \end{cases}$$

and the minimum value of the performance index $J(u^*) = -M/4$. Use this result to explain the non-existence of the solution to the minimum fuel problem in Section 6.2.3 when $M < 4$.

(6) Very often a change of state variables (e.g. using the eigenvector matrix) can be a great help in simplifying the solution of the state and costate equations. Suggest and prove equivalent results to that of problem (6) in Chapter five for both linear cost and minimum energy or fuel problems.

(7) Given the first order dynamical system

$$\frac{dx(t)}{dt} = -x(t) + u(t), \qquad |u(t)| \leqslant M$$

show that the admissible controller $u^*(t)$ driving the initial state $x(0) = 1$ to the final state $x(1) = 1$ whilst minimizing the performance index

$$J(u) = \frac{1}{2}\int_0^1 u^2(t)dt$$

is a solution of the TPBVP

$$\frac{dx^*(t)}{dt} = -x^*(t) + u^*(t), \qquad x^*(0) = x^*(1) = 1, \qquad \frac{dp(t)}{dt} = p(t)$$

$$\tfrac{1}{2}u^*(t)^2 + p(t)u^*(t) = \min_{|u| \leqslant M} \{\tfrac{1}{2}u^2 + p(t)u\}$$

$$0 \leqslant t \leqslant 1$$

Hence show that

$$\text{(a) } u^*(t) = \frac{2}{e+1}\,e^t \quad \text{if} \quad \frac{2e}{e+1} \leqslant M, \quad \text{otherwise}$$

$$\text{(b) } u^*(t) = \begin{cases} \alpha e^t, & 0 \leqslant t \leqslant t' \\ M, & t' < t \leqslant 1 \end{cases}$$

where $\alpha e^{t'} = M$ and $1 = e^{-1} + e^{-1}\left\{\frac{\alpha}{2}(e^{2t'} - 1) + M(e - e^{t'})\right\}$ i.e.

$$\alpha^2 + 2(e - 1 - Me)\alpha + M^2 = 0 \quad \text{and} \quad t' = \log_e \alpha^{-1} M$$

(8) Repeat the minimum fuel problem of Section 6.2.3 for general data m, $t_f, L, M_1 = M_2 = M$. More precisely, solve the minimum fuel problem

$$\frac{dx(t)}{dt} = \begin{bmatrix} 0 & 1 \\ 0 & 0 \end{bmatrix} x(t) + \begin{bmatrix} 0 \\ m^{-1} \end{bmatrix} u(t), \quad x(0) = \begin{bmatrix} 0 \\ 0 \end{bmatrix}, \quad x(t_f) = \begin{bmatrix} L \\ 0 \end{bmatrix}$$

$$|u(t)| \leqslant M, \qquad J(u) = \int_0^{t_f} |u(t)|\,dt$$

In particular verify that a solution to the problem exists only if $Mt_f^2/4m \geqslant L$. Provide a physical argument to justify this conclusion.

(9) Using eigenvector transformations a linear system can be described by the model

$$\frac{dx(t)}{dt} = \begin{bmatrix} 0 & 0 \\ 0 & -1 \end{bmatrix} x(t) + \begin{bmatrix} 1 \\ 1 \end{bmatrix} u(t)$$

The system is to be controlled from an initial state $x(0) = \begin{bmatrix} 1 \\ 0 \end{bmatrix}$ to the origin $x(t_f) = \begin{bmatrix} 0 \\ 0 \end{bmatrix}$ in minimum time using controllers satisfying the constraint $|u(t)| \leqslant M$, $0 \leqslant t \leqslant t_f$. Show that the minimum time controller is bang-bang with one switch and takes the form

$$u^*(t) = \begin{cases} -M, & 0 \leqslant t \leqslant t_1 \\ +M, & t_1 < t \leqslant t_f \end{cases}$$

where the minimum time t_f^* and switching time t_1 satisfy the equations

$$1 - Mt_1 + M(t_f^* - t_1) = 0$$

$$(e^{-t_1} - 2)e^{t_1 - t_f^*} + 1 = 0$$

and

$$t_f^* > t_1 > 0$$

Hence deduce that the switching time $t_1 = -\log_e(1 - \sqrt{(1 - e^{-1/M})})$ and that the minimum time

$$t_f^* = -2\log_e(1 - \sqrt{(1 - e^{-1/M})}) - 1/M$$

Check that $t_f^* > t_1$.

Remarks and Further Reading

Chapters five and six have only skimmed the surface of the general field of optimal control. The reader is referred to the reading list at the end of Chapter five for more detailed information. It should be clear from the material presented here that the presence of constraints in the optimal control problem is a major source of difficulty in the calculation of the optimal controller via the solution of the TPBVP. It has also been seen in Section 6.2 that the presence of constraints can mean that a solution does not exist! Despite these problems the study of the TPBVP can give valuable insight into the form of the optimal controller by minimization of the Hamiltonian and a parametric study of the form of the costate.

Finally, we have seen in Sections 6.1.3, 6.1.4, 6.2.2 and 6.2.3 that the minimization of the Hamiltonian frequently gives rise to the possibility that the optimal controller is not uniquely defined by the solution of the minimization problem for certain values of the costate $p(t)$. If this occurs at only a finite number of isolated points in the interval $0 \leqslant t \leqslant t_f$, then the problem can be ignored. If, however, it occurs at an infinite number of points the problem is said to be singular. Singular control problems are *very* difficult to solve. More information can be found in Munro (1979a), Clements and Anderson (1978) or Bell and Jacobson (1975).

7. Discrete Optimal Control: an Introduction

It is generally true that the existence of a control theory for continuous systems implies the existence of an analogous control theory for discrete systems. It is also generally true that the discrete version of the problem is, in many ways, simpler than the continuous version but that it has a subtle identity of its own. In this chapter, we consider the optimal control of discrete systems in the context of the material of Chapters five and six. In many cases, the concepts will be so close to those of the continuous case that ideas will be outlined only briefly. This will leave space and time to concentrate on those features of the problem that are new.

7.1 Formulation of the Optimal Control Problem

We consider a discrete non-linear dynamical system described by the recursion relations

$$x_{k+1} = f(x_k, u_k, k) \tag{7.1}$$

with $n \times 1$ state vector x_k and $l \times 1$ input vector $u_k, k \geqslant 0$. We suppose that we are interested in the dynamics of the process over the time interval consisting of the $N+1$ sample times $0 \leqslant k \leqslant N$ and that the system is subject to a known initial state boundary condition x_0 at $k = 0$. The resulting system is very often called an *N-stage process.* We could also introduce a terminal state boundary condition $x_N = x_f$ but, for the moment, we will neglect this possibility.

Just as in Section 5.1, we will suppose that the control inputs are subjected to constraints. The existence of constraints in the problem will be indicated by the symbolism

$$u_k \in \Omega(k), \qquad 0 \leqslant k \leqslant N-1 \tag{7.2}$$

and any control sequence $\{u_0, u_1, u_2, \ldots, u_{N-1}\}$ is said to be *admissible* if its elements satisfy (7.2).

The performance of the system is assumed to be assessed by the use of a performance index $J_N(u_0, u_1, u_2, \ldots, u_{N-1})$ (written as $J_N(u)$ for notational simplicity) with the (assumed) property that $J_N(u^{(1)}) > J_N(u^{(2)})$ implies that the control sequence $u^{(2)} = \{u_0^{(2)}, u_1^{(2)}, \ldots, u_{N-1}^{(2)}\}$ is "better" than the control sequence $u^{(1)} = \{u_0^{(1)}, u_1^{(1)}, \ldots, u_{N-1}^{(1)}\}$. The optimal control problem is simply the search for the best controller. More precisely, we state the following "discrete optimal control problem": Find an admissible control sequence $u^* = \{u_0^*, u_1^*, \ldots, u_{N-1}^*\}$ that minimizes the performance index $J_N(u) = J_N(u_0, \ldots, u_{N-1})$, i.e.

$$J_N(u^*) = \min J_N(u) \tag{7.3}$$

where the minimization is performed over all input sequences $u = \{u_0, u_1, \ldots, u_{N-1}\}$ that satisfy the constraints.

The admissible control sequence u^* is called an "optimal controller" for the system.

The solution of the above problem would be, in general, very complicated if we did not impose some *structure* on the performance index. In the continuous system examples considered in previous chapters, the performance index was expressed in integral form. The parallel for the case of discrete system optimization is to write the performance index in the separable additive form

$$J_N(u) = \phi_N(x_N) + \sum_{k=0}^{N-1} L(x_k, u_k, k) \tag{7.4}$$

where $L(x, u, k)$ describes the contribution to system performance of a state x and a control u at time k, and $\phi_N(x_N)$ describes the contribution from the final state x_N. (Note: the discrete system (7.1) together with its constraints (7.2) and the separable performance index (7.4) is often called an "N-stage decision process".)

EXAMPLE 7.1.1. The continuous system

$$\frac{dx(t)}{dt} = \begin{bmatrix} 0 & 1 \\ 0 & 0 \end{bmatrix} x(t) + \begin{bmatrix} 0 \\ 1 \end{bmatrix} u(t), \qquad x(0) = \begin{bmatrix} 1 \\ 0 \end{bmatrix} \tag{7.5}$$

is to be controlled by controllers satisfying $|u(t)| \leq 1$ in such a way that the performance index

$$J(u) = \frac{1}{2}(x_1(1))^2 + \frac{1}{2}\int_0^1 \{(x_1(t))^2 + (u(t))^2\}dt \tag{7.6}$$

is minimized. If the solution were to be implemented digitally using measurements of states and piecewise constant input with sampling interval $h = 1/N$ then we may replace (7.5) *exactly* by (see Section 1.8.1)

$$x_{k+1} = \begin{bmatrix} 1 & 1/N \\ 0 & 1 \end{bmatrix} x_k + \frac{1}{N}\begin{bmatrix} \frac{1}{2N} \\ 1 \end{bmatrix} u_k, \quad x_0 = \begin{bmatrix} 1 \\ 0 \end{bmatrix}, \quad 0 \leqslant k \leqslant N-1 \tag{7.7}$$

where $x_k = x(kh)$ are the sampled state vectors and $u(t) = u_k$, $kh \leqslant t < (k+1)h$, $k \geqslant 0$. We may also *approximate* the performance index by the summation

$$J_N(u_0, u_1, \ldots, u_{N-1}) = \frac{1}{2}(x_N)_1^2 + \sum_{k=0}^{N-1} \frac{1}{2N}\{(x_k)_1^2 + u_k^2\} \tag{7.8}$$

and hence obtain an approximate solution to the original continuous optimal problem by solving the *discrete* optimal control problem defined by (7.7), (7.8) and the constraints $|u_k| \leqslant 1$, $0 \leqslant k \leqslant N-1$. In terms of (7.4) we have

$$\phi_N(x_N) = \tfrac{1}{2}(x_N)_1^2, \qquad L(x, u, k) = \frac{1}{2N}\{(x)_1^2 + u^2\} \tag{7.9}$$

and it can be expected that the accuracy of the solution increases as N becomes large, i.e. $h \to 0+$.

7.2 Dynamic Programming

In essence, dynamic programming is a *computational* method for solving the discrete optimal control problem with separable, additive performance index. It is based upon the simple "Principle of Optimality" introduced below. Before this, however, it is instructive to consider the intuitively appealing, easily implementable *but entirely incorrect* methods which attempt to solve the discrete optimal control problem in a feedback manner.

Method 1. At any time $k < N$, the control inputs $u_0, u_1, \ldots, u_{k-1}$ have already been computed, applied and hence, in real time, cannot be changed. In contrast, although a measurement of the state vector x_k may be available, measurements of future states are not yet possible. If we therefore separate J_N into the form

$$\begin{aligned} J_N = &\text{ terms dependent upon past values } u_0, u_1, \ldots, u_{k-1}, x_0, x_1, \ldots, x_{k-1} \\ &+ L(x_k, u_k, k) \\ &+ \text{term dependent upon future values } x_{k+1}, \ldots, x_N, u_{k+1}, \ldots, u_{N-1} \end{aligned} \tag{7.10}$$

and note that the first terms cannot be changed and the last terms are not yet known, it is natural to attempt to minimize J_N by choosing u_k to be the admissible controller u_k^* minimizing $L(x_k, u, k)$, i.e.

$$L(x_k, u_k^*, k) = \min_{u \in \Omega(k)} L(x_k, u, k) \tag{7.11}$$

It is obvious that u_k^* depends only upon x_k and hence that the resulting control law is feedback in form. Note also that the system model is not required to evaluate the control sequence. It is certainly a pity that, in general, the procedure does not result in an optimal system response.

Method 2. If we assume that L can be separated as follows

$$L(x, u, k) = \phi_k(x) + \psi_k(u), \qquad 0 \leqslant k \leqslant N-1 \tag{7.12}$$

then we can write

$$\begin{aligned} J_N = {} & \text{terms dependent upon past values } u_0, u_1, \ldots, u_{k-1}, x_0, \ldots, x_{k-1}, x_k \\ & + \psi_k(u_k) + \phi_{k+1}(x_{k+1}) \\ & + \text{terms dependent upon future values } x_{k+2}, \ldots, x_N, u_{k+1}, \ldots, u_{N-1} \end{aligned} \tag{7.13}$$

where one of the future values (namely x_{k+1}) is separated out as its value is easily predicted if we have a model when we know that $x_{k+1} = f(x_k, u_k, k)$. The so-called "one-step-ahead" optimization method attempts to calculate the optimal controller by minimizing

$$\psi_k(u_k) + \phi_{k+1}(x_{k+1}) = \psi_k(u_k) + \phi_{k+1}(f(x_k, u_k, k)) \tag{7.14}$$

with respect to admissible controls u_k satisfying the constraints. Again, although this algorithm does have a feedback form (and hence is easily implemented on-line), it does not result in an optimal system response.

The reason that both of the above methods do not result in calculation of the optimal control sequence is that they both attempt to calculate u_k^* without taking into account the future consequences of that choice. It is useful to take an example from the stock exchange, namely, the problem of maximizing investment income over a period of N years assuming that shares are bought and sold only at regular specified times. It is well known that a policy of maximizing immediate profit will not necessarily yield the maximum income over the total period. It may well be better to take a temporary loss in order to obtain a long term gain.

7.2.1 The Principle of Optimality

The above discussion implies that the calculation of the optimal control u_k^* at time k is dependent upon all future optimal control values u_p^*, $p \geqslant k+1$ and hence that optimal controller will be most easily and naturally calculated backwards in time, starting with u_N^*, then u_{N-1}^* etc. This must, of course, be done off-line.

The mathematical nature of this "backwardness" of the solution is expressed by Bellman's "Principle of Optimality":

> An optimal controller has the property that whatever the initial state and the initial control decision, the remaining control decisions must be optimal with regard to the state resulting from the first decision.

The mathematical form of this (rather subtle) idea can be expressed as a "backwards" sequence of simple optimization problems. Let $\{u_0^*, u_1^*, \ldots, u_{N-1}^*\}$ and $\{x_0^*, x_1^*, x_2^*, \ldots, x_N^*\}$ be an optimal control sequence and its resulting state sequence satisfying

$$x_{k+1}^* = f(x_k^*, u_k^*, k), \qquad x_0^* = x_0$$
$$0 \leqslant k \leqslant N-1 \tag{7.15}$$

Let x_k be any state vector at time $k \leqslant N-1$ and let $J_{N-k}^*(x_k)$ be the smallest value of the performance index for the $N-k$ step discrete problem

$$x_{p+1} = f(x_p, u_p, p), \qquad u_p \in \Omega(p), \qquad k \leqslant p \leqslant N-1$$

$$J_{N-k}(u_k, u_{k+1}, \ldots, u_{N-1}) = \phi_N(x_N) + \sum_{p=k}^{N-1} L(x_p, u_p, p) \tag{7.16}$$

originating at x_k at time k. Also let $J_0^*(x_N) = \phi_N(x_N)$. It is clearly seen that these minimum values are functions of x_k only, $0 \leqslant k \leqslant N$.

Consider now the optimal state x_k^* at time k and let u_k be any admissible control decision (not necessarily optimal), then it follows from the definition that

$$J_{N-k}^*(x_k^*) \leqslant L(x_k^*, u_k, k) + J_{N-k-1}^*(x_{k+1}) \tag{7.17}$$

where $x_{k+1} = f(x_k^*, u_k, k)$ and hence, as the left-hand side is independent of u_k, that

$$J_{N-k}^*(x_k^*) \leqslant \min \{L(x_k^*, u_k, k) + J_{N-k-1}^*(x_{k+1})\} \tag{7.18}$$

where $x_{k+1} = f(x_k^*, u_k, k)$ and the minimizations takes place with respect to all control vectors $u_k \in \Omega(k)$. In fact equality holds in (7.18) and the controller generating the minimum is just the optimal controller u_k^*. To prove this, we prove that

$$J_{N-k}^*(x_k^*) = J_{N-k}(u_k^*, u_{k+1}^*, \ldots, u_{N-1}^*), \qquad 0 \leqslant k \leqslant N-1 \tag{7.19}$$

It follows from the definitions that $J_{N-k}^*(x_k^*) \leqslant J_{N-k}(u_k^*, \ldots, u_{N-1}^*)$. Supposing that *strict* inequality holds and let $J_{N-k}^*(x_k^*) = J_{N-k}(\hat{u}_k, \ldots, \hat{u}_{N-1})$, then

$$
\begin{aligned}
J_N(u_0^*, \ldots, u_{k-1}^*, \hat{u}_k, \ldots, \hat{u}_{N-1}) &= \sum_{p=0}^{k-1} L(x_p^*, u_p^*, p) + J_{N-k}(\hat{u}_k, \ldots, \hat{u}_{N-1}) \\
&< \sum_{p=0}^{k-1} L(x_p^*, u_p^*, p) + J_{N-k}(u_k^*, \ldots, u_{N-1}^*) \\
&= J_N(u_0^*, \ldots, u_{N-1}^*) \qquad (7.20)
\end{aligned}
$$

which contradicts the optimality of the control sequence $\{u_0^*, \ldots, u_{N-1}^*\}$. This proves (7.19). It follows that

$$
\begin{aligned}
J_{N-k}(u_k^*, \ldots, u_{N-1}^*) &= J_{N-k}^*(x_k^*) \\
&\leqslant \min \{L(x_k^*, u_k, k) + J_{N-k-1}^*(x_{k+1})\} \\
&\leqslant L(x_k^*, u_k^*, k) + J_{N-k-1}^*(x_{k+1}^*) \\
&= L(x_k^*, u_k^*, k) + J_{N-k-1}(u_{k+1}^*, \ldots, u_{N-1}^*) \\
&= J_{N-k}(u_k^*, \ldots, u_{N-1}^*) \qquad (7.21)
\end{aligned}
$$

The only way that this can be possible is if *equality* holds in (7.18) and if the minimum is generated by the optimal control u_k^*.

In summary the above analysis has proved the following mathematical version of the "Principle of Optimality":

If $\{u_0^*, u_1^*, \ldots, u_{N-1}^*\}$ is an optimal control sequence generating the optimal state sequence $\{x_0^*, x_1^*, \ldots, x_N^*\}$ satisfying (7.15), then

$$
\begin{aligned}
J_{N-k}^*(x_k^*) &= \min \{L(x_k^*, u_k, k) + J_{N-k-1}^*(x_{k+1})\} \\
&0 \leqslant k \leqslant N-1 \qquad (7.22)
\end{aligned}
$$

where $x_{k+1} = f(x_k^*, u_k, k)$ and the minimization is performed with respect to all u_k satisfying the constraints $u_k \in \Omega(k)$. Also

$$
\begin{aligned}
J_{N-k}^*(x_k^*) &= J_{N-k}(u_k^*, u_{k+1}^*, \ldots, u_{N-1}^*) \\
&0 \leqslant k \leqslant N-1 \qquad (7.23)
\end{aligned}
$$

and the minimum in (7.22) is generated by $u_k = u_k^*$, $0 \leqslant k \leqslant N-1$. (Note: equations (7.22) are frequently called the "backward recursion relations" of dynamic programming.)

EXERCISE 7.2.1. Verify that the one-step-ahead optimization method is a valid technique for calculating u_k^* on the last step (i.e. $k = N-1$) of an N-stage decision process.

7.2.2 *A computational method*

The backwards recursion relations (7.22) could be used directly to generate the optimal control sequence on-line in a *feedback* manner if all of the functions $J^*_{N-k}(x)$ are known as a function of x, $0 \leqslant k \leqslant N$. If this situation holds then, at time $k = 0$, $x_0^* = x_0$ is known and, from the principle of optimality, the optimal control u_0^* can be evaluated by minimizing

$$L(x_0, u_0, 0) + J^*_{N-1}(x_1) = L(x_0, u_0, 0) + J^*_{N-1}(f(x_0, u_0, 0)) \quad (7.24)$$

with respect to admissible controls $u_0 \in \Omega(0)$. In a similar manner at time k, the state x_k^* will be known and u_k^* can be evaluated by minimizing

$$L(x_k^*, u_k, k) + J^*_{N-k-1}(x_{k+1}) = L(x_k^*, u_k, k) + J^*_{N-k-1}(f(x_k^*, u_k, k)) \quad (7.25)$$

with respect to admissible controls $u_k \in \Omega(k)$. In fact, if all the functions $J^*_{N-k}(x)$ are known, the only problem in the implementation of the optimal control algorithm in real time is to ensure that the calculation of u_k^* can be performed quickly enough.

EXERCISE 7.2.2. A discrete system is to be optimally controlled to minimize the performance index (7.4) with

$$\phi_N(x) = \tfrac{1}{2}x^TFx, \qquad L(x, u, k) = \tfrac{1}{2}x^TQx + \tfrac{1}{2}u^TRu \quad (7.26)$$

with $F = F^T \geqslant 0$, $Q = Q^T \geqslant 0$ and $R = R^T > 0$. Assuming that it is known that

$$J^*_{N-k}(x) = \tfrac{1}{2}x^TK_kx, \qquad 0 \leqslant k \leqslant N \quad (7.27)$$

where $K_k = K_k^T \geqslant 0$, $0 \leqslant k \leqslant N$, show that the optimal controller at time k can be computed by solving the minimization problem: Minimize

$$L(x_k^*, u_k, k) + J^*_{N-k-1}(f(x_k^*, u_k, k)) = \tfrac{1}{2}x_k^{*T}Qx_k^* + \tfrac{1}{2}u_k^TRu_k + \tfrac{1}{2}f^T(x_k^*, u_k, k)K_{k+1}f(x_k^*, u_k, k) \quad (7.28)$$

subject to the constraints $u_k \in \Omega(k)$.

In the case when $f(x, u, k) = \Phi x + \Delta u$ where Φ and Δ are real $n \times n$ and $n \times l$ matrices respectively and also the control inputs are unconstrained, verify that

$$Ru_k^* + \Delta^TK_{k+1}\Phi x_k^* + \Delta^TK_{k+1}\Delta u_k^* = 0 \quad (7.29)$$

(hint: as the control is unconstrained (7.28) can be solved by differentiation) and hence that

$$u_k^* = -(R + \Delta^TK_{k+1}\Delta)^{-1}\Delta^TK_{k+1}\Phi x_k^* \quad (7.30)$$

expresses an exact linear (feedback) relationship between current state and

EXERCISE 7.2.2 contd.
current optimal controller. Show that the states satisfy the closed-loop equation

$$x_{k+1}^* = (I - \Delta(R + \Delta^T K_{k+1} \Delta)^{-1} \Delta^T K_{k+1})\Phi x_k^*, \quad k \geqslant 0 \quad (7.31)$$

and hence that the K_k, $0 \leqslant k \leqslant N$, are related by the backwards recursion relations

$$\begin{aligned} K_N &= F \\ K_k &= Q + \Phi^T K_{k+1} \Delta(R + \Delta^T K_{k+1} \Delta)^{-1} R (R + \Delta^T K_{k+1} \Delta)^{-1} \Delta^T K_{k+1} \Phi \\ &\quad + \Phi^T (I - \Delta(R + \Delta^T K_{k+1} \Delta)^{-1} \Delta^T K_{k+1})^T K_{k+1} \times \\ &\quad (I - \Delta(R + \Delta^T K_{k+1} \Delta)^{-1} \Delta^T K_{k+1})\Phi \\ &\quad 0 \leqslant k \leqslant N - 1 \end{aligned} \quad (7.32)$$

In reality, of course, the functional dependence of $J_{N-k}^*(x)$ is not known in general. An approximate solution can, however, be implemented but it requires excessive computer storage if the system state dimension n is larger than two or three. The basic idea can be illustrated by considering the situation when $l = n = 1$, i.e. the state and control vectors are scalar. We first guess the range $x_{\min} \leqslant x \leqslant x_{\max}$ over which the true optimal trajectory will move and divide this into a number ($N_S + 1$ say) of sample points equispaced with distance $h_0 = (x_{\max} - x_{\min})/N_S$. The state and time samples can then be regarded as axes for a two-dimensional table as illustrated in Fig. 81, the entry at the point (x, k) being an

x \ k	0	1		$N-1$	N
$x_{\min}$	$J_N^*(x_{\min})$	$J_{N-1}^*(x_{\min})$		$J_1^*(x_{\min})$	$J_0^*(x_{\min})$
$x_{\min} + h_0$	$J_N^*(x_{\min}+h_0)$	$J_{N-1}^*(x_{\min}+h_0)$		$J_1^*(x_{\min}+h_0)$	$J_0^*(x_{\min}+h_0)$
$x_{\min}+(N_S-2)h_0$					
$x_{\min}+(N_S-1)h_0$					
$x_{\max}$	$J_N^*(x_{\max})$	$J_{N-1}^*(x_{\max})$			$J_0^*(x_{\max})$

Fig. 81. "Look-up" table for dynamic programming.

estimate of $J^*_{N-k}(x)$. The table is filled in by working from right to left, starting with column N with the easily computed values of $J^*_0(x_{\min} + rh_0) = \phi_N(x_{\min} + rh_0)$. We then move to column $N-1$ and set

$$J^*_1(x_{\min} + rh_0) = \min \{L(x_{\min} + rh_0, u, N-1) + J^*_0(f(x_{\min} + rh_0, u, N-1))\}, \qquad 0 \leqslant r \leqslant N_S \tag{7.33}$$

where the minimization is performed with respect to $u \in \Omega(N-1)$. If $f(x_{\min} + rh_0, u, N-1)$ lies outside the range $x_{\min} \leqslant x \leqslant x_{\max}$ then the corresponding u is ignored. Alternatively, if it lies within the range the value of J^*_0 is estimated by interpolating between the values in column N. The procedure is then repeated for column $N-2$, then $N-3$ etc. In general we deduce the entries in column $k \leqslant N-1$ from column $k+1$ by

$$J^*_{N-k}(x_{\min} + rh_0) = \min \{L(x_{\min} + rh_0, u, k) + J^*_{N-k-1}(f(x_{\min} + rh_0, u, k))\}, \qquad 0 \leqslant r \leqslant N_S \tag{7.34}$$

the minimization being performed with respect to $u \in \Omega(k)$. Again, if $f(x_{\min} + rh_0, u, k)$ lies outside the range $x_{\min} \leqslant x \leqslant x_{\max}$ then the corresponding u is ignored and if it lies within this range the value of J^*_{N-k-1} is estimated by interpolation using column $k+1$.

Having completed the "*look-up table*" it can be stored and used, in principle, as the basis of a *feedback* computer control scheme with the following structure:

Step 1: await the first sample time and set $k = 0$

Step 2: measure *state* x^*_k

Step 3: find an admissible controller u^*_k solving the minimization problem

$$\min \{L(x^*_k, u, k) + J^*_{N-k-1}(f(x^*_k, u, k))\} \tag{7.35}$$

minimizing with respect to admissible controllers $u \in \Omega(k)$ and using interpolation on column $k+1$ of the look-up table to estimate the required values of J^*_{N-k-1}. (This calculation should be performed in a time much less than the sample interval if the control is computed on-line.)

Step 4: Subject the system to the calculated input u^*_k

Step 5: Set $k = k+1$. If $k \geqslant N$ stop. Otherwise await the next sample time and return to step 2.

(Note: a similar algorithm can be used for off-line calculation of the optimal control.)

EXAMPLE 7.2.1. Consider the solution of the discrete problem of minimizing the performance index

$$J_2(u) = \sum_{k=0}^{1} (|x_k| + |u_k|) + |x_2| \tag{7.36}$$

for the system $x_{k+1} = -x_k + u_k$, $x_0 = 3$ if the input u_k can only take one of two values, namely 0 or 1. We see that $N = 2$ and

$$\phi_2(x) = |x|, \; L(x,u,k) = |x| + |u|, \; f(x,u,k) = -x + u \tag{7.37}$$

We begin our (approximate) calculation of the look-up table shown in Fig. 82 by choosing $x_{min} = -x_{max} = -3$ and $N_S = 6$ and placing the appropriate values of $\phi_2(x) = |x|$ in column $k = 2$. We now fill in column $k = 1$ by applying (7.34) with $N = 2$ and $k = 1$. To illustrate this procedure consider the entry for $x = 3$, $k = 1$, namely

$$J_1^*(3) = \min_{u=0,1} \{|3| + |u| + J_0^*(-3+u)\} \tag{7.38}$$

x \ k	0	1	2
-3	9	6	3
-2	6	4	2
-1	3	2	1
0	0	0	0
1	2	2	1
2	5	4	2
3	8	6	3

Fig. 82.

For $u = 0$, the expression in brackets is $3 + J_0^*(-3) = 3 + 3 = 6$ as $J_0^*(-3) = 3$ is the element in the $x = -3$, $k = 2$ position. For $u = 1$, we obtain $|3| + |1| + J_0^*(-2) = 6$. It follows that the entry for $x = 3$, $k = 1$ is $J_1^*(3) = 6$. The rest of column $k = 1$ is filled in a similar manner. Consider now the filling of column $k = 0$. In particular consider the entry for $x = 3$ and $k = 0$, i.e.

$$\begin{aligned} J_0^*(3) &= \min_{u=0,1} \{|3| + |u| + J_1^*(-3+u)\} \\ &= \min\{|3| + J_1^*(-3), \; |3| + |1| + J_1^*(-2)\} \\ &= \min\{9, 8\} = 8 \end{aligned} \tag{7.39}$$

The rest of the column is filled in a similar manner.

The final calculation of the optimal control sequence $u^* = \{u_0^*, u_1^*\}$ proceeds using the table. We note immediately that the minimum value of the performance index for the initial condition $x_0 = 3$ is, by (7.23),

$$J_2(u^*) = J_2^*(3) = 8 \qquad (7.40)$$

We compute u_0^* as the admissible controller minimizing $|3| + |u| + J_1^*(-3+u)$. From the table, for $u=0$, $|3| + |u| + J_1^*(-3+u) = 9$ and $u=1$, $|3| + |u| + J_1^*(-3+u) = 8$ i.e. $u_0^* = 1$ and $x_1^* = -3+1 = -2$. To compute u_1^*, we must minimize $|2| + |u| + J_0^*(2+u)$, i.e. using the look-up table, this expression takes the value of 4 for $u=0$ and 6 for $u=1$ yielding $u_1^* = 0$, and hence $x_2^* = 2$.

EXERCISE 7.2.3. For the problem given in Example 7.2.1 with $x_0 = -2$, verify that the optimal control sequence is $u_0^* = 0$ and $u_1^* = 0$ *or* 1, i.e. the optimal control is not unique.

The above procedure can be applied *in principle* to n dimensional problems by simply estimating the range over which *each* state variable will vary and dividing it into $N_S + 1$ (say) sample points. The look-up table is then a multi-dimensional object (this causes no problem for computers) containing $(N+1)N_S^n$ entries. It is here that the problem occurs. Taking the (not unreasonable) case of $N_S = 100$ for an accurate solution, $N = 9$ for a nine-step process and state dimension $n = 6$, then the look-up table contains 10^{13} entries. It is immediately obvious that the computer storage requirements are immense. For this reason, it is generally true that dynamic programming can only be sensibly applied to problems of *low* state dimension $n \leqslant 3$. The one important exception to this rule is described in the next section.

Finally, even if available computer storage is adequate, there remains the problem of solving (7.35) fast enough to make real-time implementation viable. One way is to solve (7.35) *off-line* for all tabular values of the state vector and time and replace $J_{N-k}^*(x)$ in the look-up table by the $l \times 1$ solution vector $u_k^*(x)$. The look-up table then becomes an explicit tabular feedback mechanism enabling the necessary control input corresponding to a measured state x_k at time k to be computed by a simple interpolation between the tabulated values. The computer storage requirements are, of course, *increased* by a factor of $l =$ number of plant inputs.

7.3 The Discrete Linear Quadratic Problem

The discrete linear quadratic problem discussed in this section is probably the most important general problem for the purpose of application studies as

(a) it can be partially solved analytically, hence reducing the need to store large look-up tables and making application to high order systems feasible
(b) the optimal control is of a linear state feedback form (c.f. Section 5.2)

The general problem can be stated as follows – given an l-input linear, discrete, time-invariant system

$$x_{k+1} = \Phi x_k + \Delta u_k, \qquad 0 \leqslant k \leqslant N-1 \tag{7.41}$$

with a specified initial condition x_0, calculate the optimal control sequence $u^* = \{u_0^*, u_1^*, \ldots, u_{N-1}^*\}$ that minimizes the quadratic performance index

$$J_N(u) = \frac{1}{2} x_N^T F x_N + \sum_{k=0}^{N-1} \frac{1}{2} \{x_k^T Q x_k + u_k^T R u_k\} \tag{7.42}$$

where $Q = Q^T \geqslant 0$, $F = F^T \geqslant 0$ and $R = R^T > 0$. It is assumed that the elements of the input vectors are unconstrained.

To solve this problem we use the principle of optimality and the results of Exercise 7.2.2. More precisely, we will *prove* that

$$J_{N-k}^*(x) = \tfrac{1}{2} x^T K_k x, \qquad 0 \leqslant k \leqslant N \tag{7.43}$$

where K_k are symmetric, positive semidefinite matrices generated from the backwards recursion relations (c.f. (7.32))

$$\begin{aligned} K_N &= F \\ K_k &= Q + \Phi^T K_{k+1} \Delta (R + \Delta^T K_{k+1} \Delta)^{-1} R (R + \Delta^T K_{k+1} \Delta)^{-1} \Delta^T K_{k+1} \Phi \\ &\quad + \Phi^T (I - \Delta (R + \Delta^T K_{k+1} \Delta)^{-1} \Delta^T K_{k+1})^T K_{k+1} \times \\ &\quad (I - \Delta (R + \Delta^T K_{k+1} \Delta)^{-1} \Delta^T K_{k+1}) \Phi \\ &\quad 0 \leqslant k \leqslant N-1 \end{aligned} \tag{7.44}$$

and hence that the optimal controller is a linear, time-varying feedback controller of the form

$$u_k^* = -(R + \Delta^T K_{k+1} \Delta)^{-1} \Delta^T K_{k+1} \Phi x_k^*, \quad 0 \leqslant k \leqslant N-1 \tag{7.45}$$

Such a control scheme can, of course, be implemented by computing the sequence $K_N, K_{N-1}, K_{N-2}, \ldots$ off-line to obtain the "gain" matrices

$$P_k = (R + \Delta^T K_{k+1} \Delta)^{-1} \Delta^T K_{k+1} \Phi, \qquad 0 \leqslant k \leqslant N-1 \tag{7.46}$$

The P_k can be stored in a computer (only Nnl entries) and the controller implemented as

$$u_k^* = -P_k x_k^*, \qquad 0 \leqslant k \leqslant N-1 \tag{7.47}$$

To prove the above assertions, we see that (7.43) holds trivially for $k = N$. Using an inductive argument suppose that $J_{N-k-1}^*(x) = \frac{1}{2} x^T K_{k+1} x$ for some

k in the range $0 < k \leqslant N-1$ then, applying (7.22),

$$J^*_{N-k}(x) = \min_u \{\tfrac{1}{2}x^TQx + \tfrac{1}{2}u^TRu + J^*_{N-k-1}(\Phi x + \Delta u)\} \qquad (7.48)$$

But this minimization was undertaken in Exercise 7.2.2 yielding the minimizing controller (7.30), i.e. (7.45), and $J^*_{N-k}(x) = \frac{1}{2}x^TK_kx$ with K_k given by (7.32), i.e. (7.44). The inductive assumption therefore holds for $k-1$ and hence for all k. This completes the proof of the assertions.

EXAMPLE 7.3.1. Consider the solution of the linear quadratic problem

$$x_{k+1} = -x_k + u_k,\ x_0 = 1, \qquad J(u) = \frac{1}{2}\sum_{k=0}^{N-1}\{x_k^2 + u_k^2\} \qquad (7.49)$$

where the control is unconstrained. The relevant problem data are $\Phi = -1$, $\Delta = 1$, $F = 0$, $Q = 1$ and $R = 1$. The first problem is to solve the backward recursion relations (7.44) for the (scalar) K_k, $0 \leqslant k \leqslant N$, i.e.

$$K_N = 0$$

$$K_k = 1 + \frac{K^2_{k+1}}{(1+K_{k+1})^2} + K_{k+1}\left(1 - \frac{K_{k+1}}{(1+K_{k+1})}\right)^2$$

$$= (1 + 2K_{k+1})/(1 + K_{k+1}), \qquad 0 \leqslant k \leqslant N-1 \qquad (7.50)$$

yielding the results

k	K_k	P_k
N	0	—
$N-1$	1	0
$N-2$	1.5	-0.5
$N-3$	1.6	-0.6
$N-4$	1.615	-0.615
$N-5$	1.617	-0.617
$N-6$	1.617	-0.617

and the optimal control sequence, in the case of $N = 6$,

$$u_0^* = 0.617x_0^*, \qquad u_1^* = 0.617x_1^*, \qquad u_2^* = 0.615x_2^*,$$

$$u_3^* = 0.6x_3^*, \qquad u_4^* = 0.5x_4^*, \qquad u_5^* = 0$$

yielding $x_1^* = -x_0^* + u_0^* = -0.383x_0^* = -0.383$, $x_2^* = -x_1^* + u_1^* = -0.383x_1^* = 0.147$, $x_3^* = -0.385x_2^* = -0.0566$, $x_4^* = 0.034$, $x_5^* = -0.017, x_6^* = 0.017$.

Although the recursion relations (7.44) are easily programmed on a digital computer, the time-varying nature of the feedback controller and the consequent need to store lNn entries for the $l \times n$ matrices P_k, $0 \leqslant k \leqslant N-1$, is an unfortunate complication, particularly if the state dimension n is large and/or the time interval N is long, e.g. if $l = 2$, $n = 10$ and $N = 400$ we need $lNn = 8000$ elements to define the P_k, $0 \leqslant k \leqslant N-1$. This would be a particularly awkward problem in applications where the optimal controller is being designed to regulate state deviations without excessive control inputs over a long period of time. It is fortunate however that, as we go to the limit as $N \to +\infty$, the solution can be considerably simplified (see Section 5.2.2 for the equivalent continuous case). More precisely, if our system is controllable (a technical detail), the discrete problem

$$x_{k+1} = \Phi x_k + \Delta u_k, \; x_0 \text{ specified}, \; k \geqslant 0$$

$$J_\infty(u) = \frac{1}{2} \sum_{k=0}^{\infty} \{x_k^T Q x_k + u_k^T R u_k\}$$

$$u_k \text{ unconstrained}, \; Q = Q^T \geqslant 0, \qquad R = R^T > 0 \tag{7.51}$$

obtained from (7.41) and (7.42) by letting $N \to +\infty$ and setting $F = 0$, has the feedback form defined by

$$u_k^* = -P_\infty x_k^*, \qquad k \geqslant 0 \tag{7.52}$$

The constant $l \times n$ state feedback matrix is obtained from

$$P_\infty = (R + \Delta^T K_\infty \Delta)^{-1} \Delta^T K_\infty \Phi \tag{7.53}$$

and K_∞ is defined as the limit obtained by continuing the recursion relations (7.44) back to $k = -\infty$, i.e.

$$\lim_{k \to -\infty} K_k = K_\infty \tag{7.54}$$

Alternatively, letting $k \to -\infty$ in (7.44) K is the symmetric, positive semidefinite solution of the algebraic equation,

$$\begin{aligned} K_\infty = {} & Q + \Phi K_\infty \Delta (R + \Delta^T K_\infty \Delta)^{-1} R (R + \Delta^T K_\infty \Delta)^{-1} \Delta^T K_\infty \Phi \\ & + \Phi^T (I - \Delta (R + \Delta^T K_\infty \Delta)^{-1} \Delta^T K_\infty)^T K_\infty \times \\ & (I - \Delta (R + \Delta^T K_\infty \Delta)^{-1} \Delta^T K_\infty) \Phi \end{aligned} \tag{7.55}$$

Note that the "infinite-time" control law (7.49) requires the storage of only nl elements of P_∞.

EXERCISE 7.3.1. For the problem of Example 7.3.1 with $N = +\infty$, verify that the limit is $K_\infty = 1.618$ by solving (7.55). Check this result by using the backwards recursion relations and (7.54). Hence show that $P_\infty = -0.618$.

7.4 Least Squares Solutions, Costates and Lagrange Multipliers

The computational method of dynamic programming involves extra complexity if the state vector is also subjected to terminal state constraints. In such cases, alternative computational techniques can be used. It is the purpose of this section to derive a simple matrix method for solving discrete problems with quadratic performance index and linear dynamics. The analysis also introduces the notion of TPBVP and costates into the discrete problem and yields a relationship between costates and the familiar Lagrange multipliers.

7.4.1 *The discrete linear quadratic control problem revisited*

If we define the $N(n+l) \times 1$ vector

$$y = \begin{bmatrix} u_0 \\ x_1 \\ u_1 \\ x_2 \\ u_2 \\ \cdot \\ \cdot \\ \cdot \\ u_{N-1} \\ x_N \end{bmatrix} \quad \text{(composite state and control vector)} \tag{7.56}$$

then the discrete model in (7.41) can be written in the form

$$My = f \tag{7.57}$$

where M is an $Nn \times N(n+l)$ matrix and f is an $Nn \times 1$ matrix of the block form

$$M = \begin{bmatrix} \Delta & -I & 0 & 0 & 0 & \cdots & \cdots & 0 \\ 0 & \Phi & \Delta & -I & 0 & & & \vdots \\ 0 & 0 & 0 & \Phi & \Delta & -1 & & \vdots \\ \vdots & & & & & \ddots & & 0 \\ 0 & \cdots & \cdots & \cdots & 0 & \Phi & \Delta & -I \end{bmatrix}$$

$$f = \begin{bmatrix} -\Phi x_0 \\ 0 \\ \vdots \\ 0 \end{bmatrix} \tag{7.58}$$

The performance index (7.42) takes the form

$$J_N(u) = \tfrac{1}{2} y^T L y + \tfrac{1}{2} x_0^T Q x_0 \tag{7.59}$$

where the $N(n + l) \times N(n + l)$ symmetric matrix

$$L = \begin{bmatrix} R & 0 & \cdots & \cdots & \cdots & 0 \\ 0 & Q & & & & \vdots \\ \vdots & & R & & & \vdots \\ \vdots & & & Q & & \vdots \\ \vdots & & & & \ddots\; R & 0 \\ 0 & \cdots & \cdots & \cdots & 0 & F \end{bmatrix} \tag{7.60}$$

is positive semi-definite and hence the discrete linear quadratic problem can be expressed in the simple form:

$$\text{Minimize } \tfrac{1}{2} y^T L y \text{ subject to } My = f \tag{7.61}$$

This *matrix* problem can be approached by using Lagrange multipliers. More precisely, we will introduce Nn Legrange multipliers each corresponding to one of the equality constraints in (7.57). Problem (7.61) can then be solved by computing the stationary points of the Lagrangian

$$L(y, \lambda) = \tfrac{1}{2} y^T L y + \lambda^T (My - f) \tag{7.62}$$

where

$$\lambda = \begin{bmatrix} p_1 \\ p_2 \\ \cdot \\ \cdot \\ \cdot \\ p_N \end{bmatrix} \tag{7.63}$$

is an $Nn \times 1$ matrix of Lagrange multipliers and p_k, $1 \leqslant k \leqslant N$ are $n \times 1$ costate vectors. Differentiating (7.62) leads to the following conditions for the solution vector y^* for (7.61)

$$Ly^* + M^T\lambda = 0, \qquad My^* = f \tag{7.64}$$

The equation $Ly^* + M^T\lambda = 0$ can be written in the form

$$\begin{aligned} Ru_k^* + \Delta^T p_{k+1} &= 0, \qquad 0 \leqslant k \leqslant N-1 \\ Qx_k^* - p_k + \Phi^T p_{k+1} &= 0, \qquad 1 \leqslant k \leqslant N-1 \\ Fx_N^* - p_N &= 0 \end{aligned} \tag{7.65}$$

whilst the equation $My^* = f$ is simply the state equations

$$x_{k+1}^* = \Phi x_k^* + \Delta u_k^*, \qquad x_0^* = x_0 \tag{7.66}$$

More precisely, the optimal control sequence is the solution of the discrete TPBVP

$$\begin{aligned} x_{k+1}^* &= \Phi x_k^* + \Delta u_k^* \qquad x_0^* = x_0 \\ u_k^* &= -R^{-1}\Delta^T p_{k+1} \qquad 0 \leqslant k \leqslant N-1 \\ p_k &= \Phi^T p_{k+1} + Qx_k^*, \qquad 1 \leqslant k \leqslant N-1 \\ p_N &= Fx_N^* \end{aligned} \tag{7.67}$$

which should be compared with the results of Chapter five. Note that the costates are simply the Lagrange multipliers for the problem (7.61).

The TPBVP can, in principle, be used as the basis for computation. Bearing in mind, however, our knowledge that the optimal controller has a direct, state feedback form, we will attempt a solution satisfying

$$p_k = \widetilde{P}_k x_{k-1}^*, \qquad 1 \leqslant k \leqslant N \tag{7.68}$$

and hence the optimal controller

$$u_k^* = -R^{-1}\Delta^T\widetilde{P}_{k+1}x_k^*, \qquad 0 \leqslant k \leqslant N-1 \tag{7.69}$$

To find the backwards recurrence relations defining the $\widetilde{P}_k$, substitute (7.68) into the state equations to give

$$x_{k+1}^* = (\Phi - \Delta R^{-1}\Delta^T\widetilde{P}_{k+1})x_k^*, \qquad 0 \leqslant k \leqslant N-1 \tag{7.70}$$

from which the costate equations become, $1 \leqslant k \leqslant N-1$

$$\begin{aligned}\tilde{P}_k x_{k-1}^* &= (\Phi^T \tilde{P}_{k+1} + Q)x_k^* \\ &= (\Phi^T \tilde{P}_{k+1} + Q)(\Phi - \Delta R^{-1} \Delta^T \tilde{P}_k)x_{k-1}^* \end{aligned} \tag{7.71}$$

But, as the feedback gains are independent of the state, (7.71) must hold for all x_{k-1}^*, i.e.

$$\tilde{P}_k = (\Phi^T \tilde{P}_{k+1} + Q)(\Phi - \Delta R^{-1} \Delta^T \tilde{P}_k), \qquad 1 \leqslant k \leqslant N-1 \tag{7.72}$$

with terminal boundary condition derived from

$$F x_N^* = F(\Phi - \Delta R^{-1} \Delta^T \tilde{P}_N) x_{N-1}^* = \tilde{P}_N x_{N-1}^*$$

or, after a little manipulation,

$$\tilde{P}_N = (I_n + F \Delta R^{-1} \Delta^T)^{-1} F \Phi \tag{7.73}$$

EXERCISE 7.4.1. Compare and contrast the feedback solutions of Section 7.3 with the above. In particular consider both solutions when $N \to +\infty$ in the case of $F = 0$.

7.4.2 *The discrete minimum energy problem*

The problem considered here is the choice of unconstrained control sequence $u^* = \{u_0^*, u_1^*, \ldots, u_{N-1}^*\}$ for the linear, discrete system (7.41) to transfer a specified initial state x_0 to a specified final state $x_N = x_f$ at the specified time N, whilst minimizing the performance index

$$J_N(u) = \frac{1}{2} \sum_{k=0}^{N-1} u_k^T R u_k \tag{7.74}$$

representing the total amount of control energy used to complete the task. As usual we will take the case of $R = R^T > 0$.

Following the technique of Section 7.4.1 define the $Nl \times 1$ vector

$$y = \begin{bmatrix} u_0 \\ u_1 \\ \cdot \\ \cdot \\ \cdot \\ u_{N-1} \end{bmatrix} \quad \text{(composite input vector)} \tag{7.75}$$

and the $Nl \times Nl$ symmetric positive-definite matrix

$$L = \begin{bmatrix} R & 0 & \ldots & 0 \\ 0 & R & & \vdots \\ \vdots & & & \vdots \\ 0 & \ldots & \ldots & R \end{bmatrix} \tag{7.76}$$

It is easily shown that the performance index takes the quadratic form (c.f. (7.59))

$$J_N(u) = \tfrac{1}{2} y^T L y$$

Also writing the state equations in the form

$$\Delta u_{N-1} + \Phi \Delta u_{N-2} + \ldots + \Phi^{N-1} \Delta u_0 = x_f - \Phi^N x_0 \tag{7.77}$$

and defining the $n \times nl$ and $n \times 1$ matrices

$$M = [\Phi^{N-1}\Delta, \Phi^{N-2}\Delta, \ldots, \Phi\Delta, \Delta], \qquad f = x_f - \Phi^N x_0, \tag{7.78}$$

we obtain the compact system representation

$$My = f \tag{7.79}$$

and deduce that the discrete minimum energy problem can be expressed in the form of (7.61).

Introducing an $n \times 1$ vector λ of Lagrange multipliers and constructing the Lagrangian (7.62), it is seen that (7.64) still defines the optimal solution, but that the equation $Ly^* + M^T\lambda = 0$ takes the form

$$Ru_k^* + \Delta^T(\Phi^T)^{N-1-k}\lambda = 0, \qquad 0 \leqslant k \leqslant N-1 \tag{7.80}$$

The equation $My^* = f$ is simply the state equations, so, defining costates $p_k = (\Phi^T)^{N-k}\lambda$, we see that $p_k = \Phi^T p_{k+1}$, $1 \leqslant k \leqslant N-1$, and $p_N = \lambda$ and hence the optimal control sequence is the solution of the discrete TPBVP

$$\begin{aligned} x_{k+1}^* &= \Phi x_k^* + \Delta u_k^*, \qquad x_0^* = x_0, \qquad x_N^* = x_f \\ u_k^* &= -R^{-1}\Delta^T p_{k+1} \\ p_k &= \Phi^T p_{k+1}, \qquad 1 \leqslant k \leqslant N-1 \\ p_N &= \lambda \quad \text{(unknown)} \end{aligned} \tag{7.81}$$

This TPBVP has a parallel structure to that obtained in Section 5.3 for the unconstrained continuous minimum energy controller. In particular, we see that the costate terminal boundary conditions are not known and are equal to the Lagrange multiplier vector for the problem (7.61).

Although the TPBVP (7.81) could be the basis of a computation method, we will restrict our attention to solving the problem by direct solution of the defining relations (7.64). In particular, as L is nonsingular, the equation $Ly^* + M^T\lambda = 0$ yields

$$y^* = -L^{-1}M^T\lambda \tag{7.82}$$

Substituting into $My^* = f$ yields the Lagrange multiplier

$$\lambda = -(ML^{-1}M^T)^{-1}f \tag{7.83}$$

and hence the solution

$$y^* = L^{-1}M^T(ML^{-1}M^T)^{-1}f \tag{7.84}$$

The optimal control sequence is obtained by examination of the elements of y^* bearing in mind the definition (7.75).

EXERCISE 7.4.2. Verify that L is nonsingular and that $ML^{-1}M^T$ is then nonsingular only if the system (7.41) is controllable. (Hint: look at the form of M.)

EXAMPLE 7.4.1. The three-stage minimum energy problem

$$x_{k+1} = \begin{bmatrix} 1 & 1 \\ 0 & 1 \end{bmatrix} x_k + \begin{bmatrix} 1 \\ 1 \end{bmatrix} u_k, \qquad 0 \leqslant k \leqslant 2, \qquad x_0 = \begin{bmatrix} 1 \\ 1 \end{bmatrix},$$

$$x_3 = \begin{bmatrix} 0 \\ 0 \end{bmatrix}, \qquad J_3(u) = \tfrac{1}{2}(u_1^2 + u_2^2 + u_3^2) \tag{7.85}$$

with unconstrained control input can be solved by noting that

$$\Phi^k = \begin{bmatrix} 1 & k \\ 0 & 1 \end{bmatrix}, \qquad \Phi^k \Delta = \begin{bmatrix} k+1 \\ 1 \end{bmatrix}, \qquad k \geqslant 0 \tag{7.86}$$

and hence that

$$M = \begin{bmatrix} 3 & 2 & 1 \\ 1 & 1 & 1 \end{bmatrix}, \qquad f = \begin{bmatrix} -4 \\ -1 \end{bmatrix} \tag{7.87}$$

Also $R = 1$ and hence L is just the 3×3 unit matrix. This yields

$$ML^{-1}M^T = \begin{bmatrix} 14 & 6 \\ 6 & 3 \end{bmatrix} \tag{7.88}$$

and hence the solution

$$y^* = \begin{bmatrix} u_0^* \\ u_1^* \\ u_2^* \end{bmatrix} = L^{-1}M^T(ML^{-1}M^T)^{-1}f = \begin{bmatrix} -\frac{4}{3} \\ -\frac{1}{3} \\ \frac{2}{3} \end{bmatrix} \tag{7.89}$$

and the optimal control sequence $u_0^* = -\frac{4}{3}$, $u_1^* = -\frac{1}{3}$, $u_2^* = \frac{2}{3}$. The Lagrange multiplier vector

$$\lambda = -(ML^{-1}M^T)^{-1}f = \begin{bmatrix} 1 \\ -\frac{5}{3} \end{bmatrix} \tag{7.90}$$

EXERCISE 7.4.3. Show that the minimum value of the cost is

$$\begin{aligned} J_N(u^*) &= \tfrac{1}{2}y^{*T}Ly^* = \tfrac{1}{2}f^T(ML^{-1}M^T)^{-1}f \\ &= \tfrac{1}{2}\lambda^T ML^{-1}M^T\lambda \end{aligned} \tag{7.91}$$

and hence that the minimum value of the cost can be computed directly from u^*, from the initial data, or from the Lagrange multiplier vector. Verify this for the problem of Example 7.4.1.

Problems

(1) The continuous system $dx(t)/dt = f(x(t), u(t), t)$, $x(0) = x_0$ is to be controlled by controllers satisfying $u(t) \in \Omega(t)$ in such a manner that the performance index

$$J(u) = \phi(x(t_f)) + \int_0^{t_f} L(x(t), u(t), t)dt$$

is minimized. If the solution were to be implemented digitally using synchronized measurements of states and piecewise constant input with sampling interval $h = t_f/N$, consider how an approximate solution can be calculated by solving the discrete problem

$$\begin{aligned} x_{k+1} &= x_k + hf(x_k, u_k, kh) \\ u_k &\in \Omega(kh), \qquad 0 \leqslant k \leqslant N-1 \end{aligned}$$

$$J_N(u) = \phi(x_N) + h\sum_{k=0}^{N-1} L(x_k, u_k, kh)$$

if h is small enough.

(2) A discrete system $x_{k+1} = f(x_k, u_k, k)$ is controlled by inputs satisfying the constraints $u_k \in \Omega(k)$, $k \geqslant 0$. Show that an initial state x_0 can be controlled to a final state $x_N = x_f$ for some finite $N \geqslant 1$ if, and only if, the minimum value of the performance index $J_N(u) = \frac{1}{2}(x_N - x_f)^T(x_N - x_f)$ is zero. Hence use dynamic programming to show that the linear system

$$x_{k+1} = \begin{bmatrix} 1 & \frac{1}{2} \\ 0 & 1 \end{bmatrix} x_k + \begin{bmatrix} \frac{1}{8} \\ \frac{1}{2} \end{bmatrix} u_k, \qquad x_0 = \begin{bmatrix} 1 \\ 0 \end{bmatrix}$$

can be transfered to the final state $x_f = \begin{bmatrix} 0 \\ 0 \end{bmatrix}$ in *two* steps using controls $u_k = \pm 4$, $k \geqslant 0$. (Hint: as we are only interested in finding controls such that $J_2(u^*) = 0$, work backwards by finding all points x such that $J_1^*(x) = 0$ and then $J_2^*(x) = 0$ using the recursion relations of dynamic programming. Is x_0 one of these points?)

(3) Verify that the linear quadratic problem

$$x_{k+1} = -x_k + u_k, \qquad x_0 = 1, \qquad J(u) = \frac{1}{2}x_N^2 + \frac{1}{2}\sum_{k=0}^{N-1} u_k^2$$

with unconstrained control input has the solution $u_k^* = -P_k x_k^*$, $0 \leqslant k \leqslant N-1$, where

$$P_k = (1 + K_{k+1})^{-1}(-1)K_{k+1}, \; 0 \leqslant k \leqslant N-1, \; K_N = 1$$

and

$$K_k = \frac{K_{k+1}^2}{(1+K_{k+1})^2} + K_{k+1}\left(1 - \frac{K_{k+1}}{(1+K_{k+1})}\right)^2$$

$$= \frac{K_{k+1}}{1+K_{k+1}} > 0, \qquad 0 \leqslant k \leqslant N-1$$

Noting that the sequence $\{K_N, K_{N-1}, K_{N-2}, \ldots\}$ takes the form $\{1, \frac{1}{2}, \frac{1}{3}, \frac{1}{4}, \ldots\}$ and tends to zero, deduce that the gains P_k are small for k small and hence that most of the control action takes place in the later stages of the time interval $0 \leqslant k \leqslant N-1$ of interest.

(4) Show that the solution to problem (3) can be obtained by solving the discrete TPBVP

$$x_{k+1}^* = -x_k^* + u_k^*, \qquad x_0^* = x_0 = 1$$

$$u_k^* = -p_{k+1}, \qquad 0 \leqslant k \leqslant N-1$$

$$p_k = -p_{k+1}, \qquad 1 \leqslant k \leqslant N-1$$

$$p_N = x_N^*$$

Writing $p_k = \widetilde{P}_k x_{k-1}^*$, $1 \leqslant k \leqslant N$, show that the feedback law $u_k^* = -\widetilde{P}_{k+1} x_k^*$ is a solution to the problem if the gains $\widetilde{P}_{k+1}$ satisfy,

$$\tilde{P}_N = -\tfrac{1}{2}$$

$$\tilde{P}_k = \tilde{P}_{k+1}(1+\tilde{P}_k), \qquad 1 \leqslant k \leqslant N-1$$

(5) Using the techniques of problem (1), show that the solution to the unconstrained continuous problem

$$\frac{\mathrm{d}x(t)}{\mathrm{d}t} = Ax(t)+Bu(t), \qquad x(0) = x_0$$

$$J(u) = \frac{1}{2}x^T(t_f)Fx(t_f)+\frac{1}{2}\int_0^{t_f}\{x^T(t)Qx(t)+u^T(t)Ru(t)\}\,\mathrm{d}t$$

$$F = F^T \geqslant 0, \qquad Q = Q^T \geqslant 0, \qquad R = R^T > 0$$

can be approximated by solving the discrete problem

$$x_{k+1} = (I_n+Ah)x_k+hBu_k, \qquad x_0 \text{ specified}$$

$$u_k \text{ unconstrained}$$

$$J_N(u) = \frac{1}{2}x_N^TFx_N+\frac{h}{2}\sum_{k=0}^{N-1}\{x_kQx_k+u_kRu_k\}$$

where x_k and u_k are regarded as approximations to $x(kh)$ and $u(kh)$ respectively, and $t_f = Nh$. Hence show that the solution can be computed from the discrete TPBVP

$$x_{k+1}^* = (I_n+Ah)x_k^*+hBu_k^*, \qquad x_0^* = x_0$$

$$u_k^* = -R^{-1}B^Tp_{k+1}$$

$$p_k = (I_n+A^Th)p_{k+1}+Qhx_k^*$$

$$p_N = Fx_N^*$$

Writing this TPBVP in the form

$$\frac{x_{k+1}^*-x_k^*}{h} = Ax_k^*+Bu_k^*, \qquad x_0^* = x_0$$

$$u_k^* = -R^{-1}B^Tp_{k+1}$$

$$\frac{p_{k+1}-p_k}{h} = -A^Tp_{k+1}-Qx_k^*, \qquad p_N = Fx_N^*$$

and interpreting $(x_{k+1}^*-x_k^*)/h$ and $(p_{k+1}-p_k)/h$ as approximations to derivatives, deduce that the discrete TPBVP is an approximation to the continuous TPBVP derived in Section 5.2.

(6) Show that the minimum energy problem

$$x_{k+1} = -x_k + u_k, \qquad x_0 = 1, \qquad x_N = 0$$

$$J(u) = \frac{1}{2}\sum_{k=0}^{N-1} u_k^2$$

u_k unconstrained

is equivalent to the quadratic programming problem: Minimize $\frac{1}{2}y^T y$ subject to $My = f$ where $y^T = (u_0, u_1, \ldots, u_{N-1})$, $M = ((-1)^{N-1}, (-1)^{N-2}, \ldots, -1, 1)$ and $f = (-1)^{N+1}$. Hence show that the solution is $y^* = M^T(MM^T)^{-1}f$ which yields the control sequence

$$u_k^* = (-1)^k/N, \qquad 0 \leqslant k \leqslant N-1$$

Remarks and Further Reading

Bellmann's original treatment of dynamic programming can be found in Bellmann (1957) with applications in Bellmann and Dreyfus (1962). Other useful texts include Burley (1974), Layton (1976), Greensite (1970) and several others from the reading list of Chapter five. The use of dynamic programming methods for continuous systems in the form of the Hamilton-Jacobi approach is described in Munro (1979), Bryson and Ho (1969), Ogata (1967) and its use as the basis of the numerical method of "differential dynamic programming" can be found in Jacobson and Mayne (1970).

References

Adby, P.R. and Dempster, M.A.H. (1974). "Introduction to Optimization Methods". Chapman and Hall, London.

Anderson, B.D.O. and Moore, J.M. (1971). "Linear Optimal Control". Prentice-Hall, Hemel Hempstead.

Athans, M. and Falb, P.L. (1966). "Optimal Control". McGraw-Hill, New York.

Barnett, S. (1975). "Introduction to Mathematical Control Theory". Clarendon Press, Oxford.

Barnett, S. (1971). "Matrices in Control Theory". Van Nostrand Reinhold, New York.

Bell, D.J. and Jacobson, D.H. (1975). "Singular Optimal Control Problems". Academic Press, London and New York.

Bellman R. (1957). "Dynamic Programming". Princeton University Press.

Bellman, R. (1970). "Introduction to Matrix Analysis". McGraw-Hill, New York.

Bellman, R. and Dreyfus, S.E. (1962). "Applied Dynamic Programming". Princeton University Press.

Bishop, A.B. (1975). "Introduction to Discrete Linear Controls". Academic Press, New York and London.

Boksenbrom, A.S. and Hood, R. (1949). "General Algebraic Method Applied to Control Analysis of Complex Engine Types". National Advisory Committee for Aeronautics, NCA-TR-980, Washington D.C.

Brockett, R.W. (1970) "Finite-dimensional Linear Systems". John Wiley, New York.

Brogan, W.L. (1974). "Modern Control Theory". Quantum, New York.

Bryon, A.E. and Ho, Y.C. (1969). "Applied Optimal Control". Ginn, London.

Burley, D. (1974). "Studies in Optimization". Intertext Books.

Cadzow, J.A. (1973). "Discrete-Time and Computer Control Systems". Prentice-Hall, Hemel Hempstead.

Cappellini, V., Constantinides, A.G. and Emiliano, P. (1978). "Digital Filters and their Applications". Academic Press, London and New York.

Chen, C.T. (1970). "Introduction to Linear Systems Theory". Holt, Rinehart and Winston, New York.

Citron, S.J. (1969). "Elements of Optimal Control". Holt, Rinehart and Winston, New York.

Clements, D.J. and Anderson, B.D.O. (1978). "Singular Optimal Control: the Linear Quadratic Problem". Springer-Verlag, Lecture Notes in Control and Information Sciences, Vol. 5.

Craven B.D. (1978). "Mathematical Programming and Control Theory". Chapman and Hall, London.
Curtain, R.F. and Pritchard, A.J. (1978). "Infinite Dimensional Linear Systems Theory". Springer-Verlag, Lecture Notes in Control and Information Sciences Vol. 8.
Dorf, R.C. (1974). "Modern Control Systems". Addison-Wesley, Reading, Massachusetts.
Edwards, J.B. and Owens, D.H. (1977). First-order models for multivariable process control. *Proc. IEE.* **124**, 1083–1088.
Fallside F. (1977). "Control Systems Design by Pole-zero Placement". Academic Press, London and New York.
Fossard, A. (1977). "Multivariable System Control". North-Holland Pub. Co., Amsterdam and New York.
Fox, L. (1964). "An Introduction to Numerical Linear Algebra". Oxford University Press, Oxford.
Gantmacher, F.R. (1959). "The Theory of Matrices". Chelsea Publishing, New York.
Goult, R.J., Hoskins, R.F., Milner, J.A. and Pratt, R.J. (1974). "Computational Methods of Linear Algebra". Stanley Thornes, London.
Greensite, A.L. (1970). "Elements of Modern Control Theory". Spartan Books, New York.
Harris, C.J. and Owens, D.H. (1979)(eds). Special issue of the Proceedings IEE Control and Science Record on "Multivariable Systems". *Proc. I.E.E.* **126**, 6, 537–648.
Holtzmann, J.M. (1970) "Nonlinear System Theory: A Functional Analysis Approach". Prentice-Hall, Hemel Hempstead.
Jacobson, D.H. and Mayne, D.Q. (1970). "Differential Dynamic Programming". American Elsevier Publishing, New York.
Jury, E.I. (1964). "Theory and Application of the *z*-transform Method". John Wiley, New York.
Kailath, T. (1980). "Linear Systems". Prentice-Hall, Hemel Hempstead.
Kalman, R.E. (1964). "When is a Linear Control System Optimal?" *J. Basic Eng. ser. D.* **86**, 51–60.
Kirk, D. (1970). "Optimal Control Theory", Prentice-Hall, Hemel Hempstead.
Kouvaritakis, B. and MacFarlane, A.G.J. (1976). Geometric approach to the analysis and synthesis of system zeros. *Int. J. Control* **23**, 149–166 and 167–181.
Kwakernaak, H. and Sivan, R. (1972). "Linear Optimal Control Systems". John Wiley, New York.
Layton, J.M. (1976). "Multivariable Control Theory". Peter Peregrinus, Stevenage.
Lawden, D.F. (1975). "Analytical Methods of Optimization". Scottish Academic Press, Edinburgh and London.
Leigh, J.R. (1980) "Functional Analysis and Linear Control Theory". Academic Press, London and New York.
Lindorff, D.P. (1965). "Theory of Sampled-data Control Systems" John Wiley, New York.
Luenberger, D.G. (1964). Observing the state of a linear system. *IEEE Trans. MIL-8*, **74**.
Luenberger, D.G. (1966). Observers for multivariable systems. *IEEE Trans. Aut. Contr., AC-11*, **190**.

Luenberger, D.G. (1969). "Optimization by Vector Space Methods". John Wiley, New York.

MacFarlane, A.G.J. (1970a). Commutative controller: a new technique for the design of multivariable control systems. *Electron. Lett.* **6**, 121–123.

MacFarlane, A.G.J. (1970b). Return-difference and return-ratio matrices and their use in the analysis and design of multivariable feedback control systems. *Proc. IEE.* **117**, 2037–2049.

MacFarlane, A.G.J. (1970c). Two necessary conditions in the frequency domain for the optimality of a multiple-input linear control system. *Proc. IEE.* **117**, 464–466.

MacFarlane, A.G.J. (1975). Relationships between recent developments in linear control theory and classical design techniques. *Measurement and Control* **8**, 179–187, 219–223, 278–284, 319–323 and 371–375.

MacFarlane, A.G.J. (1979). The development of frequency-response methods in automatic control. *IEEE. Trans Aut. Cont., AC-24*, 250–265.

MacFarlane, A.G.J. (1980). "Frequency Response Methods in Control Systems". IEEE Press Selected Reprints Series.

MacFarlane, A.G.J. and Karcanias, N. (1976). Poles and zeros of linear multivariable systems: a survey of the algebraic, geometric and complex-variable theory. *Int. J. Control* **24**, 33–74.

MacFarlane, A.G.J., Kouvaritakis, B. and Edmunds, J.M. (1978). "Complex Variable Methods for Multivariable Feedback Systems Analysis and Design". Alternatives for Linear Multivariable Control, National Engineering Consortium Inc., Chicago.

Marshall, S.A. (1978). "Introduction to Control Theory". Macmillan, London.

Melsa, J.L. (1970). "Computer Programs for Computational Assistance in Control Theory". McGraw-Hill, New York.

Mirsky, L. (1963). "An Introduction to Linear Algebra". Clarendon Press, Oxford.

Munro, N. (1979a). "Modern Approaches to Control Systems Design". Peter Peregrinus, Stevenage.

Munro, N. (1979a). "Modern Approaches to Control Systems Design". Peter Peregrinus, Stevenage.

Munro, N. (1979b). Pole assignment. *Proc. IEE* **126**, 549–554.

Noton, A.R.M. (1965). "Variational Methods in Control Engineering". Pergamon Press, Oxford.

Ogata, K. (1967). "State Space Analysis of Control Systems" Prentice-Hall, Hemel Hempstead.

Ortega, J.M. and Rheinboldt, W.C. (1970). "Iterative Solution of Nonlinear Equations in Several Variables". Academic Press, New York and London.

Owens, D.H. (1978). "Feedback and Multivariable Systems". Peter Peregrinus, Stevenage.

Owens, D.H. (1979a). Discrete first-order models for multivariable process control. *Proc. IEE* **126**, 525–530.

Owens, D.H. (1979b). Compensation theory for multivariable root-loci. *Proc. IEE* **126**, 538–541.

Owens, D.H. (1979c). On the computation and characterization of the zeros of linear multivariable systems. *Proc. IEE* **126**, 1335–7.

Owens, D.H. (1979d). On the manipulation of optimal system asymptotic root-loci. University of Sheffield, Department of Control Engineering, Research Report No. 101.

Owens, D.H. (1979e). Some unifying concepts in multivariable feedback design. IFAC Symposium "Computer-aided-design of Control Systems", Zurich, 349–354.

Owens, D.H. (1980). On the computation of optimal system asymptotic root-loci. *IEEE Trans. Aut. Contr. AC-25*, 100–102.

Owens, D.H. (1981). Multivariable root-loci: an emerging design tool? Proceedings of the IEE Int. Conf. "Control and Its Applications" Warwick, 2–7.

Owens, D.H. and Chotai, A. (1980). Robust control of unknown or largescale multivariable systems using transient data only. University of Sheffield., Department of Control Engineering, Research Report No. 134.

Owens, D.H. and Chotai, A. (1981). Simple models for robust control of unknown or badly-defined multivariable systems. *In* "Self-tuning and Adaptive Control: Theory and Applications" (Eds S.A. Billings and C.J. Harris). Peter Peregrinus, Stevenage.

Polak, E. (1971). "Computational Methods in Optimization". Academic Press, New York and London.

Porter, B. and Crossley, R. (1972). "Modal Control: Theory and Applications". Taylor and Francis, London.

Porter, W.A. (1966). "Modern Foundations of Systems Engineering". Macmillan, New York.

Postlethwaite, I. (1978). A note on the characteristic frequency loci of multivariable linear optimal regulators. *IEEE Trans. Aut. Conts., AC-23,* 757–760.

Postlethwaite, I. and MacFarlane, A.G.J. (1979). "Complex Variable Approach to the Analysis of Linear Multivariable Feedback Systems". Springer Verlag Lecture Notes in Control and Information Sciences, Vol. 12.

Power, H.M. and Simpson, R. (1978). "Introduction to Dynamics and Control". McGraw-Hill, London.

Prime, H.A. (1969). "Modern Concepts in Control Engineering". McGraw-Hill, London.

Raven, F.H. (1978). "Automatic Control Engineering". McGraw-Hill, New York.

Rosenbrock, H.H. (1970). "State-space and Multivariable Theory". Nelson, London.

Rosenbrock, H.H. (1974). "Computer-aided-design of Control Systems". Academic Press, London and New York.

Rugh, W.J. (1975). "Mathematical Description of Linear Systems". Marcel Dekker, New York.

Sagan, H. (1969). "Introduction to the Calculus of Variations". McGraw-Hill, New York.

Sage, A.P. (1968). "Optimum Systems Control". Prentice-Hall, Hemel Hempstead.

Shinners, S.M. (1978). "Modern Control Systems Theory and Application". Addison-Wesley, Reading, Massachusetts.

Singh, M.G. (1977). "Dynamical Hierarchical Control". North-Holland Publishing, New York.

Stein, G. (1979). Generalized quadratic weights for asymptotic regulator properties. *IEEE Trans. Aut. Contr., AC-24,* 559–565.

Swisher, G.M. (1976). "Linear Systems Analysis". Matrix Publishing, New York.

Takahashi, Y., Rabins, M.J. and Auslander, D.M. (1972). "Control and Dynamic Systems". Addison-Wesley, Reading, Massachusetts.

Tropper, A.M. (1969). "Linear Algebra". Nelson, London.

Vidal, P. (1969). "Nonlinear Sampled-data Systems". Gordon and Breach. New York and London.

Willems, J.L. (1970). "Stability Theory of Dynamical Systems". Nelson, London.

Wolovich, W.A. (1974). "Linear Multivariable Systems". Springer-Verlag Lecture Notes in Mathematical and Economic Systems.

Wonham, W.M. (1967). On pole-assignment of multi-input controllable linear systems. *IEEE Trans. Aut. Contr., AC-12*, 660–665.

Wonham, W.M. (1974). Linear Multivariable Control: A Geometric Approach". Springer-Verlag Lecture Notes in Mathematical and Economic Systems.

Wonham, W.M. (1978). Geometric state-space theory in linear multivariable control: a status report. Proceedings of the 7th Triennial IFAC World Congress, Helsinki, 43B.

Zadeh, L.A. and Desoer, C.A. (1963). "Linear Systems Theory". McGraw-Hill, New York.

Index